# MANAGEMENT OF THE ELECTRIC ENERGY BUSINESS

By Edwin Vennard

Consultant and an associate of
Overseas Advisory Associates, Inc.

This authoritative book covers the fundamentals of the electric energy business from both a technical and a business viewpoint. It examines the many problems involved in supplying electric energy at a reasonable price and describes techniques for financially sound management of an electric energy system. Above all, the book emphasizes the key role of management, at all levels, in the operation of an electric energy enterprise at a time when there is a growing demand for the production of safe, reliable, and plentiful electric energy.

In easy-to-understand language, the author describes how management meets its objectives to . . .

- build and operate an electric energy system
- offer electricity at a price consistent with good service
- meet the electric energy needs of all its customers
- pay good wages to its employees
- earn a fair return for the utility's investors
- build public trust in the utility
- plan to meet anticipated energy demands in the future
- minimize any detrimental effects on the environment

The author uses a hypothetical company —the Edison Power Company—to illustrate the problems and techniques of effective management. This enables the author to define the principal terms used in the industry and demonstrate management's role in planning, building, and operating an electric utility. Although the Edison Power Company

(continued on back flap)

(continued from front flap)

represents an American utility, the principles a _____ _____ _____ ____ to any electric _____ _____ _____ me very _____ art from _____ her util- _____ cipal as- _____ ic utility and discusses in detail the ___ ____le of the management team. The emphasis throughout is on ways to earn a proper return on capital in order to encourage investors to put their savings into the electric energy business.

Whether you're interested in designing a new electric utility or building public confidence in an existing company, analyzing a utility's expenses or establishing a long-range energy policy, this book gives you an overall view of the management of an electric energy business. The book will also serve the needs of many people beyond the management group for which it is primarily intended—engineers, supervisors, regulators, suppliers, government officials, educators, environmentalists, consumers, an electric utility's own employees, and students in training courses in electric utility management.

## About the Author

EDWIN VENNARD, former president of Commonwealth Management Consultants, president of Middle West Service Company, and managing director of Edison Electric Institute, is currently a consultant and an associate of Overseas Advisory Associates, Inc. He has written hundreds of articles for professional journals and has lectured extensively in the United States and abroad on the electric energy business. He is the author of *Government in the Power Buisness* and *The Electric Power Business* (both McGraw-Hill). Mr. Vennard is a member of the Institute of Electrical and Electronics Engineers and a seven-time winner of awards from the Freedoms Foundation.

C

# MANAGEMENT OF THE
# ELECTRIC ENERGY BUSINESS

# MANAGEMENT OF THE ELECTRIC ENERGY BUSINESS

**EDWIN VENNARD**
Consultant

Sponsored by
Overseas Advisory Associates, Inc.
Walker L. Cisler, President

McGRAW-HILL BOOK COMPANY

New York   St. Louis   San Francisco   Auckland
Bogotá   Düsseldorf   Johannesburg   London   Madrid
Mexico   Montreal   New Delhi   Panama   Paris
São Paulo   Singapore   Sydney   Tokyo   Toronto

**Library of Congress Cataloging in Publication Data**

Vennard, Edwin, date.
Management of the electric energy business.

"Sponsored by Overseas Advisory Associates, Inc."
Includes index.
1.  Electric utilities–United States.     2.   Elec-
tric utilities–Management.     I.   Title.
HD9685.U5V43     658′.91′363620973     79-696
ISBN 0-07-067402-7

1234567890   KPKP   7865432109


The editor for this book was Tyler G. Hicks,
the designer was William E. Frost, and the production supervisor
was Sally Fliess. It was set in Baskerville
by Progressive Typographers.
Printed and bound by The Kingsport Press.

**THIS BOOK IS DEDICATED TO
THE CENTENNIAL OF LIGHT AND TO THE
RAISING OF LIVING STANDARDS EVERYWHERE**

# CONTENTS

PREFACE                                                                          ix

CHAPTER ONE
Setting the Scene                                                                 1

CHAPTER TWO
Trade Terms                                                                       4

CHAPTER THREE
The Evolution of Electric Energy Supply and the
    Edison Power Company                                                         24

CHAPTER FOUR
Economic Characteristics, Economic Trends, and the
    Changed Economic Climate                                                     90

CHAPTER FIVE

The Management Profession                                    115

CHAPTER SIX

How Energy Is Used in Improving Living Standards             127

CHAPTER SEVEN

Planning, Building, Operating, Pricing, and Financing an
    Electric Utility                                        181

CHAPTER EIGHT

Cost Analyses, Cost Allocation, and Rate Design             245

CHAPTER NINE

Regulation                                                  287

CHAPTER TEN

Public and Employee Relations                               310

CHAPTER ELEVEN

Research                                                    323

CHAPTER TWELVE

Energy and the Environment                                  345

CHAPTER THIRTEEN

The Energy Crisis and the Need for a Long-Term
    Energy Policy                                           365

INDEX                                                       389

# PREFACE

For most of the 1970s the impact of events and efforts to solve a multi-faceted array of energy problems have served to stimulate public recognition of the need to reduce dependence on uncertain supplies of oil and natural gas. As a consequence, increased attention has been focused on the role of electric power in meeting the energy requirements of the United States and of other nations around the world.

In the United States inflation, financing, rising electric rates, environmental regulations, delays in building generating facilities, and adequacy of fuel supplies are among the concerns engaging the attention of those with responsibilities for electric power supply. All these factors as well as other important ones affecting the reliability and price of electric service have been and continue to be subjected to intense scrutiny, debate, and action inside and outside of government. In many respects, the electric utility business is in the spotlight as never before.

This book provides an overview of the electric utility business and its problems. It describes the process by which electric power supply reached its present state of development, its essentiality to the progress of the American people, what management has done to solve problems, and how management seeks to deal with the challenges of today and tomorrow.

The readers for whom the book is primarily intended are people now employed in electric utility supervisory or management capacities or who wish to be so engaged. However, because of the breadth of concern about power supply in the United States and the variety of interests associated with providing electric service, this book may also be of assistance to those in manufacturing firms, consulting organizations, governmental agencies, and regulatory bodies and to others who work in, think about, or wish to be informed of the electric utility industry—a technologically advanced and vital part of American civilization which will continue to affect the economic progress, life-style, and well-being of everyone in the United States.

Most of the examples and case studies in this text refer to electric utilities in the United States. But this does not mean that the book is addressed solely to electric power companies in this country. Experience abroad has indicated that practically all electric utilities develop according to certain patterns, based on gross national product per capita, income per capita, and the use of energy per capita. By examining the long-term trends in the United States, utilities abroad can see where they stand in their overall energy development and the path they can take if they employ the principles and tools available to them. Furthermore, the nations now developing can benefit from the research and development that has been carried on in the industrialized nations for years and from the mistakes made in the more advanced countries.

To aid in understanding the various principles in operation, a hypothetical company has been created. In general, this hypothetical company represents approximately the investor-owned electric utility industry in the United States divided by 100. As local conditions vary widely throughout the country, the hypothetical utility cannot be used for comparison with or for judging the efficiency of any individual company. The examples, diagrams, and charts illustrate ideas or principles rather than actual situations.

The author is indebted to his many former associates and friends over the years who have helped in the preparation of this book. There have been so many that it would be difficult to name them all without omissions. Mention should be made of one in particular, however: Tom Burbank, formerly Senior Vice President of Economics, Accounting, Finance and Statistics of the Edison Electric Institute. Also, thanks are extended to Prall Culviner, Vice President of the Institute, who read the text and gave valuable comments. Most of the factual information on energy and utilities comes from the records of the Institute. Where opinions are expressed, they are those of the author.

EDWIN  VENNARD

# CHAPTER 1

# SETTING THE SCENE

Prior to the 1970s, there was a general sense of security about energy in the United States—if, indeed, it was given much thought at all by those who were not involved with its supply. Where electricity was concerned, the price of service to consumers had been declining as the result of increased customer use, gains in equipment efficiency, higher load factors, and continuing progress in management skills and techniques. Fuel prices were relatively stable because there were adequate supplies of coal, oil, and natural gas and also a healthy condition of competition among them. As a consequence, even though the price of most other commodities was rising at an inflation rate of $2\frac{1}{2}$ to 3 percent per year, the average price of electricity was able to maintain a downward trend.

At the start of the 1970s, however, inflationary pressures began to mount. Although improved efficiencies in providing electric service continued, they were unable to offset the higher cost of investment and increased operating and fuel costs caused by the new economic climate. Furthermore, although research and development was moving ahead,

1

progress in deriving more electric energy from basic fuel sources became harder to accomplish as technical limitations caused advances in efficiency to slow down.

Government regulations began to proliferate and become more burdensome, particularly in the areas of environmental controls and nuclear development. But the real impact came with the Middle East oil embargo in late 1973. The price of fuel to the utilities skyrocketed, and the inflation rate moved up sharply. The cost of money climbed, a particularly onerous problem for the capital-intensive electric utility industry, which must raise in the financial markets more than half of the capital it needs to build power supply facilities to meet the public's demand for electric energy.

All of these factors made it necessary for the utilities to obtain higher rates if they were to be able to meet their responsibilities for providing reliable electric service. This, in turn, produced consternation among the customers, long accustomed to declining electricity prices. The regulators faced an entirely new set of challenges with the need to keep the utilities financially viable and at the same time assuage the outrage of consumers.

Public recognition of the so-called energy crisis in 1973 and 1974 only provided a broader awareness of what energy experts had long known—that oil and natural gas are depletable fuels and will run out in the not-too-distant future. But there still is an abundance of coal and nuclear energy to ease the transition to such alternative sources as solar power and nuclear fusion. The problem is, however, that none of these sources will be able to meet any significant portion of United States or world energy needs until well after the turn of the century.

Transitions in the mix of sources to meet human energy needs are nothing new. About 100,000 years ago, wood became a primary fuel. Animal power was added to human muscle around 5000 B.C. By A.D. 1400, some coal was being used, as were waterwheels and windmills. In 1698, with the advent of the first working steam engine, the industrial revolution had its genesis. During the succeeding years, there were rapid changes—first to coal as the predominant fuel, then to oil and natural gas.

Electricity has an inherent advantage over other forms of energy because it can be derived from a variety of sources. Thus, with the exhaustion of oil and natural gas supplies in prospect, the use of coal and nuclear energy means even greater reliance in the future on electric power to provide the energy needed for jobs, for the realization of the economic and social aspirations of the disadvantaged, for continued progress in the overall standard of living, and for the security and well-being of all Americans—and, indeed, of everyone in the world. Which fuel resources to use for what purpose and how they should be employed are questions at the heart of the ongoing national and international debate over energy policy.

In the United States, despite governmental recognition of the vital necessity to use coal and nuclear power to meet future energy needs and reduce dependence on overseas oil, the way to their expanded use has not been made easier. Legislation enacted in the past few years and regulatory delay are adversely affecting the timely installation of nuclear and coal-fired generating facilities, which, in turn, could mean that the nation may suffer electricity shortages beginning in the early 1980s. The National Energy Act which was passed in the fall of 1978 did not provide what was needed in terms of a comprehensive and effective national energy policy. Much has to be done to remove the many inhibiting factors holding back the development and utilization of energy resources.

In essence, then, electric utility managements are working today in an atmosphere of increasing challenge to their ability to provide reliable electric power service, and, at the same time, with the full knowledge that electric energy is becoming even more essential to the future of people everywhere. Modern management has behind it a long history of successful solutions to problems in the electric utility business. New tools, such as the computer, are available and are being used in many ways to identify and deal with present problems and those to come. Through expanded research and development, answers are being sought to questions in a multitude of subject areas and new methods for power supply are being brought closer to fruition.

In many respects, the prospects for civilization depend on the wise decisions of electric utility managements, of those who regulate the electric power business, and of those who legislate and administrate with respect to matters involving the supply of electricity. Above all is the need for adequate understanding by the public of what is at stake and of the potential effects of proposals from various quarters. Only on the basis of sound information, properly understood, can sound judgments be made. The chapters that follow are intended to be of assistance to all who share the burden of responsibility for the future of electric energy.

CHAPTER

# 2

# TRADE TERMS

Like any other field, the electric energy industry has its trade terms. The people who work in the business have developed a vocabulary peculiar to their industry. Understanding these terms is one of the keys to understanding the electric energy business. Some of the principal terms of the industry will be defined in this chapter. Each new term introduced will be shown in italics. Other terms will be defined throughout the text of the book, under the subject heading to which they relate.

## THE KILOWATT AND THE KILOWATTHOUR

One of the distinguishing characteristics of electricity is that it cannot be stored economically in large quantities. It must be generated as it is used. For this reason it is important to distinguish between the quantity of electricity used and generated and the rate at which it is used and generated.

4

# MEASURES OF QUANTITY

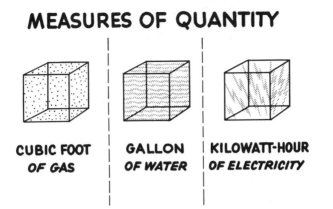

| CUBIC FOOT | GALLON | KILOWATT-HOUR |
| OF GAS | OF WATER | OF ELECTRICITY |

CHART 2.1

The *kilowatthour*, for example, is the measure of the quantity of electricity generated or consumed. The kilowatthour (abbreviated kWh) corresponds to other units of measure of quantity, as shown in Chart 2.1. A cubic foot of gas, a gallon of water, and a kilowatthour of electricity are all measures of quantity. The phrase "10 gallons of water" does not indicate whether the water is being used or whether it is flowing through a pipe; it merely identifies a quantity. The kilowatthour bears the same relationship to electricity. It is simply a measure of quantity.

## WATER ANALOGY

When talking about how fast a person is using water, both the quantity—gallons (abbreviated gal)—and the time during which the quantity is used become important. Chart 2.2 shows how a person might use water during the day. All the faucets were turned off during the night, and no

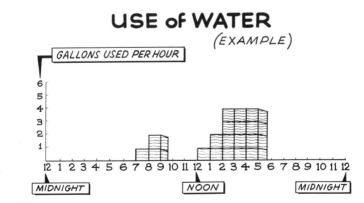

CHART 2.2

water was used until seven o'clock in the morning. Between 7 and 8, 1 gal was consumed, and the chart accordingly shows a cube representing a gallon of water used during that time.

Between 8 and 9, consumption was 2 gal, and two cubes are shown in the space representing that hour. The cubes on the chart indicate how much water was used during each of the remaining hours of the day. Counting the number of cubes shows a total usage of 18 gal of water. The chart also shows how fast it was used. From 7 to 8, the rate was 1 gal/h; from 8 to 9, it was 2 gal/h. The maximum rate of use occurred between 2 and 5 P.M.: 4 gal/h.

## ELECTRICITY EXAMPLE

In the water system, water flows through a pipe. In the electric system, energy, in the form of electricity, in effect flows through a wire to produce light and heat and to run motors. The quantity and rate of use of electricity are measured as in the water example, the only difference being in the terms used to describe the units. The unit of measurement of quantity of electricity is the kilowatthour.

Chart 2.3 shows this same household's use of electricity during a day. In this chart, the cubes represent kilowatthours of electricity. No electricity is used during the night. Cooking breakfast requires 1 kWh between 8 and 9 A.M., and a cube is shown for that hour. No further use of electricity occurs until 11 A.M., when laundry is being washed and dried. At noon, for cooking lunch and perhaps baking a cake, consumption amounts to 4 kWh of electricity.

There is no further use until evening, when dinner preparations begin. Between 5 and 6, usage is 3 kWh; between 6 and 7, it is 5 kWh as the lights go on and television watching adds to the consumption. After the cooking is finished, use of electricity drops to only 2 kWh between 7 and 8.

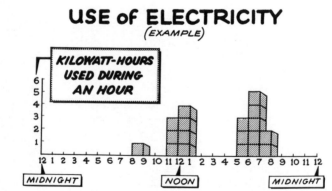

USE of ELECTRICITY
(EXAMPLE)

CHART 2.3

At 8, the whole family may leave to spend the night at a relative's home, so no electricity is used for the remainder of the day.

Counting the cubes tells how many kilowatthours this family used during the day. The total is 18. To find out how fast this family uses electricity, the number of kilowatthours used during any hour can be tallied to find the rate of use. From eight to nine in the morning, for example, 1 kWh was used. The highest rate of use occurred between 6 and 7 P.M.

*Power* means the rate at which energy is delivered or used and is analogous to the terms "gallons per minute" or "gallons per hour." The power of an automobile engine is rated in *horsepower* (abbreviated hp). This rating indicates the engine's capacity to deliver the energy required to run the automobile.

The *kilowatt* (abbreviated kW) is the same kind of unit as the horsepower. It measures the rate at which electricity is being generated or consumed.

Electric appliances are rated in *watts* and *kilowatts*. The *watt* (W) is $^{1}/_{1000}$ kW; in other words, the kilowatt is equal to 1000 W. The watt is used to rate appliances using small amounts of electricity.

The *ampere* (A) is sometimes used to indicate the rating of appliances, fuses, wires, etc., in the home. The ampere is the measure of the flow of current, as distinguished from *voltage*, which is pressure.

If an appliance is rated at 6 A and the house voltage is 115 *volts* (V), the watts are 690 (115 × 6). At unity *power factor* the watt may be referred to as a *voltampere* (VA). Also, a kilowatt is a *kilovoltampere* (kVA), a *kilovolt* (kV) being 1000 V. (Power factor is discussed later in this chapter.)

A light bulb may be rated at 100 W, which is equal to $^{1}/_{10}$ kW. This rating means that the lamp will use $^{1}/_{10}$ kWh of electricity during each hour it is used. If such a lamp is burned steadily for 60 h, it will have used 6 kWh ($^{1}/_{10}$ × 60).

Chart 2.4 shows use of electricity again, but this time the rate of use is given in kilowatts. During the hour between 6 and 7 P.M., this family used 5 kWh of electricity. During that hour, electricity was used at the rate of 5 kW (meaning at the rate of 5 kWh/h).

It is important to the power company to know how much electricity people are likely to use during any period so that the company can have enough generators to make that electricity when it is needed. The company's generators are rated according to the number of kilowatthours that can be delivered each hour (h). A generator may have a rating of 500,000 kW, which means that this machine if run for 1 h can make 500,000 kWh of electricity. If it ran for 2 h, it would make 1 million kWh. If it ran for ½ h, it would make 250,000 kWh. Also, the power company must have neighborhood transformers and power lines large

# USE of ELECTRICITY
### (EXAMPLE)

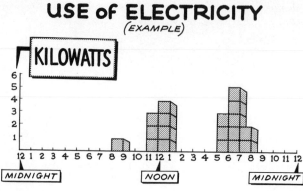

CHART 2.4

enough to take care of the customer's highest rate of use of electricity.

The power company which serves the family in the example has to have enough generating capacity to take care of that family's needs and the needs of all the other customers it serves. As shown in Chart 2.4, to meet this family's requirements between 6 and 7 P.M., the power company must be able to send out 5 kWh during that hour for this customer.

The rating of an appliance in watts and kilowatts enables a determination of how many kilowatthours of electricity the appliance is using. To find out how much electricity is used, the length of time the bulb or appliance is in use at that rate must be known. If ten 100-W bulbs using electricity at the rate of 1 kW are burned for 1 h, they will use 1 kWh of electricity. If 20 such bulbs are burned for 1 h, they will use 2 kWh.

In summary, then, the *kilowatthour* is the basic measuring unit for telling *how much* electricity has been delivered or used; the *kilowatt* tells *how fast* these units were used. Table 2.1 shows how these terms compare

TABLE 2.1   TERMS OF QUANTITY AND RATE

| *Item* | *Quantity* | *Rate* |
|---|---|---|
| Water | Gallons (gal) | Gallons per hour (gal/h) |
| Travel | Miles (mi) | Miles per hour (mi/h) |
| Oil production | Barrels (bbl) | Barrels per day (bbl/d) |
| Printing press | Impressions | Impressions per hour |
| Gas | Cubic feet (ft³) | Cubic feet per hour (ft³/h) |
| Factory machine | Units | Units per hour |
| Milk | Gallons (gal) | Gallons per day (gal/d) |
| Electricity | Kilowatthours (kWh) | Kilowatts* (kW) |

* Note that the term "kilowatt" includes a time element. It could also be stated as "kilowatthours per hour," telling how fast these units were made or used.

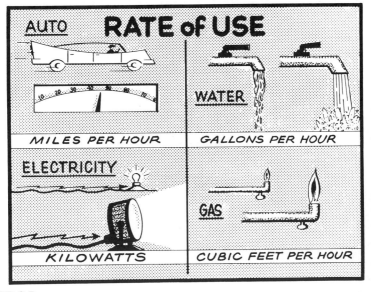

**CHART 2.5**

with other measures of quantity and rate, and Chart 2.5 shows some examples of measurement of rate of use.

## THE ELECTRIC SYSTEM

### WATER ANALOGY

Because people cannot see, smell, or hear electricity, it is hard to describe how it is made and delivered without using an analogy. Since the flow of electricity through a wire is like the flow of water through a pipe, electrical principles can be shown using water as an illustration. Some of the terms used to describe the water system are also used in the electric business.

Chart 2.6 shows a water system serving three customers. Some water systems have water-storage facilities to meet the customers' demands when the pump is not running. Other systems have no storage facilities, and the pump must be running at all times. A system with no storage facilities has been used in this illustration to compare with electric systems because storage is not practical in electric systems. The water pump, with the *capacity* to pump water at the rate of 30 gal/h, takes the water from the reservoir and delivers it to the three customers as they want to use it.

When any one of the customers turns on a spigot, water will come

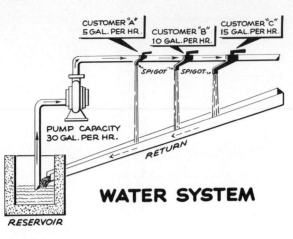

**WATER SYSTEM**

**CHART 2.6**

out at the *pressure* that is maintained in the pipe. The pump must be kept running at all times to keep up that pressure so that it will be there whenever the customers choose to use the water. The power needed to run the pump when no water is being used is merely enough to overcome the loss caused by leakage and friction in the pump.

If customer A wants water, the spigot is turned on, releasing water at the rate of 5 gal/h. A *current* of water is said to be *flowing* through the pipe. Immediately, the pump delivers water at the rate of 5 gal/h.

Assume now that customer B turns on the water at the same time A is taking water. B's spigot takes 10 gal/h, so the two of them together use 15 gal/h and the pump delivers water at this rate. It may be said that customer B makes a *demand* for water on the water system at the rate of 10 gal/h. Customer A demands water at 5 gal/h. The combined demand is 15 gal/h. There is then said to be a *load* on the water pump of 15 gal/h. The terms "load" and "demand" are synonymous.

Now customer C begins using water at the same time. C's larger spigot uses water at the rate of 15 gal/h. With all three customers using water at the same time, the load on the pump or plant is 30 gal/h, equal to the entire capacity of the pump. The pump and pipes must therefore be big enough (i.e., have sufficient capacity) to meet the maximum combined requirements of all customers at any one time.

## OPERATION

Chart 2.7 shows an electric system similar to the water system in Chart 2.6. Fuel for the system is taken from the fuel reservoir, just as water was taken from the water reservoir. The fuel may be coal, oil, gas, lignite, or nuclear fuel. (Electricity is also generated using the energy in falling water. This is discussed in Chapter 3.)

# ELECTRIC SYSTEM

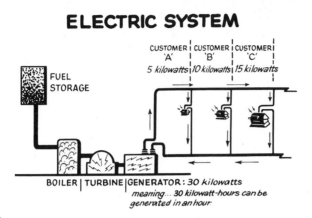

CHART 2.7

The electric *generating plant* converts the energy in the fuel to *electric energy*. The fuel is burned under a *boiler*, where the energy in the fuel is converted to heat and pressure in the steam. The steam pressure spins a *steam turbine*, which is directly connected to the *electric generator*, where the energy of the turbine is converted to electric energy.

If the customers are not using any electricity, the generator running at full speed merely keeps up the level of electric pressure (voltage) on the system. The generator uses only enough fuel in this case to overcome the *losses* caused by friction in the turbine and the generator and the losses in the system.

It may seem strange that a generator running at full speed may require only a small amount of fuel. The situation is, in fact, something like that of a man whose job is to push a wheelbarrow (Chart 2.8). When there is no load in the wheelbarrow, it is very easy to push. When the wheelbarrow is heavily loaded, it may take a great deal more energy to

# ENERGY AND LOAD

CHART 2.8

move it. Although the load of electricity on a generator is invisible, it is there, and it takes more energy to turn a loaded generator than one that is not loaded.

In the diagram of a simple electrical system shown in Chart 2.7, the electric generator, or power plant, has a *capacity* of 30 kW. It can deliver electric energy at the rate of 30 kW (at 30 kWh/h) during every hour it is operated. This generator is kept running day and night, ready to supply any and all demands for electricity as they are made by the customers. The term *demand* refers to the rate at which a customer takes energy from the electric system. A customer who is using machinery that takes 100 kW to operate makes a demand on the system of 100 kW; that is, 100 kWh of electricity will be used every hour the machinery is running. The *maximum demand* of a customer is the highest rate at which that customer takes energy. The customer's demand, or rate of use of energy (kilowatts), can be measured, as can the quantity of energy (kilowatt-hours) used during the month. The maximum demand on an *electric system* is the highest rate at which all customers combined take energy. Demands are measured in kilowatts.

If customer A in Chart 2.7 throws a switch to use electricity, a demand of 5 kW is created on the electric system, just as the customer in the water system created a demand for water by opening a spigot. When customer B decides to operate, electricity is used at the rate of 10 kW and B is said to have a demand of 10 kW. If customer A and customer B both use electricity at the same time, the combined demand on the electric system is 15 kW. If, while customers A and B are operating, customer C decides to operate, the *total demand* on the plant rises to 30 kW —its capacity. That plant cannot deliver energy at a faster rate than 30 kWh/h.

Electricity is generated as it is used, just as in the water analogy water is pumped only as it is used. The rate at which water is pumped is equal to the combined rate at which all the customers are taking water at any time. The rate at which generators make electricity is the sum of the combined rates at which all customers are taking electricity at the particular time.

The electric utility company must foresee the *highest demands* its customers will make. Then the company will plan to have enough plants and equipment to meet these requirements at all times. The demand on an electric system is sometimes called the *load* on that system.

The *maximum load* on an entire electric system during any particular period is sometimes called the *system peak*. For example, if the *maximum hourly demand* on the system is 600,000 kW for a particular month, the system peak for that month is said to be 600,000 kW.

In a water system, there must be enough water pumps of large

# WATER PUMPS

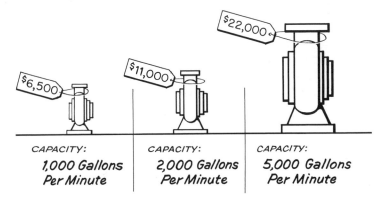

| CAPACITY: | CAPACITY: | CAPACITY: |
|---|---|---|
| *1,000 Gallons Per Minute* | *2,000 Gallons Per Minute* | *5,000 Gallons Per Minute* |

**CHART 2.9**

enough size to meet the maximum demand of all water customers. The greater the demand in gallons per hour, the larger the pumps must be. The cost of the pumps varies with the size: the larger the capacity, the lower the investment per unit of capacity (Chart 2.9).

Similarly, in an electric system, there must be enough *generators* and the generators must be large enough to meet the maximum hourly demand in kilowatts of all electric customers; the greater the demand, the larger the generator—and the more expensive (Chart 2.10). (See Chapter 8 for the effect of inflation on these costs.)

The *voltage* of an electric system is the measure of electric pressure and is analogous to water pressure in a water system. For a pipe of a given size, raising the water pressure increases the capacity of the pipe to deliver water in gallons per hour (Chart 2.11). Similarly, for a wire of a

# ELECTRIC POWER PLANTS

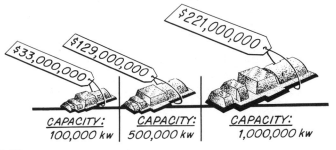

| CAPACITY: | CAPACITY: | CAPACITY: |
|---|---|---|
| *100,000 kw* | *500,000 kw* | *1,000,000 kw* |

**CHART 2.10**

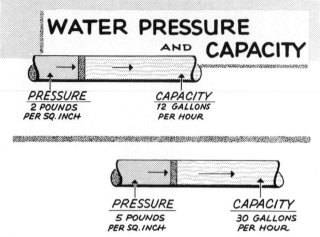

**WATER PRESSURE** AND **CAPACITY**

PRESSURE
2 POUNDS
PER SQ. INCH

CAPACITY
12 GALLONS
PER HOUR

PRESSURE
5 POUNDS
PER SQ. INCH

CAPACITY
30 GALLONS
PER HOUR

**CHART 2.11**

given size, raising the voltage (pressure) results in an increase in the capacity of that wire to deliver energy. This capacity is measured in kilowatts (Chart 2.12).

*High voltage* is used in *transmitting* power for the same reason that high pressure is used to transmit water or oil in pipes over long distances: smaller, and therefore less costly, wire can be used to carry a given amount of electricity. Bigger wires are, of course, heavier, which means that heavier, more expensive *transmission towers* would have to be built to carry the weight of the wires.

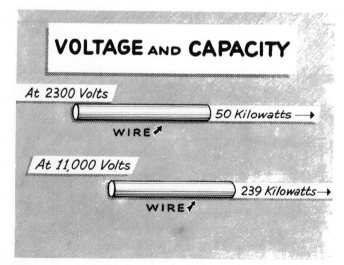

**VOLTAGE** AND **CAPACITY**

At 2300 Volts

WIRE    50 Kilowatts ⟶

At 11,000 Volts

WIRE    239 Kilowatts ⟶

**CHART 2.12**

**Power Factor** Certain electric devices have a peculiar characteristic which results in a demand for more kilowatts than are actually put to any useful purpose. The *induction motor*, the type of motor in most common use, has this characteristic when the motor is run at less than its full load. The actual work being done by the motor results in a certain kilowatt demand that can be measured by an ordinary demand meter. However, when partially loaded, the motor makes a different and useless kind of demand on the electric system. This demand is greater than the partial load on the motor and cannot be measured by the ordinary meter. This useless demand requires capacity in the electric system just the same as the useful demand.

In other words, in the operation of such a motor there is a measurable amount of *useful power* and also a certain amount of useless electric current. This useless current requires capacity in the system, and there must be investment to provide that capacity. Thus, cutting down the useless power helps cut down the cost of providing electric service. *Power factor* is an expression of the relationship between the *useful current* and the *total current* used in an electrical device.

The *watt* is the basic unit of power. There is a mathematical relationship between the watt, the volt, and the ampere, as expressed by the following formula:

$$\textbf{Watts = volts} \times \textbf{amperes}$$

or $$\textbf{Kilowatts = kilovolts} \times \textbf{amperes}$$

The term on the right-hand side is called the *kilovoltampere* (kVA). This equation is a correct one only where no useless current is in evidence. There are certain types of electrical devices that do not cause any of this useless current in the electric system. *Resistance heating devices*, such as the electric range, hot plate, toaster, and incandescent electric light (which is a heating device burning white hot), cause no useless current in the system. *Gaseous tube lights*, such as *fluorescent* and *neon lights*, cause some useless current, which can and should be corrected at the light source. When there is no such useless current in evidence, the power factor is said to be 100 percent. The above formula, therefore, is correct at a power factor of 100 percent. The formula may be expressed as follows:

$$\textbf{Kilowatts = kilovoltamperes} \times \textbf{power factor}$$

In the case of induction motors the power factor may be as low as 50 percent and is sometimes less. A 50 percent power factor means that the useful power is only half that needed to run the motor. The other half of the required power is useless. If an electric motor needs 100 kW of useful power and is operating at 50 percent power factor, it would require 200 kVA of capacity in the electric system. The electric company must

## VECTOR DIAGRAM OF POWER FACTOR
### 85% POWER FACTOR

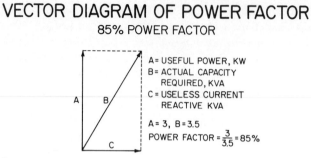

A = USEFUL POWER, KW
B = ACTUAL CAPACITY
      REQUIRED, KVA
C = USELESS CURRENT
      REACTIVE KVA

A = 3,  B = 3.5
POWER FACTOR = $\frac{3}{3.5}$ = 85%

**CHART 2.13**

provide 200 kVA of capacity in transformers, lines, and generators to serve the motor. If the power factor at that motor could be raised to 100 percent, the motor would not do any more work than it did before, but it would make a demand of only 100 kW, and only 100 kW of capacity would be needed to serve the motor.

Fortunately, there are devices called *condensers* to correct poor power factor. There are *static condensers* (those that have no moving parts) and *synchronous condensers* (those that have rotating parts). A natural place to correct the power factor is at the device causing the poor power factor. A customer who has equipment with poor power factor may correct the problem by installing a condenser. By means of this device, the power factor can be raised to any desired point. Electric utility companies usually install condensers in their electric systems to correct *system power factor.* These are synchronous condensers or *capacitors.*

The same principle may be illustrated diagrammatically by the use of *vectors*, that is, by representing quantities by length and direction of lines. Charts 2.13 and 2.14 illustrate such *vector diagrams.* The length of line *A* represents the kilowatts of useful power, sometimes called the active power. This is the power that performs useful work and the power that is measured by ordinary electric meters. Line *B* represents the kilovoltamperes, or kVA, that is, the actual capacity in the electric system

## VECTOR DIAGRAM OF POWER FACTOR
### 50% POWER FACTOR

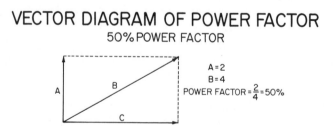

A = 2
B = 4
POWER FACTOR = $\frac{2}{4}$ = 50%

**CHART 2.14**

required to serve the particular load. Line $C$ represents what are called the *reactive kilovoltamperes*, making up the *useless current*. The term "reactive" is used to illustrate the opposite of *active power*.

The power factor of the particular load is determined by dividing the length of line $A$ by the length of line $B$. If line $A$ is 3 in and line $B$ is 3½ in, the power factor is 85 percent (3 divided by 3½). A case of a lower power factor is illustrated in Chart 2.14. In this case, the line representing active or useful kilowatts is 2 in and the kilovoltampere line is 4 in, so the power factor is 50 percent (2 divided by 4).

It is anticipated that the power factor of many customers served by the electric system will be less than unity. An electric system that is operating at 80 percent power factor is considered operating in satisfactory condition. *Transformers* and lines are usually rated in terms of their *kilovoltampere capacity* rather than their kilowatts of capacity. Electric generators are usually rated in both kilovoltamperes and kilowatts. The kilowatt rating is usually expressed at 80 percent power factor. Most electric companies have clauses in their rate schedules for large customers requiring that the customer maintain a power factor of about 80 percent.

Another illustration that has been used to help visualize power factor consists of a horse pulling a railway car on a straight track as in Chart 2.15a. The car is held to the track by the flanges on its wheels and can move only in the direction of the track. If the horse pulls straight ahead on the track, it moves the car and all of its power is effective in producing motion. In this case, it may be said to be working at 100 percent, or unity, power factor.

If the horse now turns and pulls at right angles to the track (Chart 2.15b), it may pull just as hard, but the effort serves no useful purpose because it only pulls against the rails and cannot move the car. There is no component of pull in the direction of the track. The pull is all useless and the power factor is zero.

Now suppose the horse turns halfway back and pulls at an angle of 45° to the track (Chart 2.15c). Part of its pull ($A$) is now in the direction of the track and useful in producing motion, while the other part ($C$) is at right angles to the track and useless in producing motion. In this case, the power factor would be determined by dividing the component of the pull in the direction of the track ($A$) by the total pull in the direction the horse is pulling ($B$). The part of the pull in the direction of the track ($A$) producing motion in the car corresponds to the kilowatts in an electric system, and the useless part at right angles to the track ($C$) corresponds to the reactive kilovoltamperes. The horse's pull in the diagonal direction ($B$) corresponds to kilovoltamperes.

**Load Factor**   Each customer at some time during any given month

# POWER FACTOR

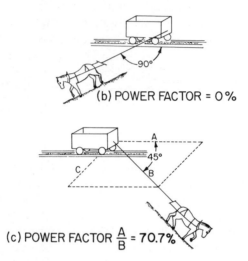

(a) POWER FACTOR = 100%

(b) POWER FACTOR = 0%

(c) POWER FACTOR $\frac{A}{B}$ = 70.7%

**CHART 2.15**

makes a *maximum demand* on the electric system. This happens when a great number of appliances are in operation at the same time, perhaps, and thus electricity is used at the most rapid rate. Very rarely is that high rate of demand maintained continuously. For example, a customer having a maximum demand of 5 kW at some time during the month does not have a demand of 5 kW during every hour of the month. During some of the hours, demand may be only 1 kW. At other times it may be only ½ kW, and in some hours at night there may be no demand at all.

As a result, the customer's use in kilowatthours is less than it would be if electricity were used at the level of maximum demand all the time. The relationship between what the customer actually uses and what would be used if electricity were consumed continuously at the customer's rate of maximum demand is referred to as *load factor*.

In Chart 2.16 each square represents 1 kWh. The chart shows how a customer who runs a small business may use electricity during a day. Between midnight and 6 A.M., no kilowatthours are consumed; between 6 and 7 A.M., 1 kWh; between 7 and 8 A.M., 2 kWh; between the hours of 11 and 12 noon, 6 kWh. Between 12 and 1, the customer's employees go to

# DAILY LOAD CURVE

## SMALL LIGHT AND POWER CUSTOMER

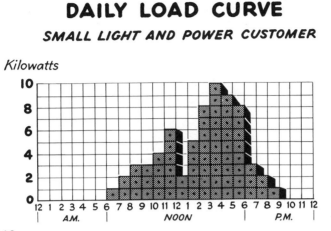

**CHART 2.16**

lunch, and use is only 2 kWh. After lunch, operations are resumed, and a *peak demand*, or a peak use, of electricity is reached between 3 and 4 P.M., when consumption is 10 kWh.

If all the dots in the squares are counted, it is found that this customer used 67 kWh during that day. The maximum rate of use was 10 kWh during the 1 h between 3 and 4 P.M. Thus, there was a maximum demand of 10 kW.

Every square under the 10-kW mark represents a kilowatthour which the customer could have used if energy had been taken uniformly during the day. If all of these squares are counted, it will be found that there are 240. If the customer had used energy uniformly at the rate of 10 kW during every hour of the day, 240 kWh would have been used, and the customer would have had a load factor of 100 percent. The customer actually used 67 kWh. The load factor is the ratio of the actual kilowatthours used to the kilowatthours that would have been used if the customer had used energy uniformly during the day at the rate of maximum use. Therefore this customer's load factor for that day was 67 divided by 240, or 28 percent.

A three-dimensional view of a customer's use for a 1-month period is illustrated in Chart 2.17. In this case the customer's use of electricity follows about the same pattern each day, with a slightly different pattern for Saturday and Sunday. From a lower level during the early morning hours, use rises to a morning peak of about 6 kW. Use drops somewhat during the noon hour, rises to an afternoon peak of 10 kW, and drops to only 1 kW after the evening activity. On Saturdays, the customer uses less, and on Sundays, almost none.

# MONTHLY LOAD
## SMALL LIGHT AND POWER CUSTOMER

**CHART 2.17**

This customer makes a peak demand on the power system of 10 kW. This means that the power company must reserve that amount of capacity for the customer's use when it is demanded regardless of the time of day or year. The company is standing by every hour ready to supply electricity at this rate, even though that demand may be used only a few hours during the month. The customer actually used 1600 kWh during the month. Consumption of 7300 kWh would have been possible had the customer used the service uniformly during the month at the rate of 10 kW. The average number of hours in a month is $365 \times 24/12 = 730$. This customer is said to have a load factor of 22 percent for the month. This percentage is obtained by dividing the 1600 kWh acutally used by the 7300 kWh that might have been used if the customer had operated uniformly at the peak rate of use. Customer load factors may vary from as low as 0 to almost 100 percent, depending upon the kind of business in which the energy is used.

In a similar fashion, the load factor for an entire electric power company can be determined (Chart 2.18). The load factor on a power system for any month is the ratio of the kilowatthours actually delivered to the number of kilowatthours that could have been delivered if the maximum demand had prevailed every hour of every day in the month. For example, assume an electric system delivered 500 million kWh during a certain month and that the maximum rate of delivery at any one time was 1.2 million kW (this being the maximum demand on the system). If that system had delivered 1.2 million kWh every hour for the whole 730 h in the month, it would have delivered 876 million kWh. The load

# A COMPANY MONTHLY LOAD CURVE

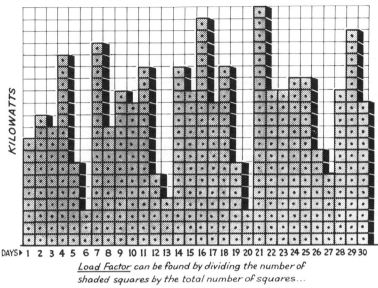

DAYS▶ 1  2  3  4  5  6  7  8  9  10 11 12 13 14 15 16 17 18 19 20 21 22 23 24 25 26 27 28 29 30

*Load Factor* can be found by dividing the number of
shaded squares by the total number of squares...

**CHART 2.18**

factor for the month was 500 million (kWh actually used) divided by 876 million, or 57 percent.

The annual load factor is found by dividing the kilowatthours delivered during the year by the kilowatthours that would have been delivered if the maximum demand during any one hour had prevailed every hour, every day of the year.

## DIVERSITY

To illustrate *diversity*, consider again the diagrammatic and hypothetical electric system illustrated in Chart 2.7. Customers A, B, and C, have maximum demands of 5, 10, and 15 kW, respectively. If all three customers choose to make their maximum demand on the electric system at the same instant, there results a combined maximum demand of 30 kW on that system. However, in practice, the habits of customers vary so that each does not make the maximum demand at the same instant as a neighbor. Consequently, during the hour that customer A is demanding 5 kW, customer B may be demanding only 7 kW instead of the 10 kW demanded during the maximum hour.

The *coincident demand* of customers A and B is therefore 12 kW (5 kW plus 7 kW). There is said to be a diversity between customers A and B of 1.25 (15 kW divided by 12 kW). Similarly, customer C operates according to personal habits and requirements and makes a maximum demand at a different hour than either customers A or B. During the maximum hour of use on the system, customer C may be using only 11 kW, so that the combined maximum demand of customers A, B, and C is 23 kW (5 kW plus 7 kW plus 11 kW). The diversity among the three customers is therefore 1.3 (30 kW divided by 23 kW). In practice, this means that the electric system can furnish all the requirements of customers A, B, and C with an overall plant capacity (and therefore investment) that is 23 percent less than would be required by each customer to furnish the individual needs by individual plants.

Diversity exists in an electric utility system because the requirements of individual customers differ. There is considerable variation between the instants at which individual customers use their maximum requirements.

In addition to diversity among residential customers, there is also diversity of use between lighting and other appliances within the home. Small appliances such as toasters and coffee makers are used when the maximum use of lighting is not being made. The electric refrigerator operates off and on throughout the day and night. The electric-range maximum demand does not occur at the time of the maximum demand of the lighting load. Also, there is diversity among individual ranges throughout a system because of the difference in individual cooking habits. The water-heating load also has diversity, especially in those cases where its use is restricted to off-peak hours. Similarly, there is considerable diversity among industrial power customers caused by the individual requirements in the operation of industrial motors.

Adding to these diversities among individual customers, there is further diversity among classes of customers. Residential and commercial lighting customers as a class make their maximum use of energy during the evening hours. Industrial customers as a class make their maximum use of energy during the daylight hours. The combined demand on the electric system is considerably less than the sum of the peak demands of individual classes of customers.

This diversity among classes has the effect of making the overall system load factor of the combined services higher than individual or class load factors, thus making greater use of the capital investment in the system. For example, residential service as a class may have a load factor of about 35 to 40 percent, commercial service about 22 percent, and general power service about 40 percent. The combined load factor of all these classes together could be as high as 60 percent.

*Diversity factor* is a measure of diversity and is the ratio of the sum of the noncoincident maximum demands of two or more loads to their coincident maximum demand for the same period.

In addition to the diversities among customers and among classes of service, there is additional diversity between communities. One community may be predominantly industrial, whereas another may be predominantly residential and commercial. The habits of residents of one community may differ from habits in another. This diversity among individual communities is one of the principal factors making it more economical for the electric utility company to furnish service from an interconnected transmission system served by a few large power stations than to use the alternative method of providing individual plants in each individual town.

Additional diversities are realized when one interconnected system connects with a neighboring interconnected system. (There is more on diversity in Chapters 7 and 8. Other terms are defined in the chapters where the subject is discussed.)

# THE EVOLUTION OF ELECTRIC ENERGY SUPPLY AND THE EDISON POWER COMPANY

In the definition of terms, it was brought out that energy is work. Energy manifests itself in a number of forms: heat, a moving body, a horse pulling a wagon, falling water, a man using his muscles. It is possible to transform one kind of energy to another. A moving body creates heat. Falling water can be made to move a waterwheel.

On the other hand, *power* is the expression of the capacity to do work or the rate of doing work. It represents the rate at which energy may be delivered or used. This capacity to work can be illustrated by a man, a horse, an automobile, a motor, or a steam engine. Power represents the rate of getting work done.

Electricity, as a useful energy form, came into being when it was discovered that when a wire or a conductor is passed through a magnetic field, a current of electricity flows in the wire. Chart 3.1a shows a coil of wire passing through the field in one direction. The electric energy in the conductor flows in one direction. It is called *direct current*. If a loop of wire is turned in a magnetic field, it will be found that the electric cur-

# ELECTRIC CURRENT

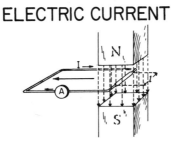

## (a) DIRECT CURRENT

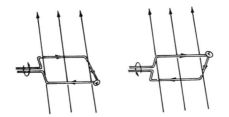

## (b) ALTERNATING CURRENT

**CHART 3.1**

rent reverses its flow with each turn (Chart 3.1b). This is called *alternating current*. Alternating current is used in practically all parts of the world where electricity is available.

Some form of energy, such as heat, is used under a boiler to generate steam at a high temperature and pressure. This steam, in turn, strikes the blades of a *steam turbine*, causing the turbine to spin, much as a windmill spins when the wind is blowing. The steam turbine is connected to a *generator* in which there is a magnetic field through which conductors rotate. If the machine rotates at a speed of some multiple of 60, the resulting alternating current will be said to have 60 *hertz* (Hz). That is, the current in the conductors alternates back and forth 60 times a second. In some countries, the standard may be 50 Hz, in which case the generators are rotated at a speed of some multiple of 50.

This process of using fuel to make steam to strike blades of the turbine so as to rotate the generator is illustrated in Chart 3.2. In most parts of the world, including the United States, various kinds of fuel and falling water are used to bring about the spinning of rotors in generators. The fuel may be wood, coal, oil, gas, lignite, or a nuclear fuel. All these fuels can be used to generate steam to run the turbines. Coal, gas, oil, and nuclear sources account for most of the fuels in the United States.

In Chapter 9 it will be noted that additional work is now being car-

# PRODUCING ELECTRICITY

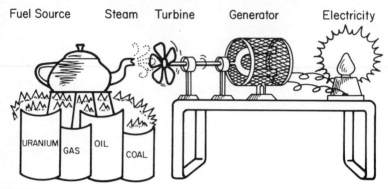

**CHART 3.2**

ried on with geothermal heat, the heat from the underground cavities in some parts of the earth. Further research is also being carried on with solar energy.

Experience has shown that the patterns in the use of various fuels change from time to time as illustrated in Chart 3.3. Before the turn of this century, wood provided most of the fuel in the United States. Bituminous coal began coming into use on a large scale during the period from about 1870 to 1910. Use of bituminous coal has declined so that it now represents 18 percent of the total energy consumption. Anthracite

**SPECIFIC ENERGY SOURCE AS PERCENTAGE OF AGGREGATE ENERGY CONSUMPTION**

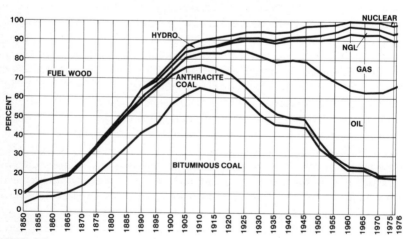

**CHART 3.3**

coal came into use beginning about 1855 and rose to a peak by 1910. It has been declining ever since. Oil became a common source in the early part of this century. Its use has risen significantly, especially beginning about 1920, because of the wide use of oil in all forms of transportation. As noted in Chapter 13, oil now accounts for 47 percent of the total energy being used in the United States.

Gas is generally produced along with the oil. Its big growth began about the same time as that of oil. Notice that in 1976 gas accounted for 27 percent of the total energy used. Gas is generally used for heating purposes and in industrial processes. *Hydroelectric* (water) power has never been a significant percentage of the total in this country, although it is significant in some other countries. Notice that nuclear fuel is the newest form of energy and now accounts for 2.7 percent of the total energy used in the United States.

The problem of the energy crisis occurs because a large percentage of the oil is imported. As became apparent in the years 1973–1974, an embargo on oil imports could cause considerable damage to the economy of the United States. The energy problem facing America is that, according to many experts, oil and gas will be the first of the major fuels to be exhausted—some time shortly after the year 2000. This has been known for more than 25 years, but only recently has it become generally recognized. Already, oil and gas are showing indications of being in short supply. In 1976, these two fuels accounted for 75 percent of the total energy used in this country. On the other hand, coal, which is in abundant supply in the United States is called upon to furnish only 18 percent of the energy needs. It is now the aim of those dealing with energy policy to bring about a change in this pattern. As will be developed elsewhere in this text, the only practical way to use coal and nuclear fuels (with presently available technology) is through the process of converting them to electricity.

Chart 3.4 shows the percentages of total U.S. energy needs met by the various sources from 1880 through 1976, with a forecast to the year 2000. Notice how the proportion for coal has dropped drastically while the percentages for crude oil and gas have increased rapidly. The aim between now and 2000 is to bring about a greater use of coal and nuclear fuel to get the useful work done. Naturally, this should be combined with practical conservation efforts.

Chart 3.5 shows the proportions of electricity generated by primary fuels and hydro in the electric utility industry. The pattern has involved using coal for about half of the power generation since 1920. The drop since 1970 is caused largely by the use of low-sulfur oil to meet environmental requirements. But since the embargo, the tendency is to move back to coal wherever practical (see Chapter 12).

## USE OF MINERAL FUEL AND
## WATER POWER ENERGY IN USA

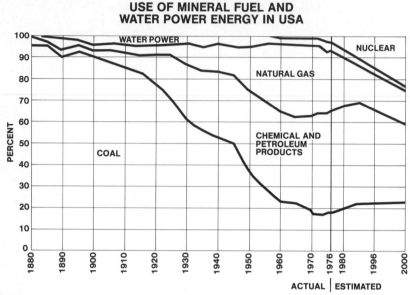

**CHART 3.4**

## TOTAL ELECTRIC UTILITY INDUSTRY PRODUCTION
## BY SOURCE IN PERCENT

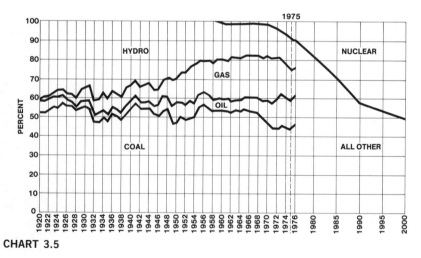

**CHART 3.5**

## EDISON POWER COMPANY

Perhaps the best way to get a clear understanding of the electric energy business is to describe a typical company engaged in the business of transforming raw fuels such as coal, gas, oil, nuclear fuel, and falling water into electric energy. Edison Power Company is a hypothetical company that will be used to demonstrate the principles that apply to the electric energy business.

The company was formed when two smaller companies merged— the Rosedale Edison Company, which served the towns of Rosedale, Sharpsville, and Crystal Lake, and the Pleasantville Electric Light and Power Company, which supplied electricity to the towns of Pleasantville and nearby Carbon City. Edison Power Company's service area includes all five of these towns. The two companies were merged to take advantage of greater economies through unified operation and the economies of scale. Chart 3.6 shows the location of some of the company's plants and the transmission lines connecting them.

## GENERATION, TRANSMISSION, AND DISTRIBUTION

Edison Power Company generates most of its electricity by means of nuclear power and oil- and coal-burning steam plants, located for the most part near the major *load centers*, places where a considerable amount of electric energy is used. In addition, the company has one hydroelectric generating station. This situation takes advantage of favorable water conditions at Fox Falls to make electricity from the power of falling water.

**Generating System**   The energy in the fuel or in the falling water is converted to electrical energy at the *generating stations*, or power plants. A power plant may have one or more *turbine generators*. The turbine generator, sometimes called a *unit* in the plant, is a steam- or water-driven turbine directly connected to an *electric generator*. The turbine is turned by steam in the steam plants and by falling water in a *hydro* plant. Generators are rated in terms of kilowatts of guaranteed continuous output.

The map shows four of Edison Power Company's generating stations. One is located at the only available water-power site. The steam plants are placed at strategic points with respect to the power requirements of the territory served. In locating these plants, the company has considered the nearness of the fuel supply and the availability of water, which is used in a steam plant for steam-condensing purposes. Also, it has considered all environmental requirements.

The power plants must have *reserve* and *standby capacity* so that the

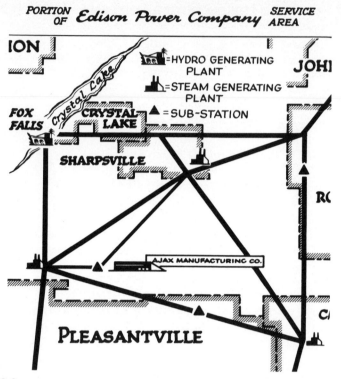

*PORTION OF* **Edison Power Company** *SERVICE AREA*

**ION**

= HYDRO GENERATING PLANT

= STEAM GENERATING PLANT

= SUB-STATION

**JOHN**

**FOX FALLS**

*Crystal Lake*

**CRYSTAL LAKE**

**SHARPSVILLE**

**R(**

**AJAX MANUFACTURING CO.**

**PLEASANTVILLE**

**C(**

**CHART 3.6**

company's customers will still have power even if a unit must be taken out of service. The reserve may be a spare generating unit, or the company may have an interconnection with a neighboring company so that it can obtain power in an emergency. Edison Power Company has both its own reserve capacity and an interconnection with a neighboring company. Edison's reserve capacity consists partly of older generators. These are kept in tip-top condition ready to go, but they stand idle most of the time because it costs more to run these older plants than it does to operate the newer ones. The capacity of the older plants is ready for use in an emergency or when people want to use more electricity than the other plants can produce. This does not happen very often.

If the reserve unit is in operation (that is, spinning at full speed), it is called a *spinning reserve*. A machine that is running can generate power almost instantaneously. The spinning reserve is always running, but often it is not generating any electricity and is using only enough fuel to overcome losses by friction. At times a boiler is kept hot, ready to operate a turbine, but the turbine may not be spinning. This is called a *hot reserve*. A reserve generator with no heat under the boiler is called a *cold reserve*. It may take an hour or more to bring the machine from a cold reserve to a spinning reserve, or to a condition where it can generate power. The *gas turbine* (to be discussed later) and hydro plant can be started very quickly.

The sum of the capacities of all units in the power plant is termed the *plant capacity* and is usually expressed in kilowatts. (Capacities, especially of larger units and plants, are frequently expressed in *megawatts*. A megawatt is 1000 kW.) The sum of all the plant capacities and the capacities which are purchased from other companies is called the *system capacity*. This is also expressed in kilowatts. The total capacity of the system minus the capacity of the largest single unit in the system is called the *firm power capacity* of the system. This *firm capacity* is the highest load the system can carry in the event that the largest unit should break down. (Some very large systems may have more than one unit as reserve.) To provide for this contingency, the reserve capacity must be at least equal to the size of the company's largest unit. Interconnections with neighbors help maintain reliable service without excess reserves.

A system with several power plants will make the greatest use of its generating units with the lowest production cost. The company will probably run its lowest-cost units all the time, day and night, to carry all the load they can carry. A unit so loaded is spoken of as a *baseload unit*. If the demand increases above the capacity of the baseload units, the next most efficient unit is put in operation.

**Transmission System**   Chart 3.7 shows a diagram of Edison Power's steam plant near Pleasantville. Right next to the power plant are *step-up transformers*. These transformers take the electricity from the generators and raise it to a higher voltage for transmission over long distances. This group of transformers is referred to as a *step-up substation*. The step-up transformers serve the same purpose as a water pump at the beginning of a water pipeline: the pump boosts the water pressure so that it can send the water over a long distance rapidly.

A line used solely for transmitting power from a generating plant to a distant center of load, or from one load center to another, is a transmission line. The group of interconnected transmission lines that carries the energy at a high voltage is termed the *transmission system*. Transmission voltages usually range from 69,000 to 765,000 V. Currently, field tests are planned using voltages of 1 million V and above. These voltages will be used to handle the increased amounts of power that the growth of the industry will soon require. They will also permit the economical transportation of power over greater distances. Edison Power, as a moderate-sized company, has found that 345,000 V is adequate for its major transmission lines.

Transmission lines emanating from power plants usually form a network looping through various communities and other power plants.

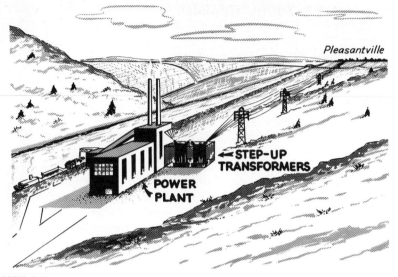

**CHART 3.7**

Chart 3.6 shows this arrangement of Edison Power's transmission system. No matter where a break in the line occurs, electricity can still be sent to any of the communities by an alternate route while the broken line is being repaired.

Chart 3.8a shows the main high-voltage transmission lines in the United States. In the Rocky Mountain region, where the population is sparse and the cities are far apart, fewer transmission lines are needed.

Transmission lines are now generally aluminum wires, called *conductors*, carried on steel towers or wood pole structures. Wood pole structures may be either a single pole or what is called an *H-frame* structure composed of two poles connected by a crossarm near the top. Some typical transmission towers are shown in Chart 3.8b.

As motors or appliances are not designed to use electricity at the high transmission voltages, it is necessary to reduce the voltage at load centers. These voltage reductions are usually made in each community served or at some other point along the transmission line to serve a very large power customer. The transformer substations used to lower the voltage from the transmission voltage are called *step-down substations*. (As a rule the term "substation" when used alone signifies a step-down substation. When a *step-up substation* is meant, it is so designated.) Chart 3.6 shows that Edison Power has a substation just outside Pleasantville, to provide power at usable voltage for the Ajax Manufacturing Company.

The *main transmission system* of bulk power supply is the system of transmission lines interconnecting a company's major power plants and the lines interconnecting with other power suppliers. Thus, the trans-

# BACKBONE TRANSMISSION MAP·1969

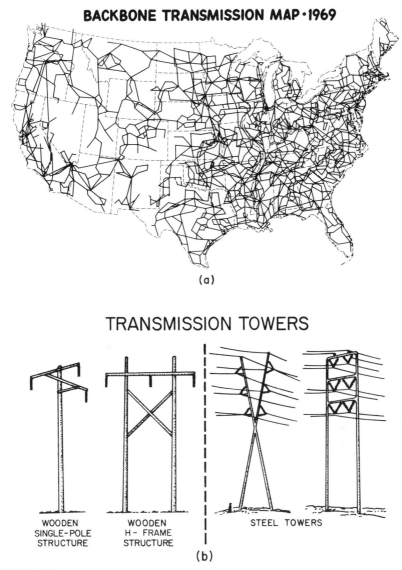

(a)

## TRANSMISSION TOWERS

WOODEN
SINGLE-POLE
STRUCTURE

WOODEN
H – FRAME
STRUCTURE

STEEL TOWERS

(b)

**CHART 3.8**

mission system for bulk power supply is distinguished from the lower-voltage transmission lines serving certain communities and industries and from the low-voltage distribution systems within the communities.

**Distribution System**   Chart 3.9 shows how a step-down substation near Rosedale lowers the voltage from 230,000 to 13,200 V. The lines carrying electricity at 13,200 V are called *primary distribution lines*, and they extend throughout the area in which electricity is to be distributed. This voltage is still too high for use in the home or factory, so it is reduced again by a *line transformer* or *distribution transformer*. The distribution transformer changes the voltage from the *primary distribution voltage* (this voltage may vary from company to company; Edison Power uses 13,200 V) to the *secondary distribution voltage,* which for use in the home is usually 120 to 240 V. For the use of motors in a factory, the voltage may be 240, 480, or 2400 V.

The lines that run to the customer's premises are called *services* or *service wires*. The combination of all the primary distribution lines, distribution transformers, secondary distribution lines, and services is called the *distribution system*.

## HOW THE INDUSTRY GREW

The first electric generating plants and distribution systems were located in the principal cities. In those days, there was no economical way to transmit electric power over any great distance. Later, as ways to transmit electricity over many miles were perfected, the electric companies were able to bring electric service to the smaller communities. The next step was to connect the power plants of one city with the power plants of another city, which permitted many more economies. As transmission lines were extended, larger electric generating plants could be built, and they could be located nearer to supplies of fuel and water. In the course of time, the electric companies built a rather complete network of transmission lines covering most of the country. Electric energy companies

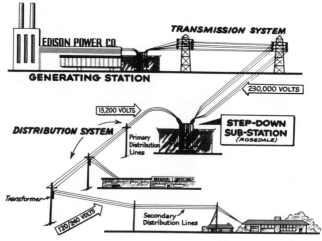

**CHART 3.9**

not only interconnected their plants within their own systems, but they also interconnected their systems with those of other companies.

After the transmission lines had been extended to the small towns and hamlets, service was further extended into rural and farm areas. Because of the distance between farms as compared with the distance between residences in the city, farm electrification developed more slowly than urban electrification.

In the late thirties, great strides were made in this final phase of building the nation's network of power lines. Today, practically every village, hamlet, city, and town in the country has electric service, and the rural electrification program was virtually complete around 1950 so far as line extension to farm customers is concerned.

## HOW THE COMPANY IS FINANCED

Edison Power Company has $1883 million invested in plants, facilities, and other assets, as shown in Table 3.1 (see Chart 3.10). The company obtained its funds from the sources shown in Table 3.2.

Money for this kind of investment is raised (1) by selling securities to the public—individuals, banks, insurance companies, pension trusts, and the like; (2) from earnings which a company is able to retain from its operations; and (3) from other internal cash sources, such as depreciation charges.

A company usually does not pay out all of its available earnings in dividends to its stockholders. It retains a portion of its earnings to provide against the possibility that it may not earn as much in some years as in others. Although this money belongs to the stockholder, he or she does not receive it, but allows the company to invest it in new facilities. The book value of the common stock is thereby increased for the stockholder, and the stockholders as a rule eventually receive an increase in their dividends as the company grows.

**TABLE 3.1**  INVESTMENT, EDISON
POWER COMPANY

| Type of Asset | Cost |
| --- | --- |
| Generating plants | $   672,990,725 |
| Transmission facilities | 270,013,701 |
| Distribution facilities | 481,545,581 |
| Miscellaneous property | 325,356,660 |
| Total plant and equipment | $1,749,906,667 |
| Other assets | 132,749,333 |
| Total investment | $1,882,656,000 |

# INVESTED CAPITAL IN PLANT AND EQUIPMENT
## *EDISON POWER COMPANY, 1976*

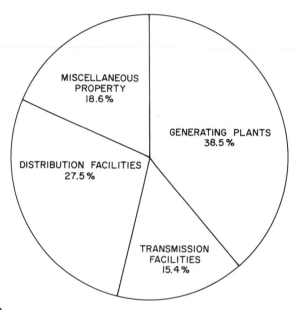

**CHART 3.10**

*Depreciation* is a charge made against a company's operations over the lifetime of a facility to provide for the anticipated replacement of the facility when it wears out. As there is no immediate need for the money represented by these charges, the cash is put to use by the company for building other facilities until it is needed for making replacements. Depreciation is further discussed in a later section of this chapter.

**Balance Sheet**  The *balance sheet* of a company shows what a company owns, what it owes, and what the stockholders' interests in the company

TABLE 3.2  SOURCE OF FUNDS, EDISON
POWER COMPANY

| Source | Receipts | Percent |
|--------|----------|---------|
| Sale of first mortgage bonds | $   767,141,225 | 40.7 |
| Sale of preferred stock | 180,417,535 | 9.6 |
| Sale of common stock | 350,810,230 | 18.6 |
| Retained earnings | 168,440,236 | 8.9 |
| Depreciation charges, etc. | 415,846,774 | 22.2 |
| Total | $1,882,656,000 | 100.0 |

**TABLE 3.3**  CONDENSED BALANCE SHEET, EDISON
POWER COMPANY
*December 31, 1976*

| Assets | | Liabilities | |
|---|---|---|---|
| Electric utility plant | $1,749,906,667 | Common stock | $   288,810,230 |
|  |  | Preferred stock | 180,417,535 |
| Reserve for |  | Capital surplus | 62,000,000 |
| depreciation | 340,306,785 | Earned surplus | 168,440,236 |
| Net electric |  | Total capital stock | $   699,668,001 |
| utility plant | $1,409,599,882 | and surplus |  |
| Net other utility | 100,022,703 | Long-term debt | 767,141,225 |
| plant |  | Total capitalization | $1,466,809,226 |
| Investments and |  |  |  |
| other assets | 36,194,085 | Current liabilities | 129,079,212 |
| Current assets | 114,611,373 | Other reserves | 64,539,605 |
| Total assets | $1,660,428,043 | Total liabilities | $1,660,428,043 |

are. A condensed balance sheet of Edison Power Company as of December 31, 1976, is shown in Table 3.3. A balance sheet is commonly set up in two columns. The left-hand column shows what the company owns. This is called *assets* and consists, in the main, of *fixed assets*—plant and equipment—and *current assets*—cash, accounts receivable, materials and supplies, etc., which usually can be turned into cash readily or are used up in the operation of the business. The right-hand column of a balance sheet shows what a company owes, called *liabilities*, and the stockholders' interest in the company, which is often referred to as the *stockholders' equity*. Liabilities consist of the *long-term liabilities*, such as mortgage bonds and other long-term debt, *current liabilities*, and *reserves*. Obligations that do not become due for over 1 year are usually considered long-term debt. Current liabilities include accounts payable, notes that are due within a year, payroll, and accrued taxes. The difference between current assets and current liabilities is sometimes referred to as *net current assets*. The relationship between current assets and current liabilities is called the *current ratio*.

*Reserve accounts* record amounts charged against earnings for future payment or contingencies, such as depreciation reserves. These accounts are, for the most part, "bookkeeping" accounts and do not necessarily represent cash reserves.

The stockholders' equity includes preferred and common stock which the company has outstanding and accounts called *capital surplus* and *earned surplus*. Preferred stock usually has a *par value* and is stated on the balance sheet at this value. This is the amount which the company is obligated to pay the preferred stockholders in the case of a voluntary dissolution.

Common stock may have a par value or it may be *no-par stock*. In the case of stock having a par value, this value is reported for the common stock outstanding, and any amounts received by the company from the sale of stock in excess of par value are reported as capital surplus. Capital surplus also includes amounts such as may have been received from the sale of assets in excess of book value and other amounts which are usually the result of corporate transactions.

*Earned surplus* represents the accumulated earnings of the company, after adjustments for such things as over- or underaccruals, less dividends paid to the stockholders.

The *stated value* of the common stock plus the capital surplus and the earned surplus represent the *common stock equity*. The sum of these is the *book value* of the common stock. Sometimes amounts in the capital and earned surplus accounts are transferred. Amounts in earned surplus can be transferred to capital surplus, or amounts in either earned surplus or capital surplus can be transferred to the stated value of common stock, as in the case when stock dividends are issued by the company. Such transfers do not change the book value of the stock, however.

As the assets of a company always equal the liabilities plus the stockholders' equity, the statement of these items is called a balance sheet.

**Stocks and Bonds**  The mortgage bondholder has first claim on the assets of the company and thus has the highest degree of security. Next in line, in order of security, after other creditors, is the *preferred stockholder*. While a bond represents, in effect, a loan to the company, which it must eventually repay, *preferred and common stock* represent shares of ownership of the company. Edison Power's preferred stock has first claim on the earnings of the company after bond and other interest is paid, so its dividend rate is comparatively secure. It also has first claim—after the bonds and other creditors, of course—on the assets of the company, so the preferred shareholder thus assumes less of a risk than the common shareholder. Table 3.4 shows how the preferred stock of Edison Power Company is distributed.

The *common stockholders* are the real owners of the company. They assume the greatest risk of loss, and if the company fails to earn enough money to pay bond interest, preferred dividends, and a return for common stock, the common stockholders suffer. On the other hand, if the company does well, the common stockholders stand to benefit and have a greater chance of benefit than holders of bonds or preferred stock. In actual practice, government regulation prevents spectacular common stock benefits in the electric energy business (see Chapter 9).

Table 3.5 shows how the common stock of Edison Power Company is held. The percentage distribution is illustrated in Chart 3.11.

**TABLE 3.4**  DISTRIBUTION OF PREFERRED STOCK, EDISON POWER COMPANY

| Owners | Number of Shares | Percent |
|---|---|---|
| Women | 2,046,000 | 34.1 |
| Men | 1,608,000 | 26.8 |
| Joint owners | 540,000 | 9.0 |
| Fraternal, charitable, religious, and educational funds | 30,000 | 0.5 |
| Trust and custodian accounts | 402,000 | .6.7 |
| Insurance companies | 978,000 | 16.3 |
| Dealers, corporations, and others | 396,000 | 6.6 |
| Total | 6,000,000 | 100.0 |

The *capitalization* of the Edison Power Company consists of its long-term debt, preferred and common stock outstanding in the hands of the public, and capital and earned surplus, as shown in Table 3.6. The manner in which a company is capitalized has an important bearing on how its securities are rated in the money market and, therefore, on the cost to the company of raising additional funds.

Securities offered on the public market are rated by professional rating agencies such as Moody's Investors Service or Standard and Poor's. The purpose of the rating is to provide the investor with a simple guide by which the relative investment qualities of bonds may be noted. These ratings are assigned to a company's securities after exhaustive studies of the company's operations, financial ratios, management, territory served, market characteristics, past performance, regulatory climate, etc. Ratings range from Aaa (Moody's) and AAA (Poor's), which are judged to be of the highest quality carrying the smallest degree of

**TABLE 3.5**  DISTRIBUTION OF COMMON STOCK, EDISON POWER COMPANY

| Owners | Number of Shares | Percent |
|---|---|---|
| Women | 5,904,072 | 30.7 |
| Men | 6,057,922 | 31.5 |
| Joint owners | 4,654,023 | 24.2 |
| Fraternal, charitable, religious, and educational funds | 76,925 | 0.4 |
| Trust and custodian accounts | 1,903,918 | 9.9 |
| Insurance | 269,241 | 1.4 |
| Dealers, corporations, and others | 365,398 | 1.9 |
| Total | 19,231,499 | 100.0 |

# DISTRIBUTION OF COMMON STOCK
## *EDISON POWER COMPANY, 1976*

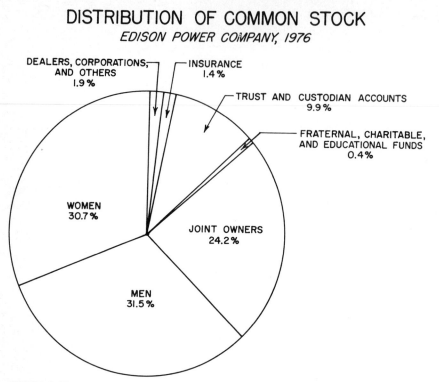

**CHART 3.11**

investment risk, to C, which is affixed to bonds having extremely poor prospects of ever attaining any real investment standing.

Most electric-utility-company first-mortgage bonds are rated A or better by the rating services. A few are triple-A and many are double-A. An A-rated bond is considered to be a high- to medium-grade obligation giving adequate security to principal and interest; a double-A rating indicates a higher quality but not quite as "gilt edge" as triple-A.

A rating generally is reflected in the relative interest cost of money;

**TABLE 3.6**   CAPITALIZATION, EDISON POWER COMPANY

| Source | Amount | Percent |
|---|---|---|
| First mortgage bonds | $   767,141,225 | 52.3 |
| Preferred stock | 180,417,535 | 12.3 |
| Common stock | 288,810,230 | 19.7 |
| Capital surplus | 62,000,000 | 4.2 |
| Earned surplus (retained earnings) | 168,440,236 | 11.5 |
| Total | $1,466,809,226 | 100.0 |

that is, in a given market, a triple-A-rated electric utility bond will generally sell at a lower yield than a double-A, and a double-A at a lower yield than an A-rated bond. Particular market conditions and terms of the contract influence the spread in yields among bonds of the different ratings.

Edison Power Company tries to see that bonds make up not much more than 60 percent of its total capital structure and that its common stock equity is between 25 and 40 percent. This helps the company to keep a good credit rating and maintains its ability to raise new funds when needed. The ratio of its equity to the total capitalization also has a bearing on how people evaluate the company's common stock.

The principal bondholders of Edison Power Company are institutional investors. They represent a group of investors or savers whose savings they invest in securities of various types of business. These institutional investors include insurance companies, investment trusts, pension funds, and banks. For example, on January 1, 1977, insurance companies held nearly half of the long-term bonds of the Edison company. Insurance companies invest part of the premiums paid by their policyholders in securities. The money earned on these securities is used to reduce the amount that policyholders have to pay in premium for their insurance. In this respect, the people who have insurance policies with the investing companies are indirect investors in the Edison Power Company.

## PRINCIPAL SOURCES OF INCOME

Edison Power Company, like most other electric utility companies, receives most of its revenue from sales at retail to residential, commercial, and industrial customers. Chart 3.12 shows the number of customers in each of these classifications and the percentage of the total each class represents. The company has 648,402 customers. Chart 3.13 shows the company's annual revenue by classes of service in percent of total revenue. The company had a total revenue from the sale of energy of about $502 million in 1976.

## EXPENSES OF OPERATION

So far, mention has been made of only one item of cost—the interest, or return, that a company must pay investors for the use of their money. There are, of course, other expenses. Those following are referred to as *operating expenses*. Chart 3.14 shows where the company's money goes.

**Production Expense**  The principal items of expense in running a power plant are fuel, labor, material, and supplies. The total of all of these is called the *production expense*. For Edison Power, the production

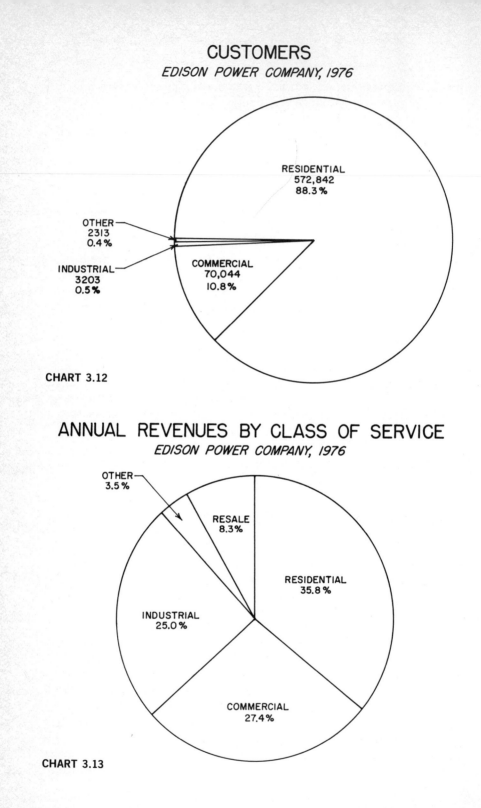

# CUSTOMERS
## *EDISON POWER COMPANY, 1976*

RESIDENTIAL
572,842
88.3%

OTHER
2313
0.4%

INDUSTRIAL
3203
0.5%

COMMERCIAL
70,044
10.8%

CHART 3.12

# ANNUAL REVENUES BY CLASS OF SERVICE
## *EDISON POWER COMPANY, 1976*

OTHER
3.5%

RESALE
8.3%

RESIDENTIAL
35.8%

INDUSTRIAL
25.0%

COMMERCIAL
27.4%

CHART 3.13

# WHERE THE MONEY GOES*
## *EDISON POWER COMPANY, 1976*

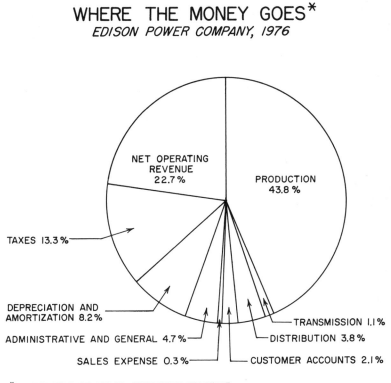

NET OPERATING
REVENUE
22.7%

PRODUCTION
43.8%

TAXES 13.3%

DEPRECIATION AND
AMORTIZATION 8.2%

ADMINISTRATIVE AND GENERAL 4.7%

SALES EXPENSE 0.3%

TRANSMISSION 1.1%

DISTRIBUTION 3.8%

CUSTOMER ACCOUNTS 2.1%

*AS PERCENT OF GROSS OPERATING REVENUE

**CHART 3.14**

expense in 1976 was $223,393,016, or 1.31 cents (13.1 mills) per kilowatthour sold.

Note that some expense items are given in cents per kilowatthour, while others are given on other bases, such as dollars per customer. These figures are given in this way here to give an idea of the size of the item. This method of expressing the data has a useful purpose, which is described in Chapter 8.

Edison Power Company uses coal, oil, and gas as fuel in its fossil-steam plants. On the average, Edison Power's plants require the equivalent of 0.95 lb of coal to make each kilowatthour. The company spent $169 million for fossil fuel in 1976, or 33 percent of the company's electric operating revenue. This cost amounted to about 10 mills per kilowatthour generated. With the coming of nuclear energy, it is contemplated that the fuel element in the overall cost of making electricity will be smaller than it is now. The company has a clause in its rate schedules providing for a corresponding increase or decrease in the rate for any increase or decrease in the cost of fuel. (Costs, pricing, and the changes in fuel prices in recent years are discussed in Chapters 4 and 8.)

**Transmission Expense**   This category includes wages of people who work on the high-voltage lines, keeping them in good order, and the expense of inspecting the transmission grid at regular intervals.

**Distribution Expense**   This expense item covers the cost of labor to operate and maintain the distribution lines, substations, and other facilities. Money paid for materials used in maintenance and the expense of patrolling, inspection, and testing is also included.

**Customer Accounts Expense**   This category contains the cost of reading meters, making out and mailing customers' bills, keeping the accounts, and collecting bills. For Edison Power Company, this expense amounted to $16.31 per customer in 1976.

**Marketing Expense**   Here are included such things as informing customers about the efficiency and safety of appliances, illustrating how to conserve energy and avoid waste, assuring customers of proper service on appliances, advising how to build load factor so as to hold down the price of service, and advising customers about any new appliances which may be available so that they can decide whether to use them (see Chapter 7).

**Administrative and General Expense**   This account covers the salaries and expenses of general officers, executives, and general office employees. It also includes insurance, employee benefits, and items of expense not specifically provided for elsewhere. This expense amounts to 5 cents per dollar of gross revenue.

## TAXES AND DEPRECIATION

**Taxes**   In addition to the operating expenses, the company must also take into account the effect of taxes and depreciation before it can arrive at the *balance for return*, the amount available to pay the investors for the use of their money.

The electric utility company is required to pay many kinds of taxes. Among these are local property taxes, franchise taxes, excise taxes, and state and federal income taxes. Edison Power Company had a total of $67,919,799 in taxes in 1976. Of this total $18,024,018 was for federal income taxes, which amounted to 3½ cents per dollar of gross revenue, or 1 percent of the total value of the company's plant investments. The total tax bill took 13.3 cents out of every dollar the company received from its customers. It amounted to almost 4 percent of the total cost of the company's plant investments.

A company has no income except the money it receives from customers for the commodity or product sold by the company. Taxes, like labor and fuel, are an expense and, like other costs, are included in the price of the commodity. Thus, in effect, it is the customer who pays the company's taxes.

**Depreciation**   A utility company must keep its property in good working condition to maintain continuous satisfactory service to the public. This means that equipment must be replaced when it wears out, becomes obsolete or inefficient, or is taken out of service for any reason. When property is taken out of service, it is retired from the company's property account. When this retirement takes place, the company must be in a position to replace the retired property with new equipment.

It is not possible to predict exactly when it will be wise or necessary to retire any particular piece of property. Each year, the company reserves from its income an amount to account for the estimated cost of property wearing out. The accumulation of these charges less the amount deducted for property retired is called the *reserve for depreciation,* or *depreciation reserve*.

Edison Power Company set aside $41,957,327 in 1976 as an accrual to the depreciation reserve for its depreciable property. This annual charge is equal to 2.4 percent of the company's total investment in physical property.

## NET OPERATING REVENUE

The *net operating revenue* or *operating income* is the amount left over from gross operating revenue after payment of all operating expenses, depreciation, and taxes. It is the amount of money left from operations to compensate the investors for the use of the funds they have entrusted to the company. Edison Power Company had some other incidental income which, when added to net operating revenue, results in *gross income*.

The company sends an annual report to its security holders and others who may be interested in the company's operations. This report details the amount of money the company received during the year from its customers and gives the amounts paid out in expenses. An extract from Edison Power Company's annual report for 1976 giving certain revenues and expenses is shown in Table 3.7, and the net operating revenue of Edison Power Company as distributed in 1976 is shown in Table 3.8.

The money that is left over after payment of bond interest and preferred dividends is the share accruing to the common stockholders. However, good management does not pay all of this in dividends; the

**TABLE 3.7** REVENUE AND EXPENSES,
EDISON POWER COMPANY
*December 31, 1976*

| Item | Amount | Percent of Electric Gross Operating Revenue |
|---|---|---|
| Electric gross operating revenue | $510,038,823 | 100.0 |
| Operating expenses: | | |
|     Production | 223,393,016 | 43.8 |
|     Transmission | 5,622,029 | 1.1 |
|     Distribution | 19,160,212 | 3.8 |
|     Customer Accounts | 10,576,812 | 2.1 |
|     Sales | 1,682,310 | 0.3 |
|     Administrative and general | 24,148,293 | 4.7 |
|     Total operating expenses* | $284,582,672 | 55.8 |
| Depreciation | 41,957,327 | 8.2 |
| Taxes | 67,919,799 | 13.3 |
| Total operating expenses, depreciation, and taxes | $394,459,798 | 77.3 |
| Electric net operating revenue | 115,579,025 | 22.7 |
| Other net operating revenue | 10,204,688 | |
| Gross income | $125,783,713 | |

* The relationship between electric operating expenses and electric gross revenue is sometimes referred to as the operating ratio. For Edison Power Company, the operating ratio is $284,582,672 divided by $510,038,823, or 55.8 percent.

company usually pays around 65 to 75 percent of it in dividends to the common stockholders. The balance is retained as a reserve for unforeseen contingencies and for the expansion and improvement of the business. This amount is reflected in earned surplus unless capitalized by issuance of stock dividends, transfers to capital surplus, or restatement of common stock stated value. It has the effect of increasing the stockholders' equity in the business. Percent return is discussed in Chapters 7 and 9.

**TABLE 3.8** DISTRIBUTION OF NET OPERATING
REVENUE, EDISON POWER COMPANY

| Distribution | Amount | Percent |
|---|---|---|
| Bond interest | $ 52,976,532 | 42.1 |
| Preferred stock dividends | 12,574,127 | 10.0 |
| Common stock dividends | 42,163,138 | 33.5 |
| Retained in the business | 18,069,916 | 14.4 |
| Total | $125,783,713 | 100.0 |

# WHAT HAPPENS TO WHAT IS LEFT
## *EDISON POWER COMPANY, 1976*

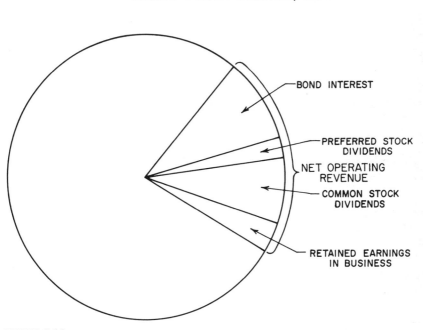

BOND INTEREST

PREFERRED STOCK DIVIDENDS

NET OPERATING REVENUE

COMMON STOCK DIVIDENDS

RETAINED EARNINGS IN BUSINESS

**CHART 3.15**

Charts 3.14 and 3.15 show the apportioning of expenses and the distribution of return to the investors in Edison Power Company. Chart 3.16 shows the company's annual sales in kilowatthours by classes of service and the relative size of each class. Total sales in 1976 amounted to about 17 billion kWh.

## HOW THE COMPANY IS OPERATED

**Loads**  Understanding the loading on the power system helps in gaining an understanding of the economics of the energy business. Starting at midnight the demand is light (Charts 8.17 and 8.18). Refrigerators are off and on all night, and certain types of water heaters are heating water. Industries are shut down. Beginning about 5 A.M., however, the load increases. Lights begin to go on, and electric ranges are being used to cook breakfast. Industries begin operating.

Later there is laundry to do and dishwashing that takes hot water. Perhaps some electricity is used in operating the television for daytime soap operas. At noon the residential demand is about 500,000 kW. Some companies that have relatively large industrial loads have a system peak

# ANNUAL KILOWATTHOUR SALES
## *EDISON POWER COMPANY, 1976*

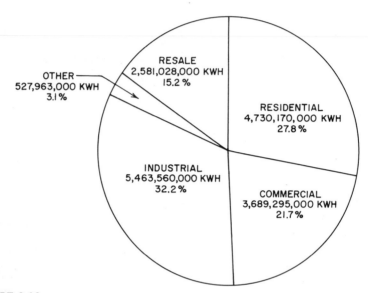

**CHART 3.16**

in the morning higher than the one in the evening. For Edison Power Company, the winter residential load rises to its highest peak of the day around 6 P.M. Most lights are on then, and evening meals are being prepared in the area. On a summer day, the system peak may occur at about 2 P.M., when air conditioning demand is at its greatest. (More load curves are discussed in Chapters 7 and 8.)

The company has a load dispatcher who is constantly watching the company's control dials that show the variation in load on the company's system. Having spent years on this job, the dispatcher can predict fairly accurately what the load will be each hour and can keep in constant touch with the plant operators, telling them which generators to start in operation. The trend is toward greater use of electronic computers in this operation.

Comparison of the load curves from day to day shows that, although the general pattern is the same, the actual load varies from day to day and from month to month. However, it is the peak, or highest, load which is most important, since this determines the amount of generating capacity the company must provide. It now takes 10 or more years to plan and build nuclear generating units. For this reason, the company

2:45 AM - The Libra
locked and lights

- 2:50 AM - Switch

- 2:55 AM - Switch

3 AM - The Library

# ANNUAL PEAK LOAD
## *EDISON POWER COMPANY, 1976*

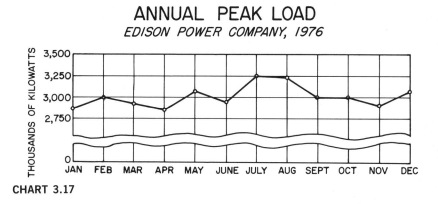

**CHART 3.17**

must plan ahead and predict the peak load for a number of years in the future. Chart 3.17 shows the highest peak load for each of the 12 months. This is called the annual load curve. The chart shows that Edison Power Company has its highest peak load in the summer months.

In order to demonstrate the peaks and valleys in the company's load curve, Chart 3.18 shows a three-dimensional view of the way Edison's customers use electric service during the course of the year. The highest peak is 3,500,000 kW. If the company's customers had used electricity at this rate during every hour of the year, the company would have generated 30 billion kWh (3,500,000 kW × 8760, the number of hours in a

# ANNUAL LOAD
## *Edison Power Company*

**CHART 3.18**

year). Allowing for system losses, the company could have sold 30 billion kWh during that time. Its load factor was 61 percent.

## USE OF ELECTRICITY INCREASING

The uses of electricity are constantly increasing. While there is less of an increase in kilowatthour use during times of depression or recession, on the whole the trend is upward and will probably be so for many years to come. These trends are discussed in Chapter 7. The load served by Edison Power Company is increasing as shown in Chart 3.19.

The chart shows Edison Power's peak load by year since 1930. Note the leveling during the Depression of the early thirties and also the increase following World War II. New generating capacity has been built to keep ahead of the load.

## UNITS IN THE PLANT

The largest unit in the company's Pleasantville power plant is a 1 million kW generator. The company can use units of this size because it is a part of a large power pool with neighboring communities (see the latter part of this chapter). Larger power companies can probably make economic use of units even larger than this.

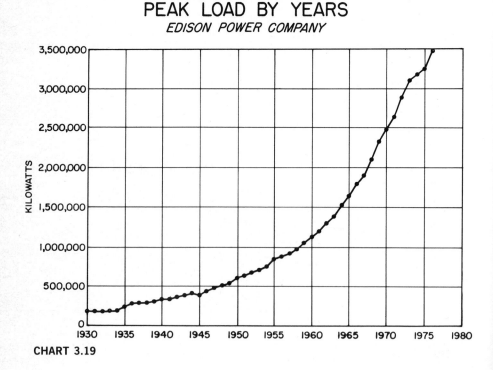

CHART 3.19

In engineering practice, several terms are used to designate the rating of a generating unit. *Capacity* or *nameplate rating* refers to the manufacturer's rating of the unit. The engineers who design the unit draw up their specifications so that the finished generator will produce electricity at a certain rate under specified conditions. The manufacturer might say of the finished product that it will generate electricity at the rate of 500,000 kW, for example. This rating will be imprinted on the nameplate attached to the generator, along with the design conditions. This rating is given in gross kilowatts, that is, without any deduction for electricity used in the generating process (such as for operation of fuel pumps and other auxiliary purposes).

**Capability**    After the company has had the 1 million-kW unit in operation for a period of time, it may find that because of favorable operating conditions the unit is actually able to deliver electricity at the rate of, say, 1,100,000 kW. This is called *capability*. Capability is given in net kilowatts, after deduction for electricity used in the generating process. After the unit is installed, it is realistic to deal in terms of capability instead of capacity. In nontechnical language, however, the word *capacity* is a generic term which includes capability. It is frequently so used in this book.

The company's capability has gone up in stair-step fashion, as shown in Chart 3.20. When the company builds a plant, it leaves ample

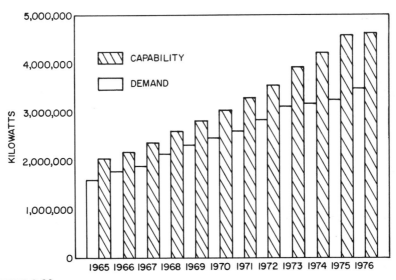

# GROWTH OF CAPABILITY AND DEMAND
## *EDISON POWER COMPANY*

**CHART 3.20**

space for additional units to be installed in future years. This tends to lower the cost per kilowatt by making more efficient use of the site, buildings, and transmission lines, and also means that the company does not need to build a new plant every time the demand rises. When the demand requires a new unit to be built, the company usually puts in the largest practical unit.

It is likely that after a new unit has been added, the company will have, for a while, more capacity than it actually needs. This excess capacity may be sold to neighboring companies until it is needed by Edison Power Company. With the development of more coordination and pooling, there frequently is joint planning with more than one company for new generating capacity. Thus maximum advantage can be taken of the economies to be derived from larger generating units.

## OTHER CAPACITY

Just as the generating capacity must be increased as the customers' demands for service increase, so also must the capacity of other equipment be increased. A step-up transformer is added at the power plant whenever a new generating unit is installed. The transmission lines, the power substations, and the distributions systems all must be increased in capacity as the rate of use of electricity increases. Always, all along the line, the customers' requirements must be anticipated and provided for in advance.

## PERSONNEL ORGANIZATION AND PAYROLL

Edison Power Company employs 4117 people. As the company has invested $1,749,906,667 in plant and equipment, this is an investment of $425,000 per employee. Because of the extensive use of machinery and the technical nature of the business, utility employees are usually more highly skilled and better paid than ordinary employees. Of the total payroll of $48 million, the salaries of all the top management amount to about $1,390,000, or 2.9 percent of the total payroll.

The stockholders, who are the owners of the company, elect the board of directors. The board selects the president, who is the company's chief executive officer and is held responsible by the board for all the company's operations. In many companies the chairman of the board of directors acts as the chief executive officer. Reporting to the president may be one or two executive vice presidents and a number of department heads, such as treasurer, chief engineer, marketing manager, director of public relations, director of employee relations, rate engineer, general counsel, and the like. Frequently many of these department heads carry the title of vice president.

Power companies vary widely as to the number of divisions they may have. The territory of the Edison Power Company is divided into four

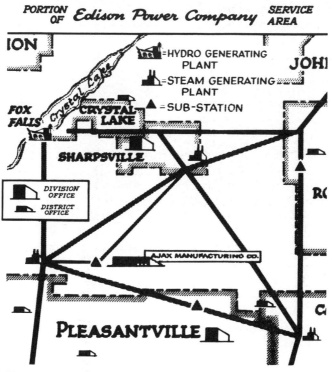

PORTION OF *Edison Power Company* SERVICE AREA

ION

=HYDRO GENERATING PLANT

=STEAM GENERATING PLANT

=SUB-STATION

JOHI

FOX FALLS

Crystal Lake

CRYSTAL LAKE

SHARPSVILLE

DIVISION OFFICE

DISTRICT OFFICE

R(

AJAX MANUFACTURING CO.

PLEASANTVILLE

C

**CHART 3.21**

divisions (see Chart 3.21). For each division there is a division manager, who is in charge of all activities except the operation of the power plants and the transmission system. These are operated from the general office. Each division is organized into about the same staff functions as the company as a whole. There is, for example, a division engineer, a division auditor, and so forth, so that each of the functions is carried on in each division. The company's customer service department establishes policy with respect to matters within its scope of operation, and the division marketing manager carries out this policy at the division level. The same is true for the other staff functions. The division manager generally reports directly to the executive vice president.

## THE EVOLUTION OF POWER SUPPLY

Looking back over the evolution of power supply as it has taken place in this country over the past 60 to 70 years, it is possible to see a number of trends. Knowledge of these trends can be combined with a forecast of what may be coming to give a preview of what may evolve between now and the year 2000. This evolutionary development may be reviewed by periods, allowing students to see something of the problems faced by electric energy management and to see how these problems have been met through innovation and change. The attempt of the industry always has been to get electricity to people where they want to use it and to render the service reliably so that customers can count on its being as continuous as possible. At the same time, the companies have striven for higher and higher efficiencies in operation through all the avenues available to them in order to allow prices to be as low as practical consistent with the cost of doing business.

Electric energy managers have been conscious of the fact that the versatile energy form they supply has an important impact on living standards. It drives the machines people use in their homes and on their jobs. It drives the machines in factories so that men and women can multiply their productivity and produce more goods for themselves and for others.

## INTERCONNECTION

In the early days of the industry, power plants were small and they were isolated. They were built where people were. These small isolated plants were much less efficient and considerably less reliable than the large interconnected systems of today.

As early as 1911, Samuel Insull demonstrated that better and more economical electric service could be provided if a number of small communities having isolated plants were interconnected with transmission lines and the electric energy brought from one larger, more efficient power plant with satisfactory reserves. His proof was the system he built in Lake County, Illinois. It was, in fact, a research project—and it worked. The country began to witness a growing practice of building transmission lines, of interconnecting power suppliers, of shutting down less efficient power plants, and of replacing them with larger and more efficient ones. All during this period of growth more and more people were able to receive the benefits of electricity. Small towns whose needs could not justify an electric generating plant were served from a transmission line passing nearby, and the process of rural electrification began.

The trend toward interconnection had advanced considerably by the start of World War II. All but the most thinly populated areas of the country were interconnected by high-voltage transmission lines. As interconnections developed and the systems grew in size, the industry was able to make use of larger and more efficient generating units. Compared to the power business, few industries have been so successful in taking advantage of the economies of scale as advancing technology brought them into being.

In all this upward sweep of unit sizes toward higher voltages and increases of interconnections, new problems had to be faced at each step along the way. Almost every new generating plant was designed to be a little larger and, if practical, to operate at a little higher temperature and at a little higher steam pressure. This presented metallurgical problems in the construction of tubing and turbine blades. Each step upward in voltage meant more research, more development work, and more experimentation to identify and solve new problems in switches, breakers, relays, and transformers—all of which had to be redesigned, tested, and proven in practice. All this research benefited not only the American industry but also the developing nations, who were freed from the need to research the size of units and the voltages they desired.

Chart 3.22 shows Edison Power Company's generating plants and transmission lines with former locations of smaller plants. As the company was organized at a satisfactory size so as to preserve the benefits of

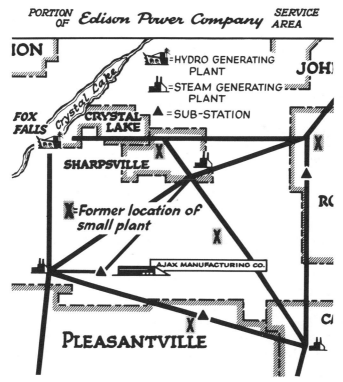

CHART 3.22

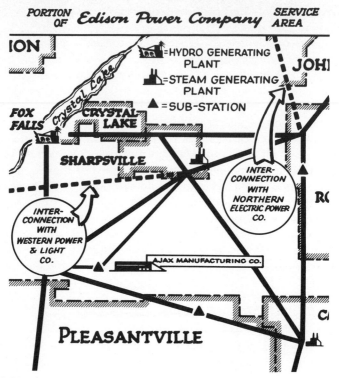

**CHART 3.23**

good local management as well as to take advantage of the benefits of scale, it was able to interconnect with neighboring power systems and bring about further benefits both in reliability and in economy.

Interconnection also helps to improve the load factors of interconnected systems because of diversity between systems. Chart 3.23 shows how Edison Power Company is connected with two other companies to the north and to the west.

Edison Power Company profits through these interconnections. The company has a peak demand of 3,474,129 kW and a total capability of 4,429,514 kW. The value of interconnections showed up strikingly in 1964. Ordinarily, the company would have to have reserve capability of about 20 percent in order to be sure it would carry the load during any kind of emergency. In that year, the company's reserve was on the order of 18 percent. Two of the company's plants, accounting for about 19 percent of total capability, were put out of commission by a spring flood. As interconnections enabled the company to call upon the facilities of its neighbors for emergency service, Edison Power's 18 percent reserve was enough to see it through, and the company was able to furnish its customers a continuous supply of electricity.

The power company maintains reserve capability so that the electricity supply will not be interrupted in case of a breakdown in one or more of the generating units. Also, the extra capability allows room for growth and possible load-forecasting errors (the load that actually develops might be more than was predicted several years before when the system additions were planned). In some instances, additional capability is required because it is necessary to take equipment out of service even on peak in view of the fact that the "valleys" between seasonal peaks are not dependable enough and long enough to permit the necessary maintenance of equipment. The company aims to keep the reserve capability as low as is practical and still maintain dependable service. The reserve capability costs money because it produces no kilowatthours and, therefore, no revenue. The company aims to keep reserve capability somewhere around 20 percent of its estimated peak load demand.

Assume that a system has a peak load of 25 million kW and that it has 60 generators, each of 500,000 kW, to carry this load.[1] The "unused" capability, or reserve of 5 million kW allows for the contingencies, possible load-forecasting error, and maintenance needs.

At any given time, the system must have generating facilities in operation with sufficient capability to pick up any reasonable amount of load instantaneously placed on the system or to replace generating capability that is lost through malfunction or failure. This capability is called *spinning*, or more properly, *operating reserve*. It must be available almost instantaneously or within a very short period of time.

To ensure dependable service, the system must be able to carry its peak load when the largest unit is out. Because of this, a company of average size cannot take advantage of the larger units with their greater efficiency unless it maintains an overly large reserve. The extra cost of carrying a big reserve often offsets the economies of the big unit. Interchange power agreements with neighboring companies can, in effect, take the place of reserve, in that the interconnecting systems can spread the risk of outage over many units. That is, the simultaneous outage of equipment on each of several interconnected systems will be less probable than an outage on a single system. This is but one of several benefits that can be derived from interconnected operations.

It is common practice for neighboring companies to contract with

---

[1] The generating units in the plants of any power company are of varying sizes and have been built over a period of years. However, it is easier to show the need for reserve capability and the value of pooling if a simplified example is used and generating units of the same size are assumed.

one another to interchange electric energy. These contracts may take a number of forms.

**Purchased Power**   Frequently one company will purchase power and energy from another company on either a short-term or a long-term basis. The purchase may be for firm power from a system's total resources or from a specific generating unit. In the former case the seller will meet its commitment to the purchaser essentially as if it were part of its system load. That is, the seller will not only supply the power under normal conditions, but it will also have adequate reserve to assure a continuous supply of power to the purchaser even under adverse conditions. This type of arrangement provides a simple approach for a small system to realize the benefits of economies of scale, lower fuel costs, and lower reserves.

Power companies also contract to sell and purchase energy from a specific generating unit. Normally this is done on a short-term basis when electric companies wish to cooperate in building a unit larger than either might wish to build alone, although this need not be the only circumstance under which arrangements are made concerning the power output from a specific unit. In this type of arrangement, company A might build a large unit and company B might agree to purchase a fixed number of kilowatts from it if and when it is available. Company B furnishes the reserve.

**Exchange of Capability**   There are many variations of energy-purchase contracts, depending on specific needs of the companies involved. One involves the exchange, rather than purchase and sale, of capability among neighboring companies. That is, company A will build a large unit and make available a certain portion of the capability to company B for a certain period of time. Then at the expiration of this time company B will make available to company A an equal number of kilowatts. In this way the electric companies will plan their system expansion together and stagger their building programs. The combined loads of the two enable them to build larger units, taking advantage of the efficiencies to be gained from increased size.

**Exchange of Economy Energy**   To keep production costs down, the electric company tries to supply its load from the most efficient generating stations as much of the time as possible. The less efficient equipment is operated when needed or kept as reserve. For example, company A may have a generating station which is not fully loaded. If the production cost of the plant is lower than that of plants of company B, company B may buy energy on a temporary basis from the plant of company A.

Company A makes no commitments as to firm power, but is willing to let company B have energy from this lower-cost plant as long as company A does not need it. Company A makes a small profit on this energy, but the price might still be less than it would cost company B to generate it. This energy is sometimes referred to as *economy energy*. Both companies benefit when between them they can keep the most efficient generating stations running steadily.

**Dump Energy**  In some hydro plants there is an abundance of water because of continuous stream flow. As the production cost of a hydro plant is low, the company naturally uses all of the hydro power that is available before using its steam plants. At times of high water a company may be able to generate more electricity from hydro power than it can sell to its customers. If the water is not run through the generating turbines, it must be wasted over the spillway.

To prevent waste, the company will use the water to generate what is called *dump energy*. The company may sell this dump energy to another company that will buy it whenever it is available. Obviously the price of this dump energy is quite low—the lowest of all types of energy. Almost any price results in a gain to the seller. The price must be lower than the production cost of any of the next available stations of the buyer, or else the buyer will not benefit and will not buy. The rates are worked out to benefit both parties in order not to waste the energy.

**Emergency Power Contracts**  Neighboring companies usually interconnect their transmission systems, even though normally they do not need to buy or sell energy. In an emergency one company can deliver energy to the other over this interconnection, and it will supply whatever energy it can to its neighbor after meeting the needs of its own customers.

**Standby Contracts**  A company may not have enough reserve capability and may want to buy a block of power from a neighbor simply as standby in case of an emergency. One company simply agrees to stand by for a certain number of kilowatts of capability for its neighbor at an agreed price.

## RECENT EVOLUTION

By the beginning of World War II, a number of company groupings had already developed, each operating in *synchronism*. A group is operating in synchronism when all the generating units on these systems are running at a precise speed of some multiple of 60 and they are producing 60-Hz electricity. They may be operating at 1800 or 3600 revolutions per minute, for example. When operating in synchronism, all these units move together like a series of pendulums swinging back and forth, in rhythm and in step, 60 times per second. If one machine gets out of step, it is out of synchronism, and relay equipment removes this unit from the system. Operating an interconnected system is a delicate process. For example, if a generating unit is to be connected to the system, it must be brought up to speed so that the frequency of the electricity it produces is the same as that of the system, and then it must be connected to the system in such a manner that it will be in phase, or synchronism, with the system. Once the machines are operating together, they tend to be held in step by magnetic force.

During World War II electric utility companies were unable to purchase new generators, as practically all the manufacturing capacity was being used to make turbine generators for ships and for other war purposes. Consequently, electric companies had to operate all their plants, both the efficient and inefficient ones. They were able to carry the war load and the civilian load because they had developed interconnections. Without these, the industry could not have met the demand. In fact, at the beginning of the war there were many in the country who predicted that it would be impossible to meet all civilian and war needs for electric energy with the existing capabilities. Practically all commodities were being rationed. It seemed reasonable to suppose that electricity would need to be rationed, too. Charles Kellogg, then president of the Edison Electric Institute, joined the federal government and studied the situation. He estimated that the growth in use of energy during the war would not vary greatly from the trend then being experienced. This especially applied to the kilowatts of demand. He pointed out that load factors would be higher with factories employing a three-shift production schedule, but energy would be used in proportion to the material and manpower available for production. This meant, for example, that machines and men would be used to make tanks and trucks instead of cars. There would be a shift in the use of energy along with increased use rather than a tremendous increase in demand. This proved to be the case.

Certain situations did need special solutions. For example, it was decided to locate a large war industry in Arkansas. No one electric company in the region had sufficient capability available to serve this new industry, nor could new capability be added. As a result, some twelve companies in the Southwest pooled their resources and together had sufficient capability to serve the industry throughout the war period. This was the beginning of what is called the Southwest Power Pool. It has been expanded in scope and size over the years and is still in existence. Other such pools were being formed in other parts of the country.

Throughout the war period the electric energy industry was able to meet war demands and civilian demands as well. J. A. Krug, Director of the Office of War Utilities, complimented the industry in an address he gave in 1943: "Power has never been 'too little or too late,'" he said. "I do not know of a single instance in which the operation of a war plant has been delayed by lack of electric power supply."

**Power Pools**    The term *power pool* generally refers to the pooling of one energy supply system with that of a neighbor or a number of neighbors. Pools may be formed for any number of reasons from a desire for information exchanges or a desire to share ownership of a large power plant to a need for doing all the things listed above as the benefits of interconnection. To obtain the maximum economies, the aggregate load of a pool should be large enough to allow the use of very large generating units. This means that with a unit rating of 1 million kW, the size of the pool should be in the neighborhood of 8 to 10 million kW or more, so that if such a large unit should fail or be shut down for overhaul, the percent of the total capability not in operation will be in line with what is considered normal reserve requirements. That is, if the size of the largest generating unit in the pool were very large in relation to the total load of the pool, an inordinate amount of reserve would be required, and this added expense would defeat the purpose of obtaining the economies of scale inherent in the large unit.

Power pools may take a number of forms, depending on the benefits sought and the resources available.

1.   A separate generating company may be formed which is owned by all companies in the pool. The companies then divide the output of the generating company. The generating company is operated for the maximum benefit to all, and the companies share the savings.

2.   Two or more companies may build a generating station jointly. In this case each company may own a part of the generating station, or one company may finance the station and own it and the other companies will contract to buy a portion of the output of the plant over a period of years.

3.   A number of companies may agree to pool all their generating plants and major transmission systems. *Holding companies* (companies which own one or more utilities) fit into this category, but the companies may be owned independently and agree by contract to build their systems as though they had common ownership and management and to operate them as one with the benefits equitably divided among them. A group of executives forms the management committee, and a committee of engineers plans the requirements of the combined systems. In this planning the idea is to build all future plants and transmission lines just as they would be built if one company owned all the facilities.

Each company may agree to pay for the construction in its territory, whether it is a power plant or a section of transmission lines, or other arrangements may be worked out for financing construction of transmission lines which are considered common to all. Each company agrees to operate and maintain the facilities in its territory even though they have been built for the benefit of the whole. A central dispatching office is established, and a central dispatcher operates all power plants in the pool to realize the maximum economy for the whole. A system of payments is worked out so that no company pays more than its share of the costs involved and each company receives its share of the savings of the enterprise. Savings can be calculated on the basis of what it would have cost each system to build and operate its own plants individually as compared with the method of building the plants and lines and operating them as an integrated system.

**Economic Loading**  Electric energy companies now use economic loading of power plants. For any large interconnected system, whether it is one company or a pool, there are a number of factors affecting efficiency.

First, there are many generating units of varying efficiencies. The efficiency of each changes as the load on it is increased or decreased. Fuel contracts vary. Then there is the factor of distance from the power plant to the load center, the place where the electricity is to be used. Power plant A may have a lower production cost than power plant B, but if power plant B is nearer the load center, the greater transmission loss of A over B may offset this difference. Assuming a given load curve, engineers can calculate the most economical loading on each generating unit for each hour of every day in the year. However, this requires an enormous number of calculations. Complicated mathematical equations are involved. It may take months to make the calculations; nevertheless, the possible savings justify them.

Today computers make the calculations for the most economical loading on each of the generators on a continuing basis. Before the advent of computers, the central dispatcher would call upon the various plant operators to meet the schedule called for by an overall schedule of operation. In the modern power pool, the computer assists the operation of the system. All data are fed into the machine, which then schedules the operation of all the power plants and the load they are to carry each hour. There may be scores of power plants covering a number of states under the central control of one computer. The central dispatcher has a schedule to follow and can take over if anything should happen to the computer. This machine calculates, automatically and almost instantaneously, which units should be operated and what load should be put

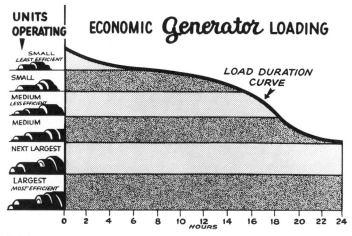

**CHART 3.24**

on each unit in order to get the most economical operation for the whole system.

Chart 3.24 shows how the various generating plants may be called into service during the day in the operation of a large power pool. It also shows the loading on a load-duration basis.

Pooling brings about the most economical construction and operation for the benefit of the companies in it. Savings are divided among all participants.

**Studies to Form a Pool**   In theory, before any pool is formed, elaborate studies are made to ensure that pooling will result in good reliability, an overall saving, and a proper division of the benefits. In practice, a pool usually is formed as an extension of interconnection agreements that already exist among the potential members of the pool. However, to understand the workings and economics of a pool it is necessary to discuss the economic studies in some detail.

First, studies are prepared as to the pool size. A pool might be so large as to be unmanageable from a practical standpoint. It also can be so small as to be impractical. Computer studies help management determine the proper size and the companies which should be included. The load characteristics of neighboring systems have some bearing on the benefits of pooling. All the systems want to obtain the maximum advantage, including any diversity.

After a determination is made as to which companies will form the pool, studies are prepared to indicate the possible savings. This requires a 10- or 20-year forecast of the kilowatthours and kilowatts for each of the companies separately. Each company is then analyzed to determine how much it would cost to build and to operate an expanded system of lines and plants to serve each of the companies separately and to operate them as separate units. To make the study, a design must be prepared for each of the companies for the selected period. A calculation must be made on economic loading of each of the systems separately for the period with the economic loading calculated by hours. Without computers such a study would take too long to be practical.

Then various plans, A, B, C, or D, may be prepared for the building of the plants and lines to serve all companies as a whole for a period of 10 or 20 years. Economic loading is applied by hours under each of the plans for the period under study. By this process the most economical of the overall plans can be determined. Thus a calculation can be made as to the savings that can be derived under any of the plans studied as compared with building plants to serve each company separately.

If the managements agree that the savings are justifiable, they enter into negotiations as to how these savings are to be divided equitably among them. These contracts have to be approved by appropriate regulatory bodies. They must be fair to all, and they must avoid any unjust discrimination. There are various ways that these pools can be formed, and there are various ways to work out the equitable division of the savings; local conditions determine how best to develop these arrangements.

**Pools to Serve Large Industries**   At times electric utility companies are called upon to serve customers with unusually large loads, as was the case with the Atomic Energy Commission (AEC; now split between the Nuclear Regulatory Commission and the Department of Energy). Two principal examples of this kind are discussed below.

At Joppa, Illinois, in the 1940s, five power companies combined their resources to finance the installation of a 960,000-kW plant for the AEC. The total plant cost over $182 million. In Ohio and Indiana a similar project was financed by 15 power companies serving parts of eight states of the greater Ohio Valley Basin. The Ohio Valley project has 2,365,000 kW of capability. It was also built to provide electricity for the AEC. The cost was over $356 million.

In these two projects, electric energy companies have about half a billion dollars invested to serve a single customer. The contracts, however, provide for cancellation charges should the customer cease operations. These charges are designed to permit the participating companies to absorb this capability on an economical basis.

**"Waste Heat" Plants**   Some types of industry develop a large amount of heat in the process of manufacture. At times, the industry may install an electric generating plant to use the heat. Experience has shown that the best way to use the waste heat is as follows:

- If the customer's generating plant is of a practical size, it is connected to the electric company's interconnected system.

- The operation of the plant is put under the control of the company's dispatch center.

- All the "waste" heat is used. The customer is given credit.

- The customer is given credit for the useful capacity of its plant with proper adjustment for reserve.

- The electric requirements, both kilowatts and kilowatthours, above the capacity of the customer's plant, are purchased from the company.

- The customer gets the benefit of having its plant and heat used, plus the benefit of the better service from the interconnected system.

**Coordination of Power Supply**   The evolution of power systems brought about the formation of a number of large interconnected and *coordinated areas* throughout the country. Before this happened, the companies in the Pennsylvania, New Jersey, and Maryland area, known as the *PJM Interconnection*, had been working together for years. The large holding companies were coordinated among themselves and were planning coordination with neighboring groups. The Northwest Power Pool had been in operation for some years. Here the investor-owned companies and the non-investor-owned systems, including the Bonneville Power Administration, had interconnected and coordinated the operation of their systems for maximum reliability and economy. The California companies were interconnecting and pooling. The Southwest Power Pool had been in existence for a number of years. The Texas companies were coordinated. Other areas were moving in the same direction.

In 1959, the electric energy companies formed a Committee on Power Capacity and Pooling in their national trade association, the Edison Electric Institute. A task force was appointed to study the interconnected and coordinated arrangements being made throughout the country so that all companies and all areas could benefit from the experience of other companies and other areas. A series of meetings was held around the country, during which these interconnection principles were discussed and the problems reviewed and studied.

When companies interconnect their systems, those systems can operate only while in synchronism. Electricity flows at the speed of light. Any mishap on one system can instantly affect neighboring systems; therefore the systems are designed to withstand almost any mishap. That is, if lightning strikes one line, the system is so set up that electric energy automatically is shifted to another line without the customer knowing it. Generating units fail from time to time, but the system is designed so that such units are automatically cut out of service and energy is supplied from other plants. On rare occasions there are even human failures, where an operator may fail to open or close a switch, which might cause a line failure or plant failure. But here again automatic devices isolate the defect and reroute the energy so that there is no interruption to the customers.

Before systems are expanded, the designs are placed in a computer so that the engineers can see how they will work. Present performance is studied as well as performance planned for years to come. This planning, in effect, is like building a model system. Engineers study this model and make sure that if any failure that might reasonably be expected occurs in one area, the automatic relays cut out that portion of the system. The remaining facilities meet the load.

Chart 3.25 shows how coordinating areas were developing during the early 1960s. They were evolving out of existing interconnection and pooling arrangements which had been created over the industry's long history of interconnection and cooperation among systems. This kind of

## COORDINATION AREAS / 1960's

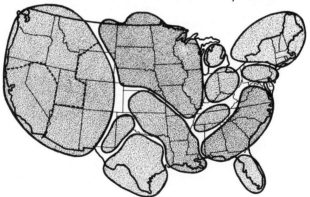

**CHART 3.25**

coordination requires that all members of the coordinated area be fully informed as to what is being done by every other supplier in the system.

**Northeast Power Interruption** On November 9, 1965, the electric energy suppliers of the Northeast were operating their systems on an interconnected basis. There were ample reserves and there was no energy shortage. But instability was triggered by the sudden loss of five transmission lines carrying power from the Sir Adam Beck Power Plant of the Hydro Electric Commission of Ontario to the Toronto area. The power delivered over those lines was higher than the backup relay settings, thus causing the circuit breakers to open. With the loss of these lines, 1,500,000 kW of power being generated at the Beck plant and at the Power Authority of the State of New York's Niagara plant, which had been serving Canada's loads, reversed and was superimposed on the lines south and east of Niagara. This quickly initiated periods of overgeneration, undergeneration, and instability which resulted in the shutdown of the systems in the Northeast. This occurrence is sometimes referred to as "cascading."

Fundamentally, here is what happened: Prior to the mishap all systems were operating in synchronism; that is, all generators were running precisely at multiples of the same speed and all "pendulums" were in step, swinging at 60 Hz. As has been noted, when in step, all machines tend to hold each other in synchronism until a disturbance or a change in generation requirements throws them out of step.

The manner in which electric energy systems behave when the energy balance between generation and load is changed is complicated by the fact that not only are there many generating plants and transmission lines in a system, but there also are many systems in an interconnected network. Fundamentally, the reaction of the interconnected systems and their facilities depends on the relative strength of the various transmission lines, the size of each generator, each generator's control equipment, and the tie-line control between the various interconnected systems.

A full explanation of these factors is beyond the scope of this book. However, the principle can be understood by considering a system with a single plant and considering only the reaction of the generating-plant equipment.

For example, as customers increase their load, the power demand on the power plant increases. Increased load will slow down the generators, with a consequent decrease in their frequency. This decrease is not appreciable with small load changes. Automatic devices called *governors* are installed on generating units to signal for increased fuel so that the plants can carry increased load and so that the generators will not slow down appreciably and thus get out of synchronism. There is a time lag between the increased load on a plant and the time the governors can react by adding fuel and steam to hold the generators at synchronous speed. If the load change is not too great, normal operations will swiftly bring the machine back to synchronous speed and there will be no appreciable or even recognizable effect on frequency. Under these conditions the system is said to be *stable*.

If a large load, such as 1 million kW, is suddenly dropped from the plant, there is not enough time for the various devices in the plant to react. The energy balance is destroyed, and the extra fuel and steam cause the generators to speed up. As they speed up, the frequency increases. In actual systems the energy-balance mismatch will differ from unit to unit depending on the factors previously mentioned. Electric energy generating units are equipped with automatic protective devices which cut the units out when their frequency goes above or below a predetermined level.

When, on November 9, 1965, the generating units serving the Toronto area were disconnected from this load, they accelerated in speed. As the speed increased, their electrical energy output increased rapidly. However, it was out of step with the other generation, and this resulted in an unstable situation.

In this way the whole Northeast experienced an interruption of energy supply in a matter of minutes. All operators then examined their energy supply and transmission equipment to determine the fault. There was no sign of equipment break or failure. Each system then began the complicated process of restarting its generators and restoring service. The restoration took longer in some areas than in others because of the intricacies of the systems.

The Northeast interruption called for a reevaluation of all coordination activities in the bulk energy supply. Consequently, the best experts in the industry, acting through the committee structure of the Edison Electric Institute, made a complete study of industry coordination practices. They prepared a report on coordination of electric energy supply which was made available to all energy suppliers. Of course, every energy supplier and coordinating area made its own reevaluation of its local energy systems.

**Coordination and Pooling**  Sometimes the terms *coordination* and *pooling* have been used interchangeably. However, there is a growing tendency to attach special meanings to each of them. For clarity, the terms, as they are used in this book, will be defined below. Generally speaking, these are the definitions now accepted in the industry.

*Area coordination* is concerned only with the reliability of bulk energy supply and not with economies available through pooling and other arrangements. *Pooling* has to do primarily with the economies of planning and operation.

Coordination areas are usually larger than pooling areas. Coordination involves a larger number of power suppliers. In fact, there may be a number of pooling areas within a single coordination area. A pooling arrangement usually involves a smaller number of suppliers. As pooling contracts among suppliers involve sharing economies of scale and possibly joint building of facilities, the negotiations leading to pooling contracts are far more complicated and time consuming than those for coordination contracts. Further, regulatory bodies need to examine the pooling contracts thoroughly to be assured that all parties share equitably in the benefits of pooling.

The coordination area groups have evolved from existing organizations and differ as to size of area and number of systems because of differences between areas, such as total population; population density; magnitude, nature, and density of electrical power loads and resources; extent of interconnection; and other factors.

The 1965 experience led to the formation in 1968 of the National Electric Reliability Council (NERC) with the stated purpose "further to augment the reliability and adequacy of bulk power supply in the electric utility systems of North America." It consists of nine Regional Reliability Councils and encompasses essentially all the energy systems of the United States and the Canadian systems in Ontario, British Columbia, Manitoba and New Brunswick. NERC has headquarters in Princeton, New Jersey.

Chart 3.26 shows the coordination areas as they existed in 1976 and 1978; and Table 3.9 shows the number of investor-owned and non-investor-owned systems in each of the areas, the size of each area in square miles, and the aggregate generating capability of each. Each coordination area is designed to do those things most fitting to meet local conditions and to provide the most reliable service. Generally speaking, the participants in each coordination area include all electric energy suppliers that make substantial contributions to the delivery of electric energy in the area, whether the system is financed in the market or through the government. These suppliers agree among themselves on the procedures, principles, and practices to be followed. Unusually there is a signed agreement among the parties. Each energy system in the area agrees to do all those things on its own system to be sure that it gives good service locally and does not operate in a way that can effect the other systems detrimentally from the standpoint of reliability. Coordination does not

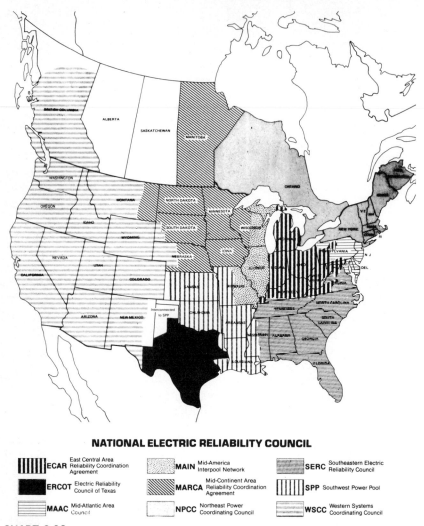

**NATIONAL ELECTRIC RELIABILITY COUNCIL**

**ECAR** East Central Area Reliability Coordination Agreement

**ERCOT** Electric Reliability Council of Texas

**MAAC** Mid-Atlantic Area Council

**MAIN** Mid-America Interpool Network

**MARCA** Mid-Continent Area Reliability Coordination Agreement

**NPCC** Northeast Power Coordinating Council

**SERC** Southeastern Electric Reliability Council

**SPP** Southwest Power Pool

**WSCC** Western Systems Coordinating Council

**CHART 3.26**

mean that an individual utility no longer plans or builds its facilities to serve its own customers. Coordination supplements this process, giving each local system more scope for improvement.

The various coordination areas have established the planning and operating criteria that can be used in simulated testing and evaluations of the performance of the combined systems. Each member discloses to the other members its plans for major generating and transmission facilities in advance of construction so that the effect of these new facilities on the reliability of the area may be evaluated and, where necessary,

**TABLE 3.9** STATISTICS ON AREA COORDINATING GROUPS

| Coordinating Group | Number of Systems | | | Year-End 1976 Capability (million kW) | Size of Area (mi²) |
| --- | --- | --- | --- | --- | --- |
| | Investor-Owned | Non-Investor-Owned | Total | | |
| East Central Area Reliability Group (ECAR) | 25 | 1 | 26 | 77.2[a] | 192,000 |
| Electric Reliability Council of Texas (ERCOT) | 8 | 3 | 11 | 34.0 | 195,000 |
| Mid-Atlantic Area Council (MAAC) | 11 | 3[b] | 14 | 41.7 | 50,000 |
| Mid-America Interpool Network (MAIN) | 10 | 6 | 16 | 38.7 | 170,000 |
| Mid-Continent Area Reliability Coordination Group (MARCA) | 11 | 13 | 24[c] | 20.3 | 400,000[d] |
| Northeast Power Coordinating Council (NPCC) | 17 | 4 | 21[c] | 73.0 | 300,000[e] |
| Southeastern Electric Reliability Council (SERC) | 15 | 13 | 28 | 104.8 | 346,000 |
| Southwest Power Pool (SPP) | 23 | 14 | 37 | 41.2 | 350,000 |
| Western Systems Coordinating Council (WSCC) | 19 | 27 | 46[c] | 97.9 | 1,600,000[f] |

[a] Does not include the liaison members.

[b] Associates representatives who participate in the activities of the region.

[c] Includes as a member a Canadian electric power system or systems.

[d] Includes southern portion of Manitoba Hydro service area.

[e] Includes the service area of the Canadian systems, which amounts to 185,000 mi².

[f] Includes the service area of the Canadian system, which amounts to 300,000 mi².

modifications may be made to assure the reliability of the bulk energy supply of the entire area.

The coordination areas also have developed procedures for the systematic interchange and review by all the participants of information pertaining to load projections, significant changes in system facilities, modes of operation, and all other matters pertinent to the reliable performance of the area systems.

Reserve capability policies are given careful consideration in each coordination area. Each area considers maintenance scheduling of major generation and transmission facilities so that such maintenance does not jeopardize service reliability. Coordination areas also agree upon essential metering, communication, and relaying facilities.

Some coordination areas have established a coordination office and maintain a staff devoted solely to coordination of bulk energy supply in the area. Through this office all information and all studies are coordinated so that all operators and planners of the coordinated systems can always be kept informed.

Most area groups have an executive or management committee to review principles and procedures on matters affecting reliability. This committee also has the responsibility to make certain the planning for generation, transmission, and other relevant matters of each of the parties is reviewed and evaluated. Assisting this top committee are one or more technical committees. In most area organizations, task forces or ad hoc committees of specialists assist the standing technical committee or committees. These technical committees generally perform studies and investigations concerning such things as overall adequacy of transmission facilities, generation reserves, operating practices, and procedures. Since 1965 there have been a number of large local *blackouts*, but the coordination program has so far prevented major cascading.

**Coordination Between Areas of Coordination**　Interconnection between power systems is not limited to those within each coordination area. Power systems in adjacent coordination areas may have interchange contracts with each other. Interarea coordination agreements supplement bilateral and multilateral interconnection arrangements. It is of interest to each coordination area to know the principal facts about adjoining areas. Consequently, procedures are established as seem appropriate to local conditions to bring about the proper exchange of all knowledge and information between areas. Also, through the industry's trade association there is a continuing exchange of information and ideas so that each area and each system can know of the practices, experience, and procedures of other areas.

For example, a committee of the Edison Electric Institute maintains a continuing file on the contractual arrangements for the purchase and interchange of energy between electric systems so that all members can review them. Through the Institute, exhaustive studies are made of diversity among all energy suppliers. These studies cover a period of years and take into account the hourly demand on individual energy systems and their neighbors. As an example of the benefits of diversity and how they can be realized, there is a contract between a group of electric energy systems in the Southwest Power Pool called the South Central Electric Companies (SCEC) and the Tennessee Valley Authority (TVA). SCEC has a heavy peak in the summer due to air conditioning. TVA has a heavy peak in the winter due to electric house heating. There is a significant annual load diversity between them. A contract calls for delivery of capability by the Southwest companies to TVA during the winter and for the delivery of an equal amount of capability by TVA to the Southwest companies during the summer. Both suppliers and their customers benefit.

## ENERGY SOURCES

Electric companies are interested in making electricity at the lowest possible cost. Prior to using steam-generated power, companies use hydroelectric power whenever it is economically available. Chart 3.27 shows the percentage of hydro and steam generation by years for the investor-owned electric utility industry since 1940. The percentage of hydro will probably continue its downward trend.

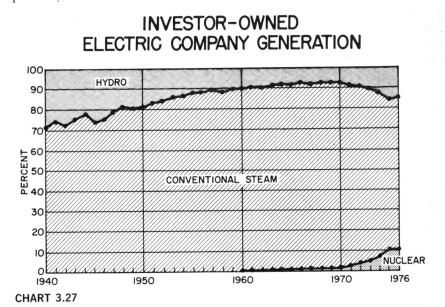

# INVESTOR-OWNED
# ELECTRIC COMPANY GENERATION

**CHART 3.27**

**Hydroelectric Power**  Hydroelectric power is made from falling water which is used to turn turbines connected to generators; the higher the fall, the more electricity can be generated from a given volume of water. If the flow of water is continuous, the hydroelectric station can make energy day and night, month in and month out. This will be possible where a dam is built on a river with a constant stream flow or where the reservoir behind the dam is large enough to store up a reserve of water to run the turbines continuously.

The number of damsites of this kind is limited. Since these are the best kind, they were the first sites to be developed. After these had been developed, engineers began to work on the next best sites for dams. Most of the practical damsites have by this time been developed. The remaining damsites are largely in places where the stream flow is limited and erratic. There may be an abundance of water during the rainy season, but there may be little, if any, during the dry season. At such sites the water is stored in an artificial lake created by the dam, where storage is possible.

When a dam is built across a river with limited stream flow, there are two principal ways that the power can be used. On rivers such as these, there is sometimes a great deal of water—even floods—and at other times the rivers are almost dry. These rivers can make a large quantity of electricity for short periods and much less electricity during the rest of the year, depending upon the flow of the water and the size of the reservoir behind the dam. If the company tried to use the generators at these dams to make electricity 24 hours a day all year long, a good portion of the capacity of the plant would be standing idle because the water would have to be stretched out over the whole year.

Sometimes the water is stored and the generators are operated only a few hours a day and a few days out of the year. Where this is done, much more electricity can be produced for every hour in which the generators are operating.[2]

The economical way to use this kind of hydroelectric power is to save up the water during the period when people are not using much electricity and then run the generators at their full capacity when people are using a great deal of electricity during the peak hours of the day. This is illustrated in Chart 3.28. At times of lower demand, steam stations take care of the load, and the supply of water is built up. Then when the peak hour arrives, the water is released through the turbines and the energy is poured into the system.

---

[2] Many dams are built by the government for flood control, navigation, irrigation, and electric energy. See Edwin Vennard, *Government in the Power Business*, McGraw-Hill, New York, 1968.

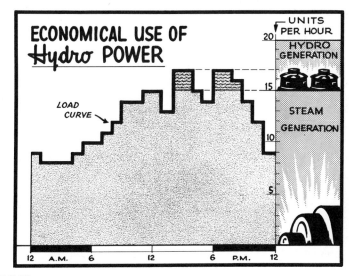

**CHART 3.28**

**Combination Dams**   At times a dam is built to serve a number of different purposes. It may be a dam for both flood control and power generation or a dam for irrigation and power generation. Navigation improvement may be one of the functions. At times the various functions operate
at cross purposes. Then it is necessary to establish the primary purpose
and secondary purpose.

Chart 3.29 shows a flood-control dam. It needs an empty reservoir
to catch the floods. Following each flood, the reservoir empties through
openings near the bottom at a rate not greater than the river channel can
take without overflowing. But, of course, no power can be obtained from
an empty reservoir, and that is the normal condition in a dam built
strictly for flood control—the reservoir is kept empty.

Chart 3.30 shows a power dam. To get power and energy, the water
must fall; the greater the fall, the greater the power. Therefore, at a
power dam the reservoir is kept as full as practical. Unlike a flood-control dam, there are no flood discharge openings near the bottom.

Chart 3.31 shows a combination dam built to provide both flood
control and electric energy. This really means two projects at the same
place—one for energy and one for flood control. A dam of this kind has
to be considerably larger and higher than either a flood-control dam or a
power dam separately. Likewise, it is more costly.

**Pumped Storage**   The *pumped-storage* process offers the opportunity for
economic development of electric energy in areas where the terrain is

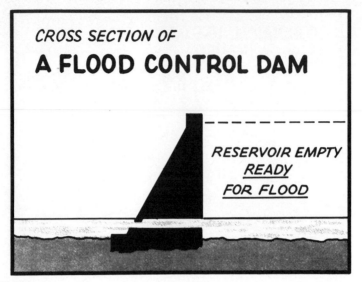

CROSS SECTION OF
**A FLOOD CONTROL DAM**

*RESERVOIR EMPTY*
*READY*
*FOR FLOOD*

CHART 3.29

suitable. Pumped storage requires that there be an elevated location where a reservoir can be constructed. There needs to be a water supply such as a river or lake nearby. A hydroelectric power plant is built at the site. It might be as large as 1 million kW. (A diagram of a pumped-storage plant is shown in Chart 3.32.)

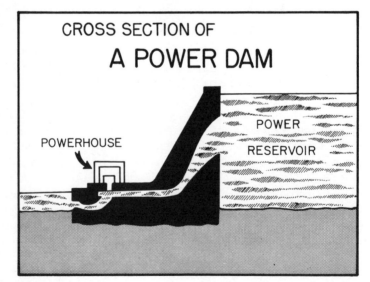

CROSS SECTION OF
A POWER DAM

POWERHOUSE

POWER

RESERVOIR

CHART 3.30

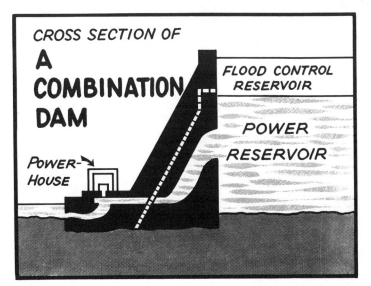

CROSS SECTION OF

# A COMBINATION DAM

FLOOD CONTROL RESERVOIR

POWER RESERVOIR

POWER-HOUSE

CHART 3.31

During the off-peak hours of the electric energy system, the hydro-electric plant is run as a motor with the energy fed to it by the electric energy system supplied by a base load steam plant. It is especially economical when a *nuclear power plant* is available to furnish this off-peak energy, as a nuclear plant has such a low production cost. The motors run the water wheels and pump the water from the river or lake up into the reservoir until the reservoir is filled. During the peak hours, the water in the reservoir is released and is run through the water wheels of

# PUMPED-STORAGE HYDROELECTRIC PLANT

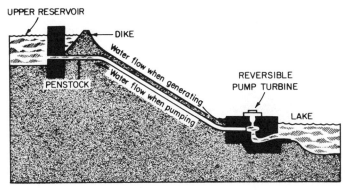

UPPER RESERVOIR

DIKE

Water flow when generating

Water flow when pumping

PENSTOCK

REVERSIBLE PUMP TURBINE

LAKE

CHART 3.32

the turbines, which then serve as generators. During this process, the project serves precisely as any hydroelectric plant would. This method is economical because, as a rule, the cost of building the pumped-storage plant and the reservoir is less than the cost of an ordinary hydroelectric station and less than the usual cost of a steam plant. When this is combined with a low-production-cost steam plant, especially a nuclear plant, the result is economical, reliable, firm power and energy.

**Conventional Fossil-Fueled Plant**   Chart 3.33 shows a diagrammatic sketch of a conventional fossil-fueled plant. Most of the plants in use today are of this type. Gas, oil, or coal is used under a boiler to make steam at high temperature and pressure. This steam is then passed through the steam turbine by stages. First, the steam enters the high-pressure portion of the turbine, where the blades are smaller. The steam expands through the turbine, where the blades get larger and larger and the temperature and pressure fall. Finally, the steam goes through a condenser which has a large number of pipes. The steam is circulated in the pipes. Cooling water from a river, reservoir, or lake is circulated around the cooling coils. The condensing water is then returned to the river, lake, or reservoir. The condensed steam is pumped back into the boiler, where it is again converted to steam at high temperature and pressure. The circulation of the steam through the turbine and the condenser and back to the boiler is a closed cycle, though from time to time small amounts of water may be added to make up for losses.

The steam turbine is directly connected to the electric generator,

# CONVENTIONAL FOSSIL-FUELED PLANT

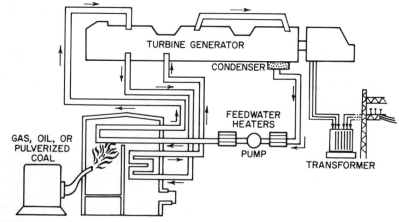

**CHART 3.33**

which generates alternating current typically at some voltage like 13,000 V. This alternating current is then passed through the step-up transformers, where the voltage is raised to the transmission line voltage which may be 220, 345, 500, or 765 kV.

Because of the economy of scale, the trend over the years has been to build generating units of larger and larger size. The temperature and pressures have been raised to higher and higher levels as, generally speaking, these higher levels result in higher efficiencies in converting the heat in the fuel to electric energy.

**Gas Turbines**   The gas turbine operates much like a jet engine on a plane. It has three characteristics:

- Low investment cost per kilowatt.

- High operating cost, as the efficiency is less than steam turbines.

- Ability for quick start.

Companies use gas turbines to carry peaks and to pick up load quickly.

**Nuclear Power**   Until recently the principal raw energy sources have been the fossil fuels, namely coal, gas, and oil. Naturally, all of these are exhaustible. Studies made about 30 years ago showed oil and gas as the two in shortest supply. Estimates then indicated these would probably be depleted early in the next century. (Coal is more abundant.) Fortunately, during the late 1940s a new source of raw energy, nuclear energy, appeared. It has enormous possibilities.

Up until 1954 the principal knowledge concerning nuclear energy was in the hands of government. In 1954 an act of Congress made this knowledge available so that businessmen and others could work toward making nuclear energy useful for peaceful purposes.

Then there followed a massive cooperative research and development effort by the electric utility industry, its equipment manufacturers and suppliers, consulting organizations, and the U.S. Atomic Energy Commission. As a member of the Edison Electric Institute, the Edison Power Company cooperated in this research and development. In 1958, the Institute appointed a Task Force on Nuclear Power composed of several of the nation's leading scientists in the field, as well as knowledgeable members of the energy industry. Its task was to study the various methods that might be utilized to convert this new source of raw energy into thermal energy that could be competitive in the production of electricity with thermal energy produced from other forms of fuel. The task force recommended a number of possibilities. No one knew whether any of the various reactor types then being considered would be economically feasible. It was necessary to do further research on the most promising concepts and first to build reactor experiments and small prototypes.

The AEC had set as its primary objective the attainment of competitive nuclear power in high-cost fuel areas by 1968—14 years after industry was given access to nuclear technology with the passage of the Atomic Energy Act of 1954.

Elaborate research and development efforts were undertaken by many organizations, with a commitment of roughly $2.5 billion for the development of this new energy source. Through 1968 the investor-owned electric power industry spent or committed almost $1 billion to nuclear research and development, with Edison Power Company paying its part.

In 1964, 4 years before the date when the AEC hoped that nuclear energy would be able to compete with fossil fuels in high-cost fuel areas, a nuclear power plant was purchased on the basis that it would prove to be competitive with a conventional plant of the same general size and at the same location.

By providing the Edison Power Company with another source of raw energy, nuclear fuel stimulates competition with and among the fossil-fuel industries—all to the benefit of the electric utility customers. As nuclear energy accounts for more than 12 percent of the electric energy now produced in the United States, its competitive effect already has been felt. For example, complete *fission* (the splitting of the atom) of 1 lb of uranium would liberate roughly the equivalent energy produced in the combustion of 2400 tons of coal. The cost of delivering nuclear fuel is insignificant, and, as a consequence, the distance from the source of the nuclear fuel supply is not an important factor in the cost of producing electric energy by this method. This means that those areas of the country with high freight costs, because of their distance from raw fuel supplies, may now especially benefit from nuclear energy.

**How a Nuclear Plant Works**   A nuclear power plant is in many respects similar to a conventional fossil-fuel-burning plant. The chief difference is the way heat is generated, controlled, and used to produce steam to turn the turbine generator.

In a nuclear power plant the furnace for burning coal, oil, or gas is replaced by a *reactor*, which contains a *core* of *nuclear fuel*. Energy is produced in the reactor by a process called *fission*. In this process the center, or *nucleus*, of certain *atoms*, upon being struck by a *subatomic particle* called a *neutron*, splits into fragments called *fission products*, which fly apart at

# BOILING-WATER REACTOR

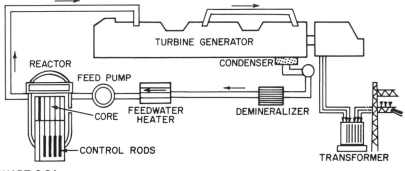

**CHART 3.34**

great speed and generate heat as they collide with surrounding matter in what is called a *chain reaction*.

The splitting of the atomic nucleus into parts is accompanied by the emission of high-energy *electromagnetic radiation* and the release of additional neutrons. The released neutrons may in turn strike other fissionable nuclei in the nuclear fuel, causing further fissions.

A nuclear reactor is a device for starting and controlling a self-sustaining fission reaction. The *nuclear core* of the reactor generally consists of *fuel elements* in some chemical form of uranium, thorium, or plutonium, depending on the type of reactor. Heat energy is produced by the fissioning of the nuclear fuel. A *coolant* is used to remove this heat energy from the reactor core so that it can be utilized in producing electricity. Fuel elements for water-cooled reactors are metal tubes containing small cylindrical *pellets* of *uranium oxide*.

There are two principal ways of getting heat from the fission of the atom in this fashion. One is the *boiling-water reactor* (BWR), which is shown diagrammatically in Chart 3.34. The nuclear reaction is controlled by *control rods*. The reaction takes place in the core. The resultant heat then boils water, producing steam which is sent through the turbine in the same manner as in the conventional power plant. There are the usual condensers enabling the return of the water to the reactor, where it can again be converted into steam.

The other type of water reactor is called the *pressurized-water reactor* (PWR), which is shown diagrammatically in Chart 3.35. This process differs from the BWR in that it has a separate loop of pressurized water going through the reactor and also through a steam generator. The steam generator generates the steam that goes through the turbine and the condenser in the normal cycle of the other plants.

# PRESSURIZED-WATER REACTOR

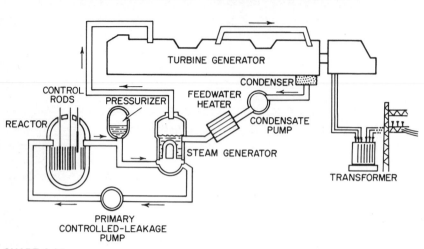

CHART 3.35

Both BWRs and PWRs are widely used throughout the world. They are sometimes referred to as *light-water reactors* (LWRs) to distinguish them from heavy-water reactors, described later in this chapter.

In the LWR, roughly 2 to 3 percent of the potential energy in uranium is obtained. The temperatures and pressures of the steam are not as high as they are in the conventional plant. Therefore, this reactor does not have the same efficiency in converting the raw fuel to electric energy. However, because of the relatively low cost of the fuel, nuclear energy generation is highly competitive with energy obtained from coal, gas, and oil. The differences in production costs are discussed in later chapters.

In the water-cooled reactor, the *nuclear fuel* which has been compacted into uniform *pellets* is placed in the tubes, or *fuel elements*. These fuel elements are then sealed at the top and bottom and arranged by *spacer devices* into *bundles* called *fuel assemblies*. The spacer devices separate the fuel elements so as to permit the coolant to flow around all of the elements in order to remove the heat produced by the fissioning uranium atoms. Scores of fuel assemblies, precisely arranged, are required to make up the core of the reactor. The geometric arrangement is necessary for several reasons. Nuclear fuel, unlike fossil fuel, has very high *energy density*. That is, tremendous quantities of heat are produced by a small amount of fuel. Because of this the fuel must be arranged in such a fashion as to permit the coolant to carry away the heat. This requires that the fuel be dispersed rather than lumped together in a large mass.

Chemical reactions between the fuel and the coolant must be avoided and, as a safety precaution, the *radioactive materials* produced must be contained. For these reasons the fuel is contained in individual tubes or fuel elements. The *cladding*, the material from which the tubes are made, must meet rigid specifications. It must have good heat-transfer characteristics, it must not react chemically with either the fuel or the coolant, and it must not unduly absorb the neutrons produced in the fissioning of the fuel to the detriment of the continuing chain reaction. The cladding material generally used for this purpose is thin-walled stainless steel or an alloy of the element zirconium.

To control the rate at which fission occurs, most reactors regulate the *population* of neutrons in the core. This is done mainly by rods which, when inserted into the core, absorb neutrons and retard the fission process. If the operator wishes to increase the energy level or reaction rate, the regulating rods are withdrawn. To shut down the reactor, the rods are fully inserted.

Neutrons liberated in the fissioning process travel at very high speeds. This is not desirable in some reactor systems where slow-moving neutrons are more effective in triggering fission than are high-velocity neutrons. To attain the desired slow-moving neutrons, a material called a *moderator* is used. It slows down the neutrons and has minimum tendency to absorb them. Materials used for this purpose include graphite or ordinary water, the latter also serving as a coolant. Most power reactors presently in operation or under construction utilize the slow-moving neutrons and are called *thermal reactors*.

After the fuel assemblies or elements are removed from the reactor, they will contain material which must be reclaimed because of the economic value. In fact, only about 1 to 3 percent of the uranium has been "burned." The other 97 to 99 percent is locked in the hundreds of fuel elements of the *spent* core.

Over the past year or two in the United States there has been a slowing down of the installation of nuclear plants. This has been caused partly by the slowdown in the economy in 1974 and 1975 (see Chapter 7 for a discussion of long-term energy trends) and partly by the lengthy delays in the building of nuclear power plants caused by the large number of hearings and approvals required by government regulatory bodies. There is further discussion of nuclear energy in Chapter 12.

It is expected that construction of nuclear installations will resume the former upward trend. As explained in Chapter 6, there is no alternative if the people of America want to continue improvements of living standards. Coal and nuclear energy must largely replace oil and gas in many applications at least until the time when some better alternative becomes available. Chart 3.36 shows the planned generation by source for

# GENERATION BY SOURCE
## *FUELS AND HYDRO*

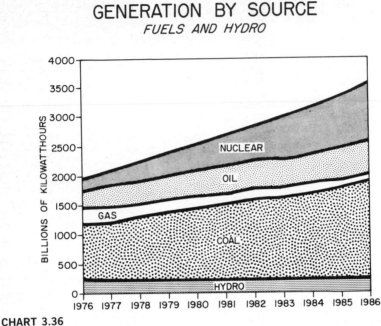

CHART 3.36

the United States from 1976 through 1986. There will be a steady rise in nuclear generation. Table 3.10 shows the number of megawatts of nuclear power by countries as of the first quarter of 1977, the percent nuclear generation for 1976, and the projection for 1985.

**Gas-Cooled Reactor**  Now under evaluation for commercial power production is the gas-cooled reactor, in which fuel elements are fabricated basically of a uranium-carbide compound and of graphite, which acts as the structural material as well as the protective enclosure for the fuel material. See Chart 3.37 for the diagrammatic sketch.

The gas-cooled reactor, which utilitizes an inert gas (helium) as the coolant, has a core structure which is different from that of the water-cooled reactor. The fuel elements are fabricated out of graphite, which acts as the structural material and the neutron moderator as well as the cladding material. The nuclear fuel, consisting of both uranium and thorium, is compressed into the center of the fuel element. As the coolant is inert, the graphite serves as adequate cladding for the nuclear fuel. The inert gas will not react with or corrode the graphite or any other structural material. Physically, the fuel elements for the gas-cooled reactor are much larger than those of the water-cooled reactor and are not bundled into fuel assemblies but are individually arranged and spaced so as

**TABLE 3.10**    WORLD NUCLEAR ENERGY

| Country | Total 1977 Operating Power, MW | Percent Generation 1976 | Percent Generation 1985 (Expected) |
|---|---|---|---|
| United States | 47,186 | 9.7 | 26.5 |
| *Outside United States* | | | |
| Argentina | 345 | 8.7 | 20.5 |
| Austria | — | | 12.5 |
| Belgium | 1,387 | 13.9 | 45.0 |
| Brazil | — | | 5.0 |
| Bulgaria | 880 | na* | na |
| Canada | 4,000 | 3.9 | 12.0 |
| China, Republic of (Taiwan) | 636 | — | 44.7 |
| Czechoslovakia | 552 | 0.81 | 16.1 |
| Denmark | — | — | 11.0 |
| Egypt | — | — | 20.6 |
| Finland | 420 | — | 20.0 |
| France | 6,535 | 5.8 | 70.0 |
| German Democratic Republic | 1,400 | na | na |
| German Federal Republic | 9,500 | 8. | na |
| India | 800 | 3.1 | na |
| Iran | — | — | na |
| Italy | 1,400 | 1.6 | na |
| Japan | 7,428 | 7.5 | 25.6 |
| Korea, Republic of South | 595 | — | 21.8 |
| Mexico | — | — | 5.0 |
| The Netherlands | 505 | 3.7 | na |
| Pakistan | 125 | 5.0 | 14.0 |
| The Philippines, Republic of | — | — | na |
| Poland | — | — | na |
| Portugal | — | — | 10.0 |
| Romania | — | — | na |
| South Africa, Republic of | — | — | 7.0 |
| Spain | 1,120 | 4.2 | 35.4 |
| Sweden | 3,760 | 13.0 | 21.0 |
| Switzerland | 1,006 | 9.0 | 30.0 |
| Turkey | — | — | 5.7 |
| Union of Soviet Socialist Republics | 7,905 | na | na |
| United Kingdom | 6,900 | 9.5 | 12.0 |
| Yugoslavia | — | — | na |

*Source:* Atomic Industrial Forum.

* Information not available.

# HIGH-TEMPERATURE GAS-COOLED REACTOR

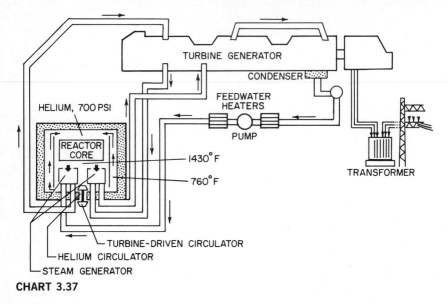

**CHART 3.37**

to allow the coolant to flow around all of the elements, removing the heat produced by the nuclear reaction. Several hundred fuel elements are required to make up the core of the reactor.

**Heavy Water Reactor—CANDU** *Heavy-water reactors* (HWRs) are now being used in Canada and elsewhere. *Heavy water* is a substance that appears in small quantities in ordinary water. The proportion is approximately 1 part heavy water to 7000 parts ordinary, or light, water.

Ordinary water is composed of 2 parts of hydrogen to 1 part oxygen ($H_2O$). Hydrogen exists in three forms, sometimes referred to as hydrogen 1, 2, and 3. Sometimes these three hydrogens have other names. As hydrogen 2 differs from hydrogen 1 by having a neutron in its atomic nucleus, it is called deuterium. Hydrogen 3 has two neutrons in its nucleus and is called tritium. The extra neutron in deuterium makes it heavier than the hydrogen atom. Hence, deuterium oxide is referred to as heavy water. It weighs about 10 percent more than ordinary water.

Heavy water is used as a *moderator* in the Canadian nuclear power reactors. Hence, the name CANDU (Canadian deuterium uranium). It is in this respect that these reactors differ from the ordinary water reactors.

The four types of moderators are ordinary water, heavy water, graphite, and beryllium. The idea is to provide a moderator that is relatively stable and does not readily absorb neutrons. Scientists measure the efficiency of the moderators by what is termed a *moderating ratio*. By this measurement, ordinary water is 60, beryllium is 150, graphite is 220, and heavy water is 1700. It is this higher efficiency that enables CANDU to use natural uranium fuel as opposed to the enriched fuel that has to be used in other major nuclear power systems.

Canadian authorities estimate that the fuel cost of CANDU will be considerably less than the corresponding fuel cost of the LWR. The investment cost per kilowatt of CANDU may be somewhat higher than the corresponding investment cost of the LWR, but the Canadians estimate that the lower fuel cost will more than offset this higher investment cost.

The CANDU concept is demonstrated by the Pickering Generating Station which has four 500,000-kW units.

**Breeder Reactors**  Much remains to be done before the full benefits of nuclear energy are obtained. Most energy experts in the United States and abroad are of the opinion that the next major step in the evolution of energy supply will be the *breeder reactor*. This reactor gives further promise of economical energy because it breeds more nuclear fuel than it uses. The breeder reactor's fuel cost should be low and relatively independent of market fluctuations in the price of uranium ore. Without breeders, according to some experts, the world may exhaust its uranium supply shortly after the year 2000.

While present commercial reactors use only 2 to 3 percent of the potential energy in the fuel, breeders offer usage of 80 percent or more of the potential energy. Chart 3.38 shows a diagrammatic sketch of the *liquid-metal fast breeder reactor* (LMFBR), the one now being given the greatest attention in the United States and in a number of foreign countries. Notice that the reactor has control rods and fuel elements like other reactors. The primary coolant is liquid sodium, which passes through the reactor and then through a *heat exchanger*, where the heat is transferred to a *sodium intermediate loop*. In turn this loop passes through the steam generator which develops the steam for the turbine. After this point, the flow of steam through the generator and back through the condenser to the generator again is much the same as in the other reactors and the conventional plants. The first experimental LMFBR of any material size was the Enrico Fermi plant located on the property of The Detroit Edison Company. The principal initiator of this research project was Walker Cisler, who was then Chairman of the Board of Detroit Edison. Through his leadership, funds were made available for the Enrico Fermi research project by many electric utility companies in the United States and abroad and by other institutions as well. The project made a significant contribution to world knowledge of the LMFBR.

# LIQUID-METAL FAST BREEDER REACTOR

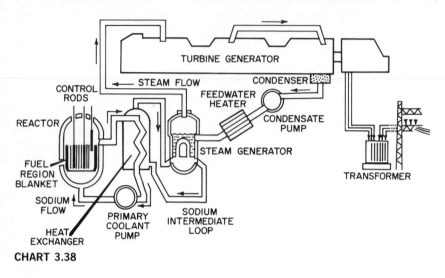

**CHART 3.38**

In the United States, research has been carried on for the development of the next large breeder reactor, to be located on the Clinch River in Tennessee. An LMFBR, it is referred to as the Clinch River Breeder Reactor. At the moment, the Carter Administration is frowning on this project, which may cause a delay in the development of a breeder reactor in the United States. Other nations are going forward with research, development, construction, and operation of breeders. More is said about the breeder in Chapter 11.

**Fusion**   The process that takes place when certain light atomic elements are joined, or fused, together with a release of heat energy is called *fusion*. Theoretically, it is now known that the energy released by fusion can be controlled and released over a period of time. Fusion is the reaction that occurs in the hydrogen bomb, just as fission is the reaction in the atomic bomb. In the case of the bomb, the enormous energy is released in a fraction of a second. By the controlled reaction process, this energy can be spread over a number of months.

The problems in the research and development of controlled fusion are far greater than those experienced in the control of fission. Fusion takes place at a temperature of millions of degrees. To harness this enormous energy, scientists must find some way to contain and control such temperatures.

Nations all over the world are devoting considerable research to the field of fusion, which makes use of elements obtainable from seawater.

When the process is finally developed, there will be available an almost unlimited supply of energy. Scientists give assurance that it can and will be accomplished. Fusion is in about the same state of development as the early research in controlled fission. It took about 25 years to make nuclear energy from fission commercially feasible. Scientists predict that fusion will be available as an energy source sometime after the year 2000.

Although research is going foward with many other forms of energy, most experts throughout the world look upon fusion as being the next most promising source after the breeder reactor. There will be more on fusion and other possible future sources of electricity in Chapter 11.

# 4

# ECONOMIC CHARACTERISTICS, ECONOMIC TRENDS, AND THE CHANGED ECONOMIC CLIMATE

## ECONOMIC CHARACTERISTICS OF POWER COMPANIES

Electric utility companies have certain economic characteristics which are unlike those of manufacturing companies. The electric energy suppliers are among the utility group, in which duplication of property is contrary to public interest. Furthermore, electric utilities have certain characteristics which make them unlike ordinary utilities.

Managers should be generally familiar with these characteristics as a requisite for wise decision making, particularly as the unique nature of the business makes electric utilities especially sensitive to any large or sudden changes in the economic climate, such as those experienced in recent years. Naturally, this familiarity should also be shared by those having regulatory responsibilities.

**INVESTMENT REQUIRED TO GET
ONE DOLLAR OF ANNUAL SALES**

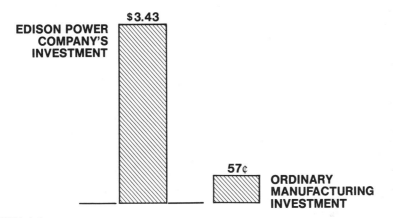

CHART 4.1

Following are some of the special economic characteristics of electric utility companies:

1.   Electric utilities, as Chart 4.1 illustrates, are the most *capital-intensive* (ratio of capital investment to annual gross revenue) of all industries. Furthermore, this large capital investment must be committed 5 to 12 years in advance. This is about the time required to build a large power plant. Because of these factors, any severe change in the rate of inflation or cost of capital has a serious effect on utility earnings and the price of electric energy.

2.   Because of the large capital investment required, most of the annual expenses are *fixed*, i.e., they do not vary with the number of kilowatthours sold (see Chapters 8 and 9).

3.   Electricity cannot be stored economically in large quantities. It must be generated (produced) the instant the customer throws a switch. The water business and the gas business are both utilities and have relatively high investment costs per dollar of revenue, but their products can be stored. Even the telephone company, which is also highly capital-intensive, can in effect, "store" its product by telling the customer that the line is busy. Because of their ability to store their product, the other utilities can have a higher load factor (the measure of the utilization of the investment). This higher load factor results in a lower unit cost in their prices.

As will be noted in Chapter 11, the electric industry is endeavoring to develop an efficient large-scale battery. When such a battery is achieved, possibly after 1990, the economics of the electric utility business will change and the customer will further benefit through price.

4. Because of the characteristics mentioned above, load factor is of primary inportance to the electric utility. Trends and management practices that have enabled the building of load factor will be presented in this chapter. In Chapter 7 management's plans for increasing load factor in the future will be considered. Load factor is important, and the pricing policies of electric utilities must take it into account, as described in Chapter 8.

5. Customers' patterns of use of electric energy vary in both volume and load factor. Chapter 8 includes a discussion of the manner in which cost studies are made and rates are designed so that they represent as nearly as possible the cost of furnishing service over this wide range of use.

6. The business is regulated. Thus, there is a fine line as to price, which must not be too high nor too low. Fluctuations in the economic climate are reflected in the unit cost of service and therefore in the price.

7. Because of the heavy, intricate machinery in power plants, especially nuclear plants, construction must be initiated 8 to 12 years in advance. Chapter 7 describes how management makes these plans.

## THE ECONOMIC CLIMATE

The wholesale price index is a good gauge of the economic climate, and one which indicates the impact on prices of each of the major wars since 1800 (Chart 4.2). Steep rises took place at the beginning of the War of 1812, the Civil War, and World War I. However, sharp declines took place following each of these wars, brought about by disarmament and a return to a normal economy. During such major wars, it is usually the practice to control prices of labor and commodities, as this is the only practical way to operate a society at war. In short, during a major war, the free society operates somewhat like a government planned economy (see Chapter 6).

A similar rise in prices was experienced during the early stages of World War II. Contrary to previous patterns, however, there was no decline in the index at the close of World War II but rather a continuation of the rise, which has become even more severe during the past few years.

From the standpoint of the economic climate, World War II never ended. Because of continued threats from without, the United States maintained the necessary equipment for defense. Therefore, defense spending continued to require a large percentage of the gross national product, although not as much as during the immediate war years.

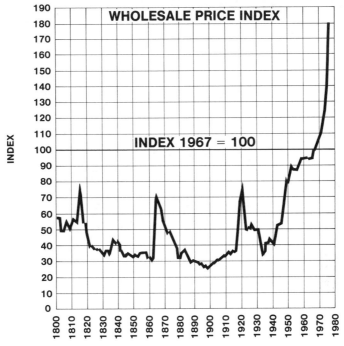

**CHART 4.2**

Furthermore, the United States adopted a policy of material assistance when called upon by other free societies which feared that their freedom was being threatened militarily by outside forces. Added to this burden was the U.S. commitment to help in the rehabilitation of allies and to help all the developing countries. Thus, following World War II, there continued to be large government expenditures that did not materially better the civilian economy. This required a continued high tax burden.

The civilian economy itself was influenced largely by two major factors: (1) the large and increasing demands on government for welfare and (2) the unusual demands for wage increases. The result has been that government has continued to operate at a deficit, spending more than its income (see Chart 4.3). A rising wholesale price index has been inevitable.

Wisely, government has decided that control of wages and prices by government would not be advisable in peacetime. A free people will gladly sacrifice a large part of their freedom during a war emergency when control of wages and prices is necessary. Otherwise, people are unwilling to make the same sacrifices.

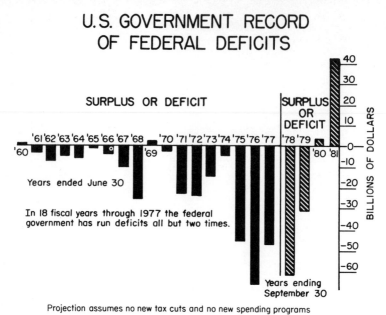

## U.S. GOVERNMENT RECORD
## OF FEDERAL DEFICITS

Projection assumes no new tax cuts and no new spending programs

**CHART 4.3**

**The Handy-Whitman Index**    The generally accepted standard reflecting the cost of equipment used by the electric utility industry is called the Handy-Whitman index. Chart 4.4 shows this index from 1950 through 1976 (actual). Despite the rise in the index from 1950 through 1965, the cost per kilowatt of steam turbines actually declined. This decline occurred for two reasons: First, the manufacturers improved their efficiency of construction and, second, the utilities purchased larger and larger turbo-generators.

Including the sharp upward changes in 1968 and again between 1973 and 1975, the index shows that the capital cost per kilowatt has more than doubled since 1968. In this situation, utility management has found itself facing problems that had not existed during prior years. Although it was clear that the cost of electricity would not take any significant percentage of the family budget or any significant percentage of the cost of manufacturing, a severe problem resulted, largely because the public had become accustomed to a constantly declining average price. Previously, the customers' electric bills rose as they used more electricity in an increasing number of applications, but the average *unit* price declined, and customers therefore accepted their higher bills.

Furthermore, government regulation under a climate allowing a constantly reduced average price is a fairly easy procedure—both tech-

# HANDY-WHITMAN INDEX
# AND COST PER KILOWATT

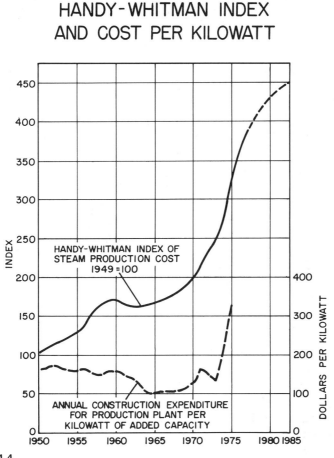

**CHART 4.4**

nically and politically. It is not difficult to explain to a regulatory body why the average price could go down with increased efficiency and with increased volume.

Today, however, regulatory bodies and the public need an enormous amount of information to help them realize that, although the improvements in efficiency will continue, the average price will have to rise if customers are to receive the electric service they need and demand 5 years hence and beyond. If poor service were to result immediately after a failure to raise the price sufficiently, the public and the regulatory body might be more conscious of the necessity for the increased price. In fact, decisions made today are affecting the nature of the service years in the future. Managements must raise money now in the free market, where people can choose whether or not to invest in the electric energy business. Furthermore, the funds are required to begin construction on needed power facilities that will come into service a decade hence.

## THE CHANGED ECONOMIC CLIMATE

Through all the years of the business, up until about 1968, the price over the long term could be reduced because improvements in operating efficiencies more than offset the inflation rates of around 2 or 3 percent. Chart 4.5 shows that (with few exceptions) the average price declined from 4 cents per kilowatthour in 1902 to about 1.5 cents in 1968. It is true that there were some rate increases during this long term, but the general price trend was downward. As will be shown later, improvements in operating efficiency have continued and can be expected to continue for some time in the future, although these improvements will be less in degree than they have been in the past. However, the improvements in efficiency now being realized and being contemplated for the immediate future cannot possibly offset the much higher inflation rate experienced since 1968 and cannot offset the other factors that have added greatly to the cost of producing electric energy.

Naturally, the increased volume of electricity consumption added materially to the ability to reduce the price. This will be brought out

## AVERAGE COST OF MAKING AND DELIVERING A KILOWATTHOUR
*EDISON POWER COMPANY*

**CHART 4.5**

more clearly in the discussion of the reasons for the decline in price in Chapter 12.

From now on in the text, the terms "cost" and "price" will be frequently used. As explained in Chapter 9, the fair return allowed does not necessarily have to be the cost of money (capital). It is true that this is the minimum return that should be allowed if the companies are to continue raising the capital needed to construct facilities, which by law they are required to build, and if the regulatory body is to meet its obligation of assuring reliable and adequate service for the future. In recent years, *cost of capital* has been given a great deal of weight in regulation. Where the return allowed is cost of capital, and where rates are designed to include this capital cost and no more, the total unit cost of service is the same as unit price.

Mention should be made of two areas on the curve in Chart 4.5 which seem to deviate from the long-term trend between 1902 and 1968. The rise in average price during the Great Depression of the early thirties was not caused by rate increases, but by a decline in the volume of industrial sales, which carry a lower-than-average price. The volumes in residential and commercial service, at higher prices, were fairly steady.

The dip in the average price which occurred during the early forties was caused by another reason. These were the war years when electric utilities could not purchase turbo-generators. Plant reserves dropped to a very low level. The price of service was based upon the plant in service, not on the plant that would have been in service if companies could have bought power plants.

Because of other higher costs some rates did have to rise in the post-war period. But if the "bump" of the early thirties and the "valley" of the early forties are ignored, it appears that the general trend in price from 1902 to 1968 was downward.

In economic terms, for the period from 1902 to 1968, it is said that the *unit incremental cost* (i.e., the added unit cost of building power plants and operating them) was less than the average *unit embedded cost*. This means the average cost of making and delivering a unit of energy had been declining.

However, since 1968, the incremental unit cost has been above the average unit embedded cost, with the resultant rise in the average unit embedded cost. This fundamental change in the economic trend suggests that management should reexamine all its plans and practices so as to adapt them to this wholly new condition—at least as far as electric utilities are concerned.

Where the change in the climate originated, or which country sparked it, is not a subject for this study. As the United States is a large contributor to the world's total "gross national product," it is likely that

# ESTIMATED INVESTMENT PER KILOWATT RELATED TO UNIT SIZE

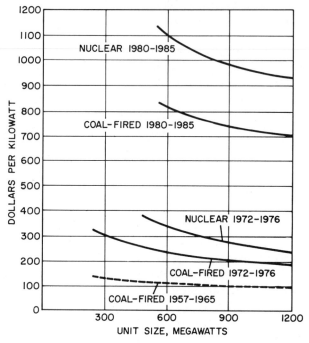

**CHART 4.6**

the United States made some contribution to the change in the world climate. Whatever the cause, or whatever the nature of the economic climate, it is the purpose of electric utility management to operate so as to earn not less than the cost of money to finance the expansion necessary for meeting customer demands.

Interestingly, the change in the basic economic trend appears to have occurred in practically all countries. The factors that produced the shift in the United States to rising electricity prices are indicated below.

1.   There was an extraordinary increase in the rate of inflation. In Chapter 6 there is a discussion of the manner in which energy is used to improve living standards, which also covers the elementary economics of any country operating as a free society with a free market for capital. Some of the primary factors that have been reflected in the higher rate of inflation in the United States are also mentioned.

Chart 4.6 shows the investment cost per kilowatt of capacity for power plants of various sizes for both the conventional steam plant and nuclear plant. Note that the investment cost per kilowatt of capacity is lower for the larger units. This was a contributing factor in reducing the price of electricity between 1902 and 1968, and the principle will continue to be applicable in the future, but to a lesser degree.

All the unit investment costs of power plants increased materially between 1957 and 1976, and the investment costs in the electric utility business, with respect to annual revenue, are the highest of all industry. Thus, substantial inflation in the unit investment cost of a power plant has a severe effect on the total cost of producing a unit of electric energy.

Also, a large part of the manufacturer's cost in making a turbo-generator is construction labor. The unusual rise in labor costs has been a major factor in causing the much higher capital cost of a power plant.

2.  The cost of capital increased sharply. This is illustrated in Chart 4.7, showing the average yield of utility bonds, preferred stock, and common stock by years from 1928 to 1977.

3.  Labor, especially construction labor, demanded and obtained increases above those justified by increased productivity. The discussions in Chapter 6 show how these increases may have contributed to inflation.

4.  Environmental problems have multiplied since 1970. As pointed out in Chapter 12, the environmental concerns of electric utilities did not begin in 1968, but rather had their beginning many years ago. Public awareness of environmental problems came with the great influx of people to the large industrial centers and the resultant pollution. Obviously, utilities as well as practically all manufacturing processes make some contribution to pollution. The environment needs to be improved, but as pointed out in Chapter 12, it is always a matter of weighing the benefits against the cost. Nevertheless, the demands for further

## AVERAGE YIELD OF UTILITY BONDS, PREFERRED STOCK, AND COMMON STOCK

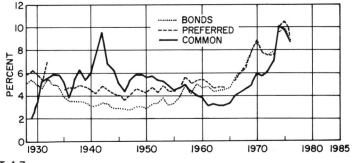

**CHART 4.7**

improvements became apparent. More government directives were issued. Utilities have tried to meet these demands within the limits of practicability. When people realize the enormous cost of providing some of these improvements, and when they realize the only way in which these costs can be paid is through the electric bill, the public demands become less. Therefore, the problem is also materially influenced by the degree of public information and the degree to which people will balance the benefits against such costs.

5.   Fuel costs, which are discussed more fully in Chapter 7, increased steeply. It is now fairly well recognized that the United States unwisely failed to establish a more specific long-range energy policy when, about 25 years ago, it first became known that oil and gas would probably be exhausted early in the twenty-first century and that coal would last much, much longer. But, until recently, little has been done to develop a constructive policy. Most of the fuel used in the United States has been oil and gas. Only a small percentage has been coal. Furthermore, as pointed out in Chapter 13, oil imports have become a necessarily high percentage and can be expected to increase to even higher proportions. The results of this dependence on oil were dramatized by the embargo and the fixing of much higher prices. There followed a tremendous increase in the cost of fuel, which was reflected directly in the price of electric energy.

6.   During the early 1970s there was an unusual rash of strikes and other disturbances causing delays in the delivery of electrical equipment and especially in the construction of power plants. Also, it was found that the completion of nuclear power plants required a longer time than anticipated. Government regulations and hearings added materially to the delays. As a consequence, the engineers of power companies decided that in the interest of reliability, it would be advisable and necessary to increase the power-plant reserves from about 15 percent to about 20 percent.

**Index of Prices**   Chart 4.8 shows the consumer price index compared with the electricity component, with 1915 representing 100. Note the long period during which the index of electricity was down. Chart 4.9 shows the same components with 1930 representing 100. Later discussions will cover more of the reasons for the decline in price until 1968.

## ECONOMIC TRENDS

**Trend Toward Electric Energy**   Almost since the beginning of electric energy use, the trend has been toward the application of more and more energy in this form as contrasted with primary fuels, such as coal, gas,

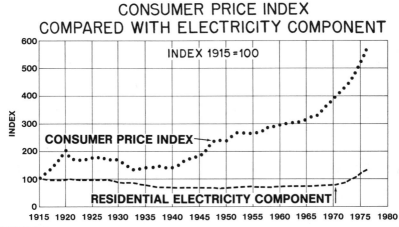

CHART 4.8

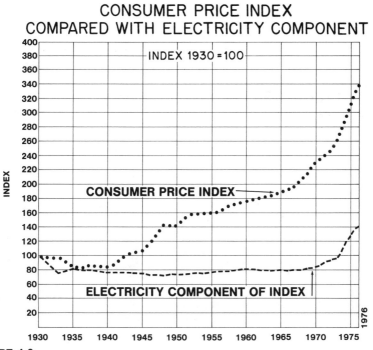

CHART 4.9

# INDICES OF ELECTRICITY GENERATION AND TOTAL ENERGY USE IN THE U.S.A. 1902-1976

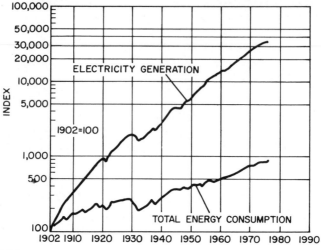

**CHART 4.10**

and oil, for end use. This is illustrated in Chart 4.10, which shows the index of growth of total energy consumption contrasted with the index of growth in electric generation. Electricity has been growing consistently at a rate of growth that is about 2½ times the rate of growth of total energy.

Why has this been so? Possibly, the primary factors affecting the customers' choice of energy are:

1. Price

2. Reliability

3. Convenience

It has already been noted that, until 1968, the average price of electricity decreased, with general economic inflation of about 2 to 3 percent. The price of primary fuels tends to rise with inflation. This is no reflection on the management of enterprises concerned with other fuels. They did not have the same opportunities for improvements that electricity offered, as will be shown.

Over the long term, the reliability of electric service has been 99.98 percent "customer-hours on." To obtain this figure, each company in the country was asked to send in its 10-year past annual record of customer-hours off. The countrywide customer-hours off per year amounted to 0.02 percent of the total customers multiplied by the total hours in a year.

Finally, most people clearly agree that electric energy is convenient.

The factors of price and reliability are especially important to industry, where the largest percentage of electricity is used. Manufacturers could make use of other primary fuels. Furthermore, they could generate their electricity at their manufacturing sites. The trend has been away from this *on-site* generation, however.

**Reasons for Price Reductions**   Why was it possible to bring about a lowering of the average price of electricity during the period 1902–1968, when most commodity prices were rising? Here are the principal reasons:

1.  *Increased volume.*   Customers wanted to use more and more electricity. As illustrated in Chapter 8, this has an important bearing on unit cost and, therefore, unit price.

2.  *Larger and larger generating units.*   Chart 4.6 shows how investment cost per kilowatt goes down with size of unit. Chart 4.11 shows the trend in the maximum size of a unit. Furthermore, generally speaking, the large units utilize higher temperatures and pressures, both of which contribute to conversion efficiency.

3.  *Higher and higher transmission voltage.*   Higher voltage allows a

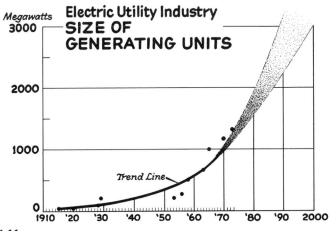

CHART 4.11

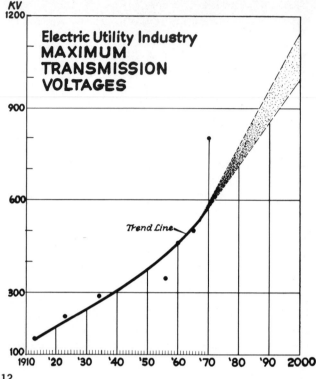

**CHART 4.12**

lowering of the cost per kilowatt in transmitting large blocks of power (Chart 4.12).

4. *More and more interconnection, coordination, and pooling.* The interconnected systems now cover the country (Chart 3.8). Most of the diversity benefit in interconnection has been obtained.

5. *Higher load factors, until recently* (Chart 4.13). The unit of cost (price) benefit of this factor is covered in Chapter 7. Also discussed in this chapter is what management did from 1931 through the period 1968–1970 to build load factor from about 50 percent to about 65 percent, as well as an explanation of why load factor has fallen since 1970.

6. *Consistent improvement in thermal efficiency.* Chart 4.14 shows the equivalent number of pounds of coal required to produce 1 kWh of electric energy from 1890 through 1976. This is the measure of *thermal efficiency* (i.e., the number of *British thermal units*, or Btu, required to make 1 kWh) of the power plants.

## PERCENT LOAD FACTOR
### *TOTAL ELECTRIC UTILITY INDUSTRY*

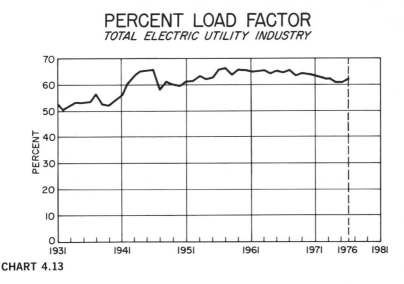

**CHART 4.13**

Chart 4.15 shows Edison Power Company's power-plant thermal efficiency.

7. *More and more diversity* (Chart 4.16).   Higher diversities result in lower unit cost (price). How management improves this is discussed in Chapters 7 and 8.

8. *Less and less self-generation,* (Chart 4.17).   With the lowering of the price and high reliability, fewer and fewer customers have

## EQUIVALENT POUNDS OF COAL REQUIRED TO PRODUCE ONE KILOWATTHOUR OF ELECTRIC ENERGY FROM FOSSIL FUELS

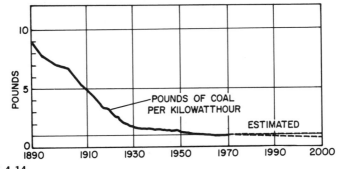

**CHART 4.14**

# THERMAL EFFICIENCY

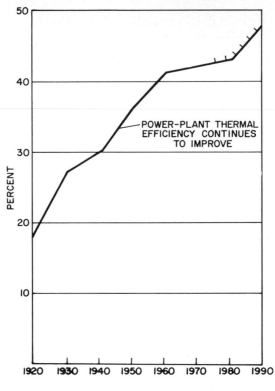

POWER-PLANT THERMAL
EFFICIENCY CONTINUES
TO IMPROVE

CHART 4.15

# EFFECT OF DIVERSITY

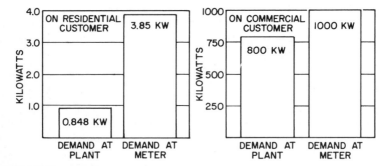

ON RESIDENTIAL CUSTOMER

0.848 KW — DEMAND AT PLANT

3.85 KW — DEMAND AT METER

ON COMMERCIAL CUSTOMER

800 KW — DEMAND AT PLANT

1000 KW — DEMAND AT METER

CHART 4.16

## TREND IN SELF-GENERATION

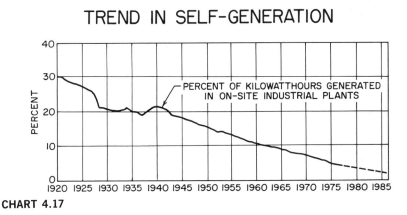

**CHART 4.17**

generated their own electricity. They prefer to purchase from the interconnected system.

9. *Increased productivity* (Chart 4.18).

10. *Electric energy has been constantly available.*  Availability and reliability have been factors affecting volume, which in turn helped to lower price.

## PRODUCTIVITY INDEX
### 1967 = 100

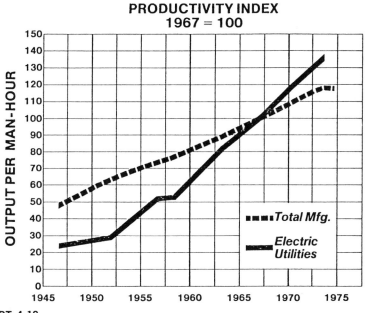

**CHART 4.18**

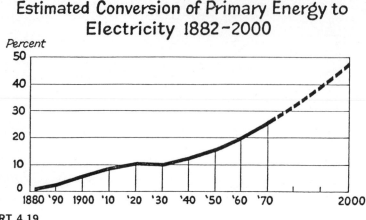

### Estimated Conversion of Primary Energy to Electricity 1882–2000

CHART 4.19

Because of the factors listed above, and others, there has been a constant increase in the percent of the primary fuels used in the conversion to electric energy (Chart 4.19). The forecast to 2000 is discussed in Chapter 7.

Before concluding the discussion on these economic trends, it should be mentioned that they did not "just happen." Management has played a role each step of the way in each factor mentioned. True, the opportunities were there, but someone had to make decisions, to make the opportunities realities.

**Other Economic Trends**   Chart 4.20 shows the trend in plant investment and gross annual revenue for the years 1930 through 1976.

## TRENDS IN PLANT INVESTMENT AND REVENUE

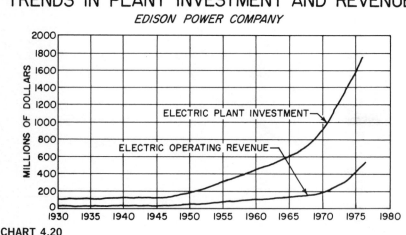

*EDISON POWER COMPANY*

CHART 4.20

# TRENDS IN REVENUE AND EXPENSES
### EDISON POWER COMPANY

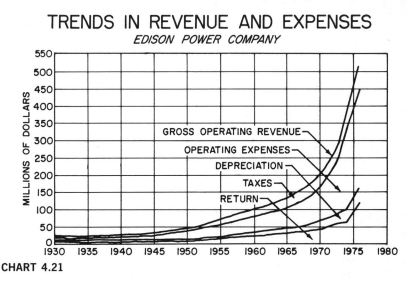

**CHART 4.21**

Chart 4.21 shows the trend in gross operating revenue, taxes, operating expenses, depreciation, and balance for return.

Chart 4.22 shows, for Edison Power Company, the amount of investment required per $1 of annual sales. This is the *investment ratio*, the ratio of plant investment to gross operating revenue. Note the rise to al-

# INVESTMENT RATIO
### EDISON POWER COMPANY

**CHART 4.22**

most $7 of investment per $1 of sales in 1933 when the company's sales were low because of the Depression. The investment ratio has come down substantially since then and is lower now than in the years 1947–1948, when the margin of reserve was at the low point.

Now that fuel costs are high, management is working hard to lower the investment ratio and must do so if the company is to earn a satisfactory return on investment.

**Production Costs**  Chart 4.14 indicates the continued improvement in the efficiency of converting primary fuel to electric energy.

Chart 4.23 shows the trend in the cost of fuels to electric utilities in the United States—coal, gal, oil, nuclear, and the composite cost.

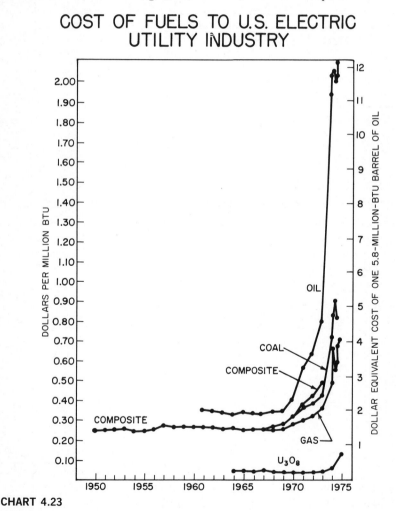

COST OF FUELS TO U.S. ELECTRIC UTILITY INDUSTRY

CHART 4.23

# COST OF FUEL FOR TOTAL ELECTRIC UTILITY INDUSTRY, 1950-1976

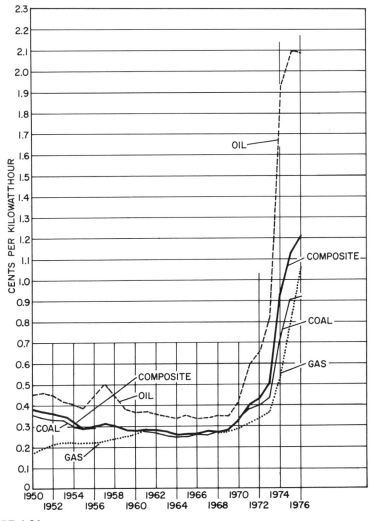

**CHART 4.24**

Chart 4.24 shows the trend in fuel costs, expressed in mills per kilo-watthour.

This chart and Chart 4.6, showing the increase in the cost per kilo-watt of power plants, show the effect on the total unit cost (price) of electricity. These prices will be evaluated in Chapters 7 and 8.

Chart 4.25 shows the taxes of Edison Power Company as a percent

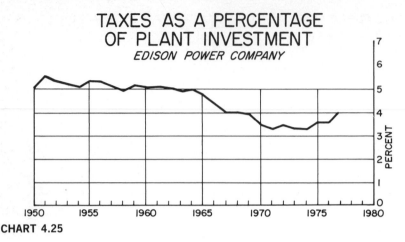

CHART 4.25

of plant investment, and Chart 4.26 shows taxes as a percent of gross revenue.

## OPERATING TRENDS BY PERIODS

In analyzing these various trends, it appears that the company has been affected by different circumstances during different periods of its growth. At a recent staff meeting, Edison Power Company's economists and top management reviewed the company's growth and its effect on these trends. They found that the company went through six significant periods, each of which produced marked changes in the nature of the company's operations.

**Period I: 1931 to 1933** These were the years of the severe slump in business caused by the Great Depression. Total sales, net income, and

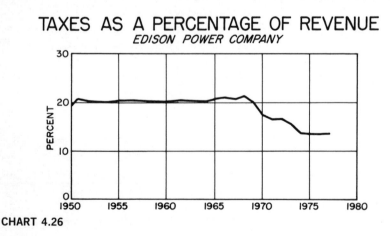

CHART 4.26

percent return on investment all dropped. Edison Power's management was doing all it could to cut expenses during this period, but despite these efforts it could not prevent a severe drop in percent return on the investment.

The peak load on the company dropped after 1930, with the result that, like most other companies in the industry, Edison had considerable excess capability during the thirties.

**Period II: 1934 to 1940** This is the period of the long climb out of the Depression. Sales of electricity began to rise as people had more money to spend and as companies stepped up their sales efforts. During this period a number of companies, including Edison Power, voluntarily lowered their rates with commission approval. They did this in the hope of increasing sales, even though earnings at the time did not justify rate reductions.

During the period 1934–1940 taxes, operating expenses, and depreciation also increased.

Largely because of the rate decreases during the period, Edison Power Company showed hardly any rise in net return during the period 1934–1940, although sales had increased considerably. This was not overly important because plant investment did not go up much in this period. As a matter of fact, investment decreased slightly because the value of old equipment taken out of service during the period was greater than the amount of new construction.

**Period III: 1941 to 1945** About 1940, Edison Power Company saw a coming need for more power capacity. Plant reserves were nearing 20 percent and dropping rapidly. As capacity was added, the investment in Edison Power showed a slight rise beginning about 1940. However, there was still no increase in net income.

The utility plant investment leveled off again beginning about 1943. The reason was that the company was unable to buy generators and other equipment; the manufacturers were busy building turbines and generators for the armed forces. The electric utility industry had to find ways to get along without new capacity and had to utilize all its old and somewhat obsolete facilities. The increased use of the older equipment resulted in higher costs.

**Period IV: 1946 to 1960** This was the period of postwar expansion. Total sales rose sharply, taxes continued to increase, and other expenses went up in proportion.

Despite the continued rise in sales, the balance for return still remained almost level until about 1947 or 1948. As a result, from 1945 through about 1948 the percent return on investment declined.

An increase in net operating revenue began about 1948, resulting from the increased operating efficiencies of the new generators coming into operation and higher rates. It was during this period that a number of companies had to reverse their longtime practice of reducing rates and ask the commissions to grant some rate increases. They were feeling the effect of those years when they had been unable to purchase newer, more efficient machines. The effect of the new machines they were then purchasing had not yet had sufficient impact on lowering production costs. However, despite these rate increases the average price of residential service continued its downward trend. This was because the sliding-scale nature of electric rates includes a built-in feature that results in automatically lowering the average price with increased use.

**Period V: 1961 to 1968**  This was a period of continued expansion. Generating-unit sizes increased, and Edison Power Company began using transmission voltages of 345,000 V and higher. Interties with neighboring systems were strengthened and pooling agreements tightened. The company began planning for nuclear plants, and electric space heating began to be a meaningful part of the company's load.

Increased efficiency of operation and increased use by customers made it possible for rate reductions to be made during this period. However, costs were going up and, particularly in the latter years of the period, were having an effect on percent return.

**Period VI: 1968 to 1976**  In 1968 Edison Power Company management could see economic problems about to develop. There was the continued pattern of larger units, higher voltage, more pooling, etc., but clouds were gathering. Construction costs began to rise with large increases in construction labor. Government was overextending and promising more. Money was getting tight. There were signs of increased inflation.

Then came the economic change mentioned earlier in this chapter. The whole economic climate changed severely, thus causing a reversal in the long-term downtrend in the price.

Much of the text that follows in this book is a description of what management is doing and should do to operate properly in the new climate.

# CHAPTER 5

# THE MANAGEMENT PROFESSION

In one way or another, most people are committed to the idea of improving standards of living, not only for themselves and their families but for others as well. Americans have developed a society and an economic system that allow initiative to prosper. There are rewards for those who are willing to risk time, effort, and capital on new ideas that bring benefits to all. Americans encourage innovation and have been able to find more ways to use machines to get their work done. Machines have multiplied the human ability to perform tasks and to produce the things which people want and need.

In an industrialized society, roughly 10 percent of the people are leaders in the community. These are the professional people, the managers and supervisors in business, the teachers, the leaders in government, the scientists, and the like. They contribute in large measure to getting work done and maintaining law and order. They contribute to the creation of jobs and improved living standards.

Their value is in proportion to their knowledge and the degree of excellence in their use of the required skills of leadership. These leaders must adapt themselves to the rapid increase in available knowledge and the more rapid resultant increase in rate of change. These things are true in particular of electric utility management.

## GETTING A JOB DONE

In any large, efficiently run company, the various levels of management from the top down to the supervisor make up the management team. Expert managerial teams operate highly complex industries and research facilities.

Today "getting the job done" is carried out on a massive scale. For example, an industrial management team might have to:

- Build a pipeline across the country
- Build a nuclear power plant
- Build a subway in a congested city
- Produce a better cereal
- Build a computer that will do the brainwork of 100,000 people
- Find a cure for cancer
- Build a model city
- Convert the hydrogen-bomb reaction to electricity production
- Plan and carry out a program to finance, build, and operate an electric utility over 10 years and earn the full cost of capital
- Develop and carry out a plan to use energy to feed the hungry

The idea of concentrating a mass of brainpower on a particular problem has been followed with considerable success. This method came into being with the advent of a new profession called management.

The matter might be simply stated in this way: All growth, progress, development, and building depend upon "getting a job done." The manager is the one who takes the required ingredients—material, supplies, knowledge, people, machines, energy, and capital—and then fashions them to accomplish a desired result.

Stated another way, the manager:

1. Begins by recognizing a problem
2. Analyzes the problem

3.  Sets up various possible hypotheses

4.  Tests the hypotheses as carefully as possible

5.  Weighs benefits against cost

6.  On the basis of the information gathered, makes a decision

7.  With the proposed solution in mind, provides the organization and the tools necessary to see that it is carried out

8.  Then, periodically, measures the results and makes the necessary adjustments

This is the basic management method, and it can be applied with only slight variations to every phase of business activity. It requires intelligence, judgment, and the willingness to take risks. Of all the professions, management is probably the youngest. Engineers have been known since the first log was thrown across a stream. Lawyers were needed when people first began living in groups. The medical profession can be traced to the dawn of history. But management, as it is known today, is not much more than 100 years old.

Two great inventions made it possible and necessary, the first being the industrial revolution. This great transformation began when a Scotsman named James Watt watched steam coming from a teakettle and realized that it represented energy which could get work done. This led to the development of machines, working for people and helping them produce more things. The machines cost money. Their work had to be organized and their output marketed in a rational way. The need for capital and better organization led to the second invention, called the *corporation*, or *company*. The idea of the corporation is so widespread that people tend to think of it as a natural resource, like air and water; but historically, it is a new idea. Its development led directly to the development of the profession of management.

It takes an enormous amount of capital to finance the required machinery in a modern industrial society. Capital comes either from savings of people or from government which gets its capital basically through taxes. Because few individuals have sufficient savings (capital) to buy all the machinery needed, the company or corporation was formed as a means through which many people could pool their savings to acquire the facilities for manufacturing the products they and others want and need.

The purpose here is to concentrate attention on the skills of management as they apply to the electric utility industry today. It is a considerable task, for it means outlining, in very brief form, the management job in a changing industry in a changing world.

Management in the business corporation is simply the act of accomplishing by and through people desirable economic results—taking into consideration the interests of the owners, the employees, the customers, and the community.

Management must be concerned with behavior patterns of people. This requires being familiar with their knowledge, attitudes, beliefs, hopes, and aspirations. Managers must understand what motivates people and be able to turn this understanding into effective action.

It is advisable, therefore, to identify the skills required of effective management in terms of objectives and in terms of specific skills.

## MANAGEMENT OBJECTIVES

**Maintaining the Economic Health of the Organization**   This is the first and most important objective. For an electric utility company the ability to operate a business at a profit or a "fair return" is of the greatest economic importance. Of course, this does not mean that the manager should be shortsighted and take a quick profit today at the expense of long-range goals. Rather, managers should view economic performance in the light of all conditions, both internal and external to the company. Every manager, no matter at what level of responsibility, is benefited when the company is successful and is adversely affected when it is not.

**Integrating the Viewpoints of People and Functions**   Each person is different, with individual ideas and viewpoints, loyalities, goals, and values based on experience. The diversity of people in an organization should be coordinated and united to form a harmonious whole.

It does not really matter what size the group is; it could be 20, 200, or 2000 people. The skill involved in motivating them toward a common purpose is the same. The accountant should realize the importance of that role, but must also realize that the business is not functioning just to maintain accounting records. An engineer should realize that the distribution department has problems which must be solved. The distribution people should realize that a sound planning program is vital to the company. This requires communication—constant communication within the company—at all levels of authority.

**Instilling the Service Motive**   The manager should be adept in the skill of instilling the service motive in other people. One way is to have all employees, especially supervisors, understand the manner in which energy serves the community's welfare—of how it helps improve living standards (see Chapter 7). Inspiring excellence in subordinates involves a combination of factors, not the least of which are setting a good example and providing incentives to stimulate above-average effort.

**Making the Organization Dynamic and Adaptable**   The economic climate is constantly changing. Good management will adjust to it. New methods and tools are being developed. Management will choose the best, test them, and, if they are deemed sound, adopt them. Top management should be able to demonstrate that it does not get "set in its ways."

**Providing Human Satisfaction in Work Output and Relations**   There are few areas in which management's attitude is so telling and so important. Ways are found to encourage the outstanding individual while keeping the whole organization moving harmoniously toward a common goal. The manager endeavors to use all the knowledge and skill being developed by social scientists to help understand how to motivate employees. Pay alone is not enough, nor is work satisfaction. Both are surely part of the answer, but there is much more to be learned in this area. One valuable approach to motivation is employee education as outlined in Chapters 6 and 10.

**Relating the Company's Affairs to the Community**   Electric utility management should take the initiative in community affairs, which may involve such considerations as cleaning up the environment, conservation, area development, jobs, and industrial growth. Energy plays a role in most of them. Leaders should know the role of the electric utility in the community. Company plans and objectives are related to those of the community. While public relations programs represent a factor, this concept basically has to do with the company's conviction that its interests and those of the community are the same.

## THE SKILLS OF MANAGEMENT
All leaders in the community, whether professionals, managers, or others, have developed certain skills that they utilize in carrying out their affairs. Just as doctors and lawyers learn the special requirements of their professions, so, too, do the managers. The more proficient they become in practicing these skills, the more effective they will be as managers. Generally speaking, such managers are given greater responsibilities by their superiors. The following skills refer particularly to the electric utility business.

**Forecasting**   It may take 12 or more years to complete the building of a nuclear power plant. It takes less lead time to build transmission lines and distribution systems. Nevertheless, the utility, whether in generation, transmission, or distribution, must plan in advance and plan wisely and accurately. Managerial success depends on how accurately the plan can be made. Planning is started by a forecast, which should be for not less than 5 years and preferably for 10 years or longer.

In the budget, there is an overall company forecast and a plan for 1 year. Chapter 7 discusses as an example a 5-year plan which begins with an accurate forecast. Wise managers of energy in any country make long-range forecasts of all energy requirements, including the primary fuels needed as well as electric energy requirements. Such a forecast should be for 25 or 30 years. From this forecast is prepared a long-range energy policy for the development and use of energy, especially as it relates to primary fuels that may be scarce. Those countries that have not established such a policy find themselves in difficulty now that it is apparent that oil and gas will probably be the first two fuels to be exhausted. This was known more than 20 years ago, but some countries did not recognize the significance of this fact because they had not made satisfactory forecasts and plans.

**Planning**  Planning is conducted across the board, from the lowest supervisor to top management. Every function is included, and all departments have input. Alternative plans are prepared, and the best is selected—after tests. Once top management adopts the overall plan, the plans for all departments and subdepartments are established. There is constant follow-up, and changes are made when desirable.

**Selecting Staff**  Modern management has a skilled personnel department. There is a computer record on each current employee, giving all pertinent information, including education, skills, experience, training, and qualifications. When someone is needed for a higher or different position, an examination is made of all qualified existing employees. Prospective employees are screened and tested to determine their skills, aptitudes, and abilities. In the interest of the employee and the company, the aim is to place each person in a position for which the individual is qualified.

**Organization**  Skill in organization comes largely through experience. No one organizational design fits every situation, and routines can vary depending somewhat upon the existing staff and whether it is functioning smoothly.

The organizational situation reflects the nature and skills of the manager of the company or the departments. There are, however, some possibly obvious rules. For example, avoid the "one over one" setup, where one person has only one to supervise. The number reporting to one supervisor should be neither too small nor too large, with the range being perhaps four to ten. There should be clear lines of authority and well-established lines of communication, up and down and across. Means should be provided for full coordination. Rules should be established for jobs requiring joint effort. Rules for delegation and decision making should be established and practiced.

**Selecting Equipment, Material, and Facilities**  Through experience, study, and practice, managers of electric utilities develop the skills needed in selecting equipment, some of which is the most complicated of all machinery. Naturally, skilled technicians assist in this process.

**Establishing and Maintaining Controls**  The efficient manager sets up a tight control system on every phase of the business: construction work in progress, inventories, financing schedules, preparation of short- and long-term plans, budgets, all studies regarding new methods and procedures, employees by categories, fuels, and maintenance schedules. The manager calls for and receives periodic progress reports.

**Reviewing and Appraising**  With respect to problems, the manager goes through a review and appraisal process, or instructs someone to do so. This involves stating the problem, analyzing it, setting up possible hypotheses and testing them, weighing the benefits against the cost, making decisions, and allocating the necessary organization, people and money. Periodic measurements of progress, along with review and appraisal of results then follow. Any errors or schedule slippages are detected and corrected.

**Establishing Rewards and Incentives**  Through many avenues and methods, employees are encouraged to perform efficiently, avoid waste, improve productivity, be courteous, practice safety, plan ahead, and strive for excellence. Competitive pay scales are important, and should provide a sufficient difference in wages between positions to encourage striving for promotion.

Besides wage and salary scales, management uses other means for recognizing excellence and unusual performance. One important incentive is to encourage the purchase of the company's securities by prospective investors. This means paying a proper return on investment.

**Communicating**  Many of the problems faced by managers of electric utilities and of all business stem from the lack of understanding on the part of employees, government agencies, and the public. A basic example is illustrated in Chapter 6. There is a strong correlation between lack of knowledge and unproductive behavior. Knowledge breeds understanding, which, in turn, breeds trust. These matters are subject to close measurement and analysis.

**Leadership**   Those in management positions probably have already displayed some degree of leadership. This is one of the primary reasons for being chosen. Some acquire the quality through heredity, some through home training or more formal training, some through self-training.

To be a leader requires the possession and display of those qualities which others admire and respect. Generally speaking, a leader has above-average intelligence, a background of knowledge and experience which enable sound decision making, and good personal character. In addition to being a hard worker, a leader has imagination and enthusiasm, an understanding of human nature, and kindliness. Desiring loyalty from others, a leader also exhibits this quality. Most people know the required characteristics. Those who develop them become the leaders.

**In Summary**   Here, then, are the skills of management:

1. Forecasting

2. Planning

3. Selecting staff

4. Organizing

5. Selecting equipment, materials, and facilities

6. Establishing and maintaining controls

7. Reviewing and appraising results

8. Establishing rewards and incentives

9. Communicating

10. Leadership

## DEVELOPMENT OF THE MANAGER

The question is frequently asked: How does one go about becoming a high-level manager? That question is not easy to answer, but it is deserving of comment.

The human race is composed of individuals having a wide range of qualities, some of which are subject to varying degrees of measurement. For example, intelligence ranges from roughly 60 to 140 on a scale where 100 is average. The bulk of the measurements fall in the range from 95 to 105. There is a wide range in honesty, in willingness to work, and in such virtues as courage, humility, magnanimity, obedience, and perseverance. The manager or leader of the community is one who strives for excellence in those qualities he or she knows instinctively to be good.

There is no lack of opportunities for people with these characteristics. Indeed, the higher the level of management, the more difficult it is to fill the positions.

Where do people get these qualities? Possibly from:

1. Heredity

2. Training by others, such as parents, teachers, or company leaders

3. Self-training

4. Environment

Studies indicate that there is much that can be done in the areas of self-discipline and training. It may well be that this could be the most important source of better quality. Reading the history of great men, their biographies and ancestry, indicates that this may be so.

Through a combination of all these qualities, there appears to be a group of people in any community, relatively small in number, who are the "doers." Generally speaking, these are people who have developed the qualities required for leadership and who "strive for excellence." If there is one definition and one only for a manager, it is this: a leader who strives for excellence in living habits, in character, in work performance, and in concentration on getting the job done well.

Anyone can see exceptions to these rules. Nevertheless, the chances of reaching the positions of leadership are much greater among those who embody the qualities described here.

**In Summary**  With notable exceptions, then, the manager is one who:

1. Strives to develop those personal characteristics that others admire

2. Is looked upon by others as a leader

3. Constantly strives for excellence

4. Is willing to make the self-sacrifices necessary to become a leader

5. Makes efficient use of time

6.  Likes to help others

7.  Continually strives for improvement through exercise and strives to build physical power, mental power, willpower

## OBJECTIVES OF ELECTRIC UTILITY MANAGEMENT

Following is a brief statement of the generally accepted overall purposes of utility management. All management takes some part in carrying out these functions, which will not be elaborated upon here as each chapter in some respects will dwell upon them.

In considering these basic functions, it becomes apparent why the company must operate as a unit, why each department depends upon the others, and why each department head and each manager should know what the other managers are attempting to do. Here is the list:

1.  To provide all the electricity required, when, where, and in the amounts desired

2.  To provide service as reliably as possible

3.  To design the electric poser system and operate it in the most economical fashion consistent with good service

4.  To achieve the most economical operation which requires, as a fundamental premise, the striving for maximum utilization of investment, which is measured by load factor

5.  To pay those wages and salaries required to attract, hold, and keep satisfied good employees

6.  To make the price to the customers as low as is practical consistent with good service and the earning of a fair return to investors

7.  To constantly earn not less than the cost of the capital required to finance needed expansion

8.  To operate the business in a way that is pleasing to employees and the public

9.  To operate in a way so as to minimize the detrimental effects on the environment and maximize the beneficial aspects

Management constantly faces the need for making sound decisions respecting complicated problems, all having to do with aspects of these purposes. It has been wisely said that a person's judgment on any subject is no better than that person's knowledge of that subject. Management's first responsibility is to accumulate appropriate knowledge of the electric power business itself, in all its phases, not merely the phase which has previously occupied the manager's attention. Furthermore, the manager should acquire knowledge of the overall purposes of the energy business, the skills needed to carry them out, and the tools available. With this knowledge, the effective manager is able to improve his or her decision-making functions.

## HOW OBJECTIVES OF ELECTRIC UTILITIES ARE CARRIED OUT

1. Management first designs a plan for:
   a. 1 year (the budget)
   b. 5 years (fairly firm, but reviewed each year)
   c. 10 years (on parts of the plan)
   d. 20 years (on energy policy and priorities)
2. Management develops the plan under the following categories:
   a. System planning
   b. Operations (expense control)
   c. Building utilization of investment (load factor)
   d. Employee and public relations
   e. Pricing

The purpose of this book is to briefly outline general principles and practices and then to specifically describe how they are applied to a case study. Computer models will be used and illustrated, and the corporate model will be discussed.

The latest application which is now sweeping management is the use of the computer in assisting top and middle management in making wise decisions. With an appropriate computer model, management is able to develop on paper and put into the computer various alternative plans for the solution of a problem. The computer then calculates the results that will be obtained from each of the trials. Management can select from many alternatives the plan that seems best suited to its needs. Management can see the impact of a decision without waiting for 5 years for the actual results to occur.

The corporate model is the latest procedure in building a model for management. Here the complete 5-year or 10-year plan of a company, with all its details, is put into the computer model. All department heads know of the decision to carry out that plan, and each deals with the appropriate portion. Subsidiary computer models of departmental functions are built. After the plan is made, management up and down the line can check all future decisions to determine what effect they have on the overall plan.

## MANAGEMENT AND HUMAN BEHAVIOR

Some managements are quite proficient in the use of their skills in dealing with the behavior of their employees and the public. Managers who know of these skills and the tools available are the ones who generally have good employee relations and good public relations.

1.   Management works with and through people.

2.   The behavior of people varies depending upon their knowledge and beliefs, both of which are subject to measurement beforehand.

3.   There are definite correlations that will be illustrated between a person's knowledge and attitude, such as (a) attitude and efficiency and (b) attitude and safety record.

4.   Attitudes and behavior patterns can be dealt with scientifically.

5.   A large body of knowledge on how to deal with problems concerning people is already available.

6.   A manager's effectiveness depends upon knowledge of how to deal with problems involved with human behavior and the skillful application of that knowledge.

If a manager believes employees are not behaving properly, the question should be asked: Are these employees properly informed as to the facts and purpose of the business? If they are not informed, why not? Part of management's responsibility is to keep employees properly informed on the affairs of the company and to improve their knowledge of what the company is doing so that they can exercise better judgment.

If the employees are behaving poorly, to what extent is it traceable to their ignorance of facts about the company which management possesses, but has failed to furnish to them? There are definite and specific skills and tools that can be used by management to build good employee and public relations. These will be illustrated in Chapters 6 and 10.

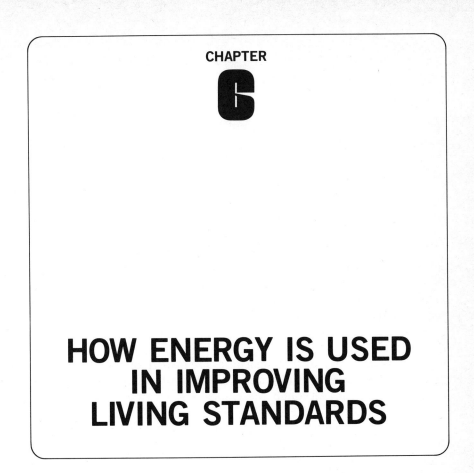

# HOW ENERGY IS USED IN IMPROVING LIVING STANDARDS

## ENERGY AND LIVING STANDARDS

This is a description of how management uses energy to run machines to increase productivity and living standards. As defined here, *good living standards* mean:

1. Sufficient food, clothing, and housing for satisfactory living, and prospects for improvement.

2. Good health, which depends on such factors as a balanced diet and good medical services.

3. Sufficient insurance to provide security and peace of mind.

4. Leisure time for social intercourse, recreation, and personal development.

127

5. Material possessions and services which reduce drudgery and enhance enjoyment, ranging from clothes washers and televisions to telephones and travel agencies.

All these "things" are spoken of by the economist as *goods and services*. All the goods and services of a country make up the *gross national product* (GNP). For better understanding of the terms as used here, see Chart 6.1 [1].*

For people to obtain these items and satisfy the wants of a growing population, it is necessary to increase the total production of all goods and services. The ingredients used are materials, people, capital, machinery, energy, and management.

Obviously, all living beings must exert energy to get food and shelter, or they will not survive. Humans, with relatively little muscular energy, made the big leap forward in building living standards with the development of machines operated by inanimate energy, which made it possible to produce a great deal more per hour of work. Productivity has been shown to be the key to the good life and to improved living standards.

The manner in which the machines are owned, operated, and managed depends to some extent upon the kind of political society under which people live. Although these societies have been given various names and although there are some variations, all of them basically fall into two major categories. One calls for government ownership, operation, and management of all the machinery of production. It is sometimes referred to as a government planned economy. The Soviet Union is the principal example today.

The other system is generally referred to as the *free society*. Under this method, people as individuals and as groups own the machinery of production, and arrange for operation and management. Government is assigned the task of regulator. Although these two systems will be briefly described, this chapter deals largely with the methods used in the free society.

It appears that roughly 80 percent of all of the jobs available in a country like the United States are with a company or corporation under some type of business management. Experience has shown that frequently there is considerable lack of knowledge among people on such matters as profits and the division of income between owners on one hand and workers on the other. This knowledge, or lack of it, has important bearing on the behavior of employees. This chapter will analyze these behavior patterns especially from the standpoint of what management can do about them.

---

* Numerals in brackets refer to source notes at the end of this chapter.

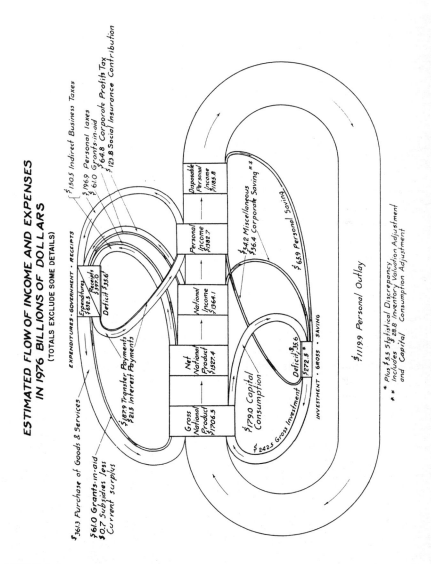

CHART 6.1

## USING MACHINERY RUN BY INANIMATE ENERGY TO GET WORK DONE

Any advanced industrialized nation could be used to illustrate these beginnings and these principles. As the United States was one of the first to apply these principles on a broad scale and to utilize modern business management and as the statistics to illustrate these principles are readily available, it will be used as the example.

This should not be interpreted as an endeavor to indicate that the United States is so much further advanced than other countries in this respect. Also, it is not intended to indicate that the American people, as a class, are in any way superior to other people. It is believed that the reasons for any advancements that may have been experienced can be traced to the principles used. The endeavor will be to illustrate the principles which can be applied elsewhere.

Charts will be shown indicating some of the long-term trends in the United States. From these it can be seen that it was not very many years ago when the living standards in America were about the same as they are now in some of the developing countries. Throughout this study, an attempt will be made to evaluate the contributions that electric energy has made and can continue to make in this overall development.

As will be illustrated, the biggest contribution of machinery is in helping to keep the prices of goods and services lower than they would be otherwise. By keeping prices low, people can buy more of the product being made by a particular manufacturer and have more money to buy other commodities. In this way, new jobs are created.

**The Beginnings of Machinery Driven by Inanimate Energy** As group living evolved among humans since the dawn of prehistory, social arrangements developed to keep pace with the needs along the way. Before the advent of machines, people improved their living standards by dividing the work, but progress was still slow because they had little energy in their own muscles and those of the animals they could harness.

Imagine 20 athletes riding one bicycle in tandem. Attached to the bicycle there is an electric generator, which indicates the amount of electric energy they can produce. When these 20 well-conditioned and strong people pump as hard as they can steadily at their maximum strength, for 1 h they will generate only 1 kWh of electric energy. Roughly, this is about the same amout of energy that is expended by one horse pulling a full load for 1 h. Today, 1 kWh can be purchased for a few cents. With modern machinery the average worker now has available tens of thousands of kilowatthours per year to help get work done.

**Elementary Illustration of a Tool—Sometimes Called a Machine** An elementary example of the principle of the machine involves the killing of a bear in a primitive society by a person whose muscular strength alone is not great enough to accomplish the task, but who makes a club to get the job done (Charts 6.2 to 6.4). In such a case the person wants something and uses human energy to kill the bear (or to build a shelter), but this energy is made more effective through the use of a tool, in this case a club. With the club, human production of food, clothing, and shelter could be increased.

Producing the club may have taken a few days, during which its manufacturer could neither hunt nor fish. This deprivation made possible greater future production through the creation of a basic machine. If this person was in the business of selling furs to the neighbors, income was reduced by the time it took to make the machine or the tool. The money (or working time) spent in developing a machine is called *capital*. People deny themselves something they want today in order to invest their time and savings in machinery to get more things tomorrow. All investments of money put into machinery to increase production are spoken of as capital.

The clubmaker exercised self-discipline and self-denial in the expectation of reward, an increase in *productivity*, or output of goods per hour. In a much later era, such a person might be in the business of manufacturing and selling machines and might be called a capitalist because of investing savings (time) in machinery.

## CONSTRUCTING A PRIMITIVE TOOL
### *THE CLUB*

CHART 6.2

# EARLY USES OF A PRIMITIVE TOOL
## *KILLING A BEAR*

**CHART 6.3**

**The Big Jump Forward**   The big jump in machinery (and living standards) came when a way was found to use steam to run engines. As was noted in Chapter 5, James Watt conceived the steam engine, which marked the beginning of the Industrial Revolution (Chart 6.5). He developed the steam engine for commercial use in 1769. No longer limited to using energy from wind, water power, animals, or humans, people could now build machines wherever fuel could be found to heat water to make steam. Charts 6.6 and 6.7 show various kinds of machines that have been built to increase productivity.

The conversion of primary fuels, such as wood, coal, gas, and oil, to electricity provided a form of energy that was more flexible, more convenient, easier to use, and easier to transmit. Furthermore, as illustrated in other chapters, electric energy could be made more reliable and more

# EARLY USES OF A PRIMITIVE TOOL
## *BUILDING A SHELTER*

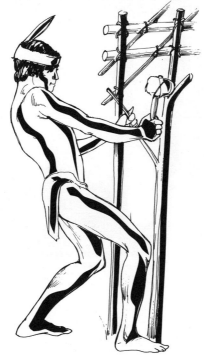

**CHART 6.4**

# THE BEGINNING OF THE INDUSTRIAL REVOLUTION

**CHART 6.5**

## EXAMPLES OF MACHINES
### TRACTOR, TRUCK, AND TRAIN

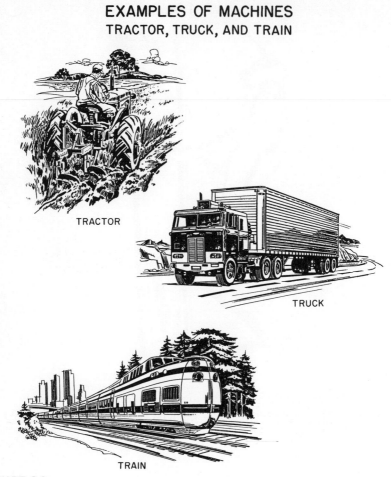

TRACTOR

TRUCK

TRAIN

**CHART 6.6**

economical than other forms of energy. As shown in Chapter 4, electricity has increased at about 2½ times the rate of growth of all primary fuels, such as coal, gas, and oil. Furthermore, almost any kind of primary fuel or falling water could be used in producing electric energy. In Chapter 13 it will be noted that there have been continuing changes in the patterns of use of various primary fuels available. The world is now going through another gradual change as it becomes more evident that gas and oil may be exhausted sometime shortly after the year 2000. But other fuels, such as coal, are still in abundance, and nuclear fuel will last a very long time, especially when used in the breeder reactor. In Chapter 11, it will be seen that fusion, and possibly solar energy (on a large scale), probably will be available after the year 2000.

## EXAMPLES OF MACHINES
### PLANE, FACTORY, AND COMPUTER

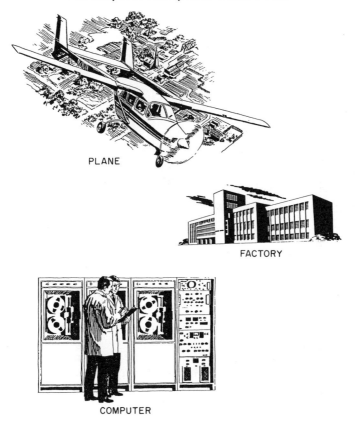

PLANE

FACTORY

COMPUTER

**CHART 6.7**

It is expected that by the year 2000 about half the consumption of primary fuels will be for generation of electric energy for end uses. Today, most factories and industrial processes are operated with electric energy. Electricity use has become a good index, not only of total industrial production, but also of productivity—the amount a person can produce in 1 h of work. Chart 6.8 shows for the United States the industrial use of electricity per worker [2]. This was 47,903 kWh in 1974 and it continues to increase.

If the 20 athletes in the previous example could pedal with the same strenuous effort for a year, they could generate 3650 kWh, each athlete producing 183 (3650 divided by 20) kWh. Note in Chart 6.8 that each worker now has at hand, not 183 kWh a year, but 47,903 kWh a year.

Ordinary people doing an ordinary day's work each produce the

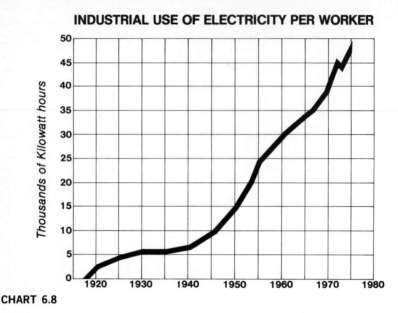

**CHART 6.8**

muscle-equivalent of 67 kWh of energy per year. With 47,903 kWh a year at hand running machinery, each worker has the equivalent of 700 helpers (Chart 6.9), thus increasing individual productivity 700 times.

Chart 6.10 illustrates how production by machinery has grown since 1850. In 1850, 65 percent of the work output was from animate labor by humans or animals. Only 35 percent was from inanimate energy. Today, about 98 percent of all work is performed by machines run by inanimate energy [3].

Note from Chart 6.11 that with machinery and energy, a country with one-twentieth of the world's population can produce 26 percent of the world's goods and services [4]. This does not illustrate any superiority of a race, but rather what might be expected through a system of getting work done by machinery to improve living standards.

## MACHINES, ENERGY, WAGES, AND JOBS

**Machines Make Jobs**  When the machine was first being developed, Karl Marx and others feared that machines designed to help humans do their work would create great unemployment. This contributed to his belief that, in the worker interest, government should own all machinery of production. Indeed, even today, surveys of public opinion indicate that some people still believe that machines put people out of work and destroy jobs. The facts may help to correct this misinformation.

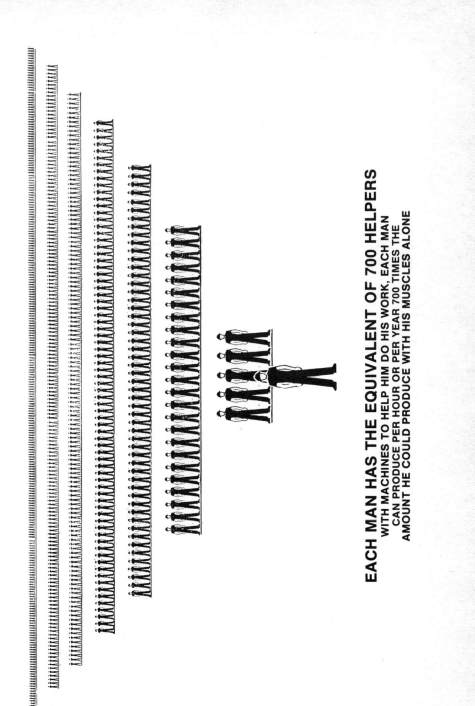

**EACH MAN HAS THE EQUIVALENT OF 700 HELPERS**
WITH MACHINES TO HELP HIM DO HIS WORK, EACH MAN
CAN PRODUCE PER HOUR OR PER YEAR 700 TIMES THE
AMOUNT HE COULD PRODUCE WITH HIS MUSCLES ALONE

CHART 6.9

## ESTIMATED WORK OUTPUT BY SOURCE OF ENERGY IN THE UNITED STATES

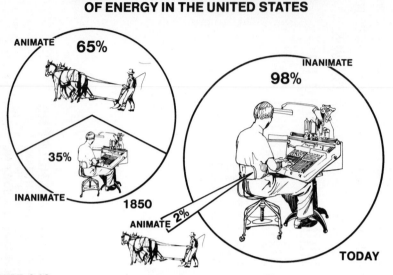

CHART 6.10

## WORLD NATIONAL INCOME and POPULATION
*NET OUTPUT OF GOODS AND SERVICES*

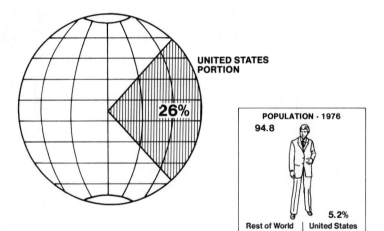

CHART 6.11

No doubt, new machines cause unemployment in some areas while creating jobs in others. It is likely that when Henry Ford developed the automobile, the few hundred workers who may have been engaged in the manufacture of buggy whips or wagons lost their jobs. But today there are 800,000 people employed directly by the automobile and trucking industry. Counting those indirectly employed by this industry, the total may be in the neighborhood of 2 million people [5].

Note from Chart 6.12 that during the period from 1870 through 1976, population increased 435 percent. Over the same period, the number of workers increased by 794 percent. Workers in mining increased only 266 percent, while workers in manufacturing increased 757 percent. In transportation, communication, and public utilities, the increase was 603 percent [6].

Where new machines are created to do a better job more economically, they may cause the displacement of some workers. However, it is the aim of modern management to retrain these people so that they can be gainfully employed elsewhere.

The process of mechanization results in greater efficiency of production. Thus the price of one product may be lower than otherwise, making it possible for people to devote some of their income to another product, thereby increasing manufacturing in that area and creating more jobs.

With the computer, an advanced machine, it is being found that further economies can be realized. It can do for mental labor what the ordinary machine has done for physical labor. It is likely that the computer

## MACHINES MAKE JOBS

### *Population Growth and Occupational Employment, 1870-1976*
### *(Per Cent Increase)*

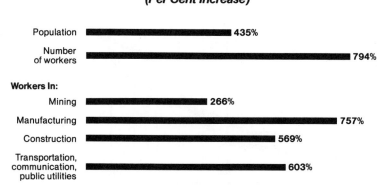

CHART 6.12

will have the same beneficial result in improving living standards as the ordinary machine. The aim should always be for employees to retrain themselves or for management to retrain them so that they can fill better positions.

**The Relationship between Use of Energy Per Hour and Income Per Hour**   Income per hour is a good index of living standards. Chart 6.13 shows how the income per hour goes up with the use of energy per hour [7]. Here the energy is reduced to *horsepower-hours per worker-hour*. (The *horsepower-hour* (hp·h) is close to the kilowatthour in energy value. A horsepower-hour is roughly equivalent to the energy a horse can exert when pulling a loaded wagon for one hour).

This has risen steadily from almost zero in 1850 to almost 8 hp h per worker-hour in 1974. Each worker can now produce each hour the equivalent of that which seven horses could produce.

Notice how income per worker-hour rose from about 40 cents per hour in 1850 to about $3.65 per hour. These values are in constant 1958 dollars. The two curves rise together. They go hand in hand in improving living standards. This is the key to America's improvement in living standards and to such improvement in any country.

Electric utility management is conscious of the manner in which this resource contributes to the overall living standards of people. Today,

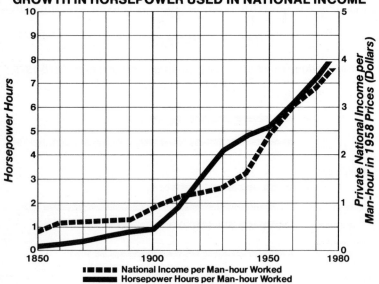

**GROWTH IN HORSEPOWER USED IN NATIONAL INCOME**

■■■■■ National Income per Man-hour Worked
■■■■ Horsepower Hours per Man-hour Worked

**CHART 6.13**

there are some who advocate a "no growth" economic policy as well as a "no growth" energy policy. Others call for a "go slow" energy policy, without proper distinction between useful energy and wasteful energy and between energy which is in short supply and that which is not.

Obviously, it is always wise to avoid waste, to operate efficiently, and to conserve valuable resources, such as declining reserves of gas and oil. However, other fuel resources, such as coal and uranium, are relatively plentiful. Even so, these also should be used wisely.

The point is that if the overall aim is to build living standards, care should be exercised to weigh all factors before instituting a "go slow" policy regarding useful (not wasteful) energy. Income per capita, jobs, productivity, and total production can be adversely affected by a slowing of growth.

Wise management tries to reach a proper balance between conservation, environmental protection, and the overall well-being of people as measured by their living standards. For sound decision making, it is first necessary to have knowledge of all the interrelated factors. Prudent management then shares this knowledge with employees, customers, and all who may have an interest. The aim of this chapter is to help provide some knowledge of the relationship between useful energy and living standards.

**Shift in Workers from Farm to Industry**  In Chart 6.14, note the decline in percentage of workers in agriculture from about 70 percent of all the workers in 1820 to about 3.5 percent today. The percentage of workers in factories and in industry has risen from less than 30 percent in 1820 to about 97 percent in 1975 [8]. In about 1880 the two groups were approximately half and half. The shift from an agricultural society to an

# MORE WORKERS ARE FINDING JOBS IN INDUSTRY

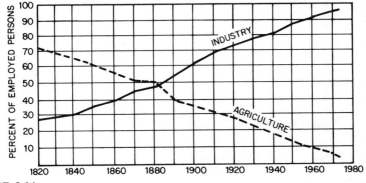

**CHART 6.14**

industrialized society is reflected in this chart. Each world society passes through these steps as it becomes industrialized. In the United States, the large farms are really large manufacturing concerns. As the farms get fewer in number, they become larger in size. This size enables them to produce more by means of machinery and to keep prices lower than otherwise. The total average investment per farm is now about $200,000 [9]. In the United States about 3.5 percent of the workers can feed the entire population—some 200 million people. In addition, the United States sends food abroad. This again illustrates the value of the use of energy and machines for getting work done. Farm productivity has increased many times over. Charts in other chapters will illustrate these improvements.

Today, electricity is available to practically all farms and small-town dwellings in America. That point was reached around 1950. As each country expands its production program and thus expands its energy use per capita, it will be found that the pattern will be somewhat similar.

**Further Evidence of Improved Living Standards with Energy**  Chart 6.15 is another example of this principle. It shows the buying power of the average hourly factory worker, based on 1960 constant prices. The real buying power of the average wage earner has risen from about 40 cents per hour in 1850 to about $2.50 today, in 1960 constant dollars [10].

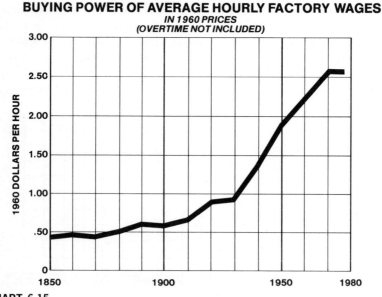

**BUYING POWER OF AVERAGE HOURLY FACTORY WAGES**
*IN 1960 PRICES*
*(OVERTIME NOT INCLUDED)*

**CHART 6.15**

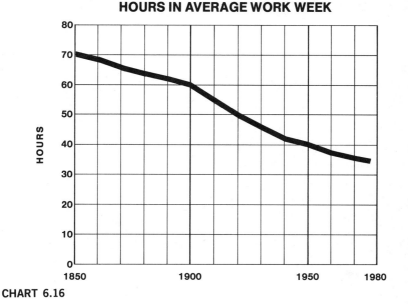

**CHART 6.16**

Another measure of improved living standard is the decline in the number of hours in the workweek of the average worker. This has dropped from about 70 h in 1850 to about 36 h today (Chart 6.16) [11].

**Investment in Tools or Machinery—The Corporation**  It requires an enormous amount of capital (savings of people) to build machines that use the energy to make the things to increase production and productivity. Later, it will be demonstrated that the electric utility business requires about one-eighth of all the total capital in society. The machines that use the energy require the other seven-eighths of the capital. The two must go hand in hand. Energy is used as it is needed. The amount of energy used will rise directly in proportion to total production and productivity.

Chart 6.17 shows the manner in which the investment in tools for each worker in industry has increased in a little over 100 years. It was under $1000 per worker in 1850. As late as 1930, it was only some $6000. Now it is about $43,000 per worker [12]. This means that some people have denied themselves something so as to save and invest $40,000 in tools for each of the workers on the machines. It is this enormous investment in machinery, using so much energy, that enables a worker to produce so much per hour.

During the early years of industrial development, the individual manager or manufacturer owned the machinery. Later it was found that

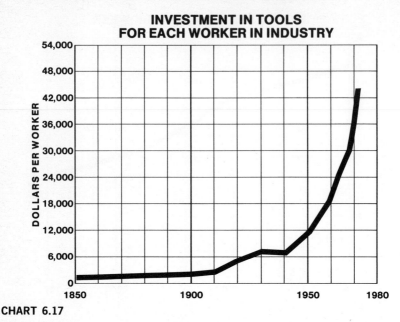

**CHART 6.17**

investment requirements were so great that few individuals had the necessary savings to assume the whole task.

The idea of the corporation or the company came into being. This was simply a means by which many people could pool their savings to provide the necessary capital. As mentioned in Chapter 5, this marked the beginning of the modern professional manager.

All capital comes from the savings of somebody. It is money not spent for food, clothing, housing, or other living expenses. To encourage people to save and to invest their savings in machinery requires that the person who saves be paid interest or dividends on the investment. Later, a discussion of the stock market and bond market will illustrate what the interest rates are.

When government provides the capital to buy machinery, it decides how much of a person's income will be placed into machinery. In the government planned economy, the government manufactures and sells all the goods and services. Government decides what to charge for these goods and services. The charges are enough to cover the cost of manufacturing plus whatever additional amount government deems necessary to provide additional machinery. Both kinds of society aim to increase productivity through the process of using more and more machinery run by more and more energy. On the one hand, the government decides how much of the workers' income will be placed into machinery. On the other, in the free society every individual is free to save or not to save and free to invest or not to invest. If there is sufficient incentive to save and to invest, a person will do so for the benefit of the return that can be obtained on the savings.

Chart 6.18 shows the investment per employee in various industries for the year 1972 [13]. Notice that the largest investment, $313,000, expressed in investment per production worker, was in the petroleum industry. The next largest was $250,000 per employee in the electric light and power industry [14]. Third was chemicals, with $89,000 per worker. The average for all manufacturing was $43,000.

Today, in America, roughly 80 percent of the workers are employees of some company or corporation, under business management. In the United States, most of the capital for industry comes from what is called the *free market*. Anyone can invest in the hundreds or thousands of industries that make the goods that people want or need. Roughly, there are about 5 million people owning shares in companies directly. Also, there are various groups known as *institutional investors*. Included are banks, which may invest their deposits, and various trust funds and groups called investment trusts. The insurance companies have their premiums invested mostly in bonds and stocks of industry. As practically all Americans have some savings in insurance policies, pension funds, or the like, it might be said that almost all Americans are investors, directly or indirectly, in industry.

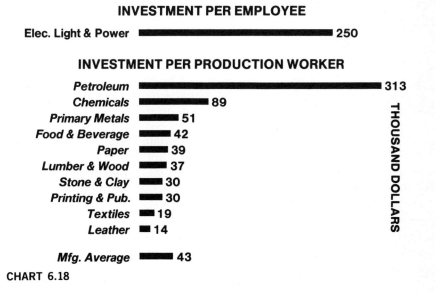

**INVESTMENT PER EMPLOYEE**

Elec. Light & Power ━━━━━━━━━━━━ 250

**INVESTMENT PER PRODUCTION WORKER**

Petroleum ━━━━━━━━━━━ 313
Chemicals ━━━━ 89
Primary Metals ━━ 51
Food & Beverage ━ 42
Paper ━ 39
Lumber & Wood ━ 37
Stone & Clay ━ 30
Printing & Pub. ━ 30
Textiles ━ 19
Leather ━ 14

Mfg. Average ━ 43

THOUSAND DOLLARS

CHART 6.18

## THE AMERICAN ECONOMIC SYSTEM

What is the system used in the United States to get work done to improve living standards and to help people improve their personal welfare? In summary, the principles of the system are these:

1. *The Right to Private Property.* A person has the right to own all the things bought from earnings. Clothes, house, and land are an individual's property. Also, investment may be made in various kinds of securities: bonds, preferred stock, or common stock of a company. An ownership share of a company can be sold anytime desired to anybody who will pay the price the stockholder is asking. In effect, a person will sell property to the highest bidder.

2. *Rewards for Hard Work and for Savings.* An individual wants to work in a personally chosen job and wants to be free to change jobs if desired. Freedom to bargain for wages and the opportunity to enhance value through higher education, to command a higher income, and to be rewarded for hard work and study are all factors in the dynamics of the U.S. economic system.

3. *Profits and Losses.* The American system recognizes the right of an individual, or a group, to make a profit on an investment in an enterprise. If a company is managed well and produces a quality product at a fair price, people will buy the product. If management controls expenses and avoids waste, a profit can be made on the business. The profits go to the stockholders or the owners. A person is free to go into any business and has the right to know that individual effort can produce a profit. It is the right to earn a profit that provides the encouragement to save and invest in machinery that will improve the productivity of all workers. But, by the same token, the system is one of risk and possible loss, as this chapter will demonstrate.

4. *Private Investment in Machinery of Production.* The American system encourages people to save and to invest (or not to invest) their savings in the machinery of production, i.e., in the companies that make goods people want and need.

5. *Competition.* Except in the case of utilities, where prices are controlled by government regulation (discussed in Chapter 9), the American system is one of competition. A person must make a quality product and sell it at a fair price or else some competitor will get the business. This competition is one of the best incentives for economy and efficiency in operation and management. Management learns that it must make a quality product at the lowest possible price, consistent with the cost of making the product, to encourage its sale, and to meet competition.

6. *Government Regulation.* The American system is one of government regulation, not government operation (management) of business.

Thus a fundamental American principle is preserved—the right of redress for wrong. When government both regulates and manages business enterprise, the right of redress is lost, and the only appeal is to the one who committed the wrong. Also, this principle is in keeping with another American principle—the separation of powers.

## DOES THE SYSTEM WORK?

The system is not perfect. Errors occur, and probably always will, because humans are not perfect. But as long as people maintain their free society, they are free to change and improve the economic system of getting work done any time they feel a change is desirable.

But the question is does this system work? Does it work to improve the living standards listed at the beginning of this chapter?

The functioning of the economic system can be measured in various ways. The following are references to a few charts that illustrate how this system improves living standards, increases jobs, and provides for the distribution of incomes.

1. Chart 6.12 illustrates how the number of jobs available has grown faster than population has increased.

2. Chart 6.15 shows the long-term increase in the buying power of average hourly factory wages.

3. Chart 6.16 shows the long-term decrease in the hours in the average workweek.

4. Increasing GNP and productivity are two of the primary aims of any economic system. Chart 6.19 [15] is unique in a number of respects. Note the relatively low rate of growth of worker-hours. This is limited by population growth and further by the decline in hours worked per week. GNP, in constant dollars, has risen steeply with increased productivity. The steep increase in output per worker-hour (productivity) reflects the increased use of machinery. As previously noted, the increase in machinery is encouraged by stimulating people to save and invest. This, in turn, is brought about by paying a satisfactory return on capital.

5. Chart 6.20 shows that in constant 1975 dollars there has been a steady increase in disposable personal income both on a total basis and on a per capita basis. The trend is expected to continue through 1980.

6. The rise in median family income is shown in Chart 6.21, both in current dollars and in constant dollars.

## GNP, PRODUCTIVITY AND MAN-HOURS

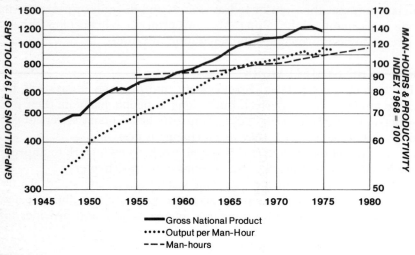

Gross National Product
Output per Man-Hour
Man-hours

**CHART 6.19**

7.   Chart 6.22 shows the increase in personal spending in total dollars, while Chart 6.23 shows per capita spending, both in current and constant dollars. These charts indicate the amount people are buying in goods and services that make up what is referred to as their living standard. Per capita consumer spending has increased from approximately $2800 in 1950 to about $4600 in 1975. It is expected that the figure will be close to $6000 per capita by 1990.

## THE TREND IN REAL DISPOSABLE INCOME
### ALL FIGURES IN 1975 DOLLARS

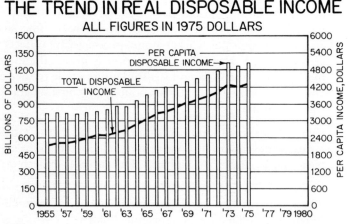

**CHART 6.20**

# MEDIAN FAMILY INCOME

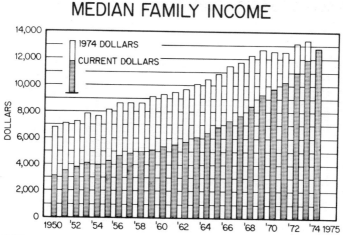

**CHART 6.21**

8. As shown in Chart 6.24, food accounted for a little over 18 percent of expenditures in 1974. Shelter was less than 16 percent of the total. Automobile transportation accounted for 12 percent.

9. Sometimes it is said: "The rich are getting richer and the poor are getting poorer." Chart 6.25 seems to indicate otherwise. The number of families making less than $5000 per year has declined. The same is true for families in the category between $5000 and $7000 per year. Even the number of families in the category between $7000 and $10,000 has declined. However, there has been an increase in the number of families earning $10,000 to $15,000, and there has been a substantial increase

# PERSONAL CONSUMPTION EXPENDITURES

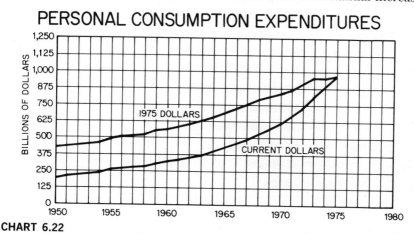

**CHART 6.22**

## THE RISE IN PER CAPITA SPENDING

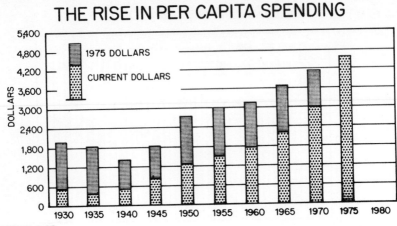

**CHART 6.23**

in the number of families earning $15,000. and over. People with low incomes are moving into the higher income categories. This pattern is expected to continue through 1990. Chart 6.26 shows the upward movement in this income scale in a different way.

10. For persons of 25 years and over, Chart 6.27 shows the level of education completed by years. There has been improvement at all levels of education.

# THE DISTRIBUTION OF CONSUMER EXPENDITURES
## TOTAL EXPENDITURES, 1974 = 100%

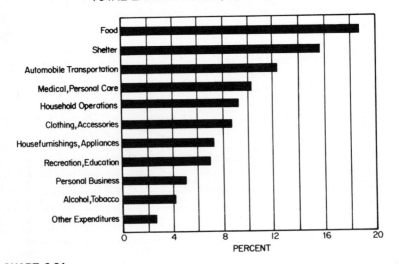

**CHART 6.24**

## FAMILIES BY INCOME CLASS
### BASED ON 1974 DOLLARS

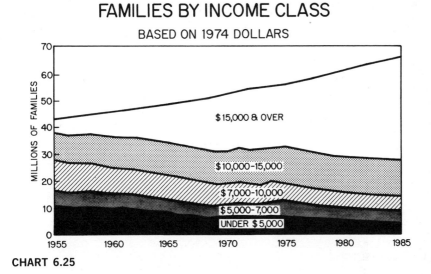

**CHART 6.25**

11. Most people like to travel. Chart 6.28 shows the expenditure for foreign travel from 1960 through 1975.

12. Security is one of the aims of the good life. People are buying more and more life insurance to help obtain it (Chart 6.29).

## THE CHANGING INCOME PYRAMID
TOTAL FAMILIES EACH YEAR = 100%; BASED ON 1974 DOLLARS

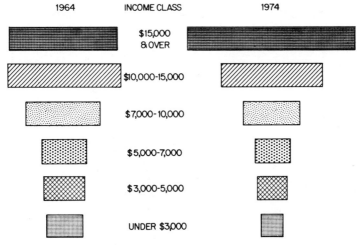

**CHART 6.26**

# EDUCATIONAL ATTAINMENT

### YEARS OF SCHOOL COMPLETED, PERSONS 25 AND OVER

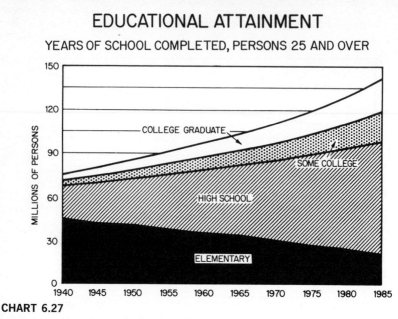

**CHART 6.27**

13. Health is important for the good life. More and more money is spent on medical care. In 1974 it accounted for 8.6 percent of consumer spending (Chart 6.30).

14. Supernumerary income is an interesting statistic (Chart 6.31). It represents the income over and above the income required

# EXPENDITURES FOR FOREIGN TRAVEL

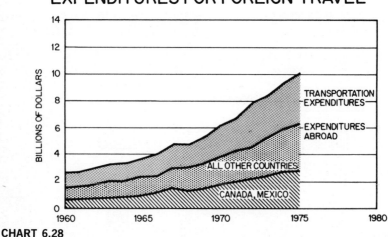

**CHART 6.28**

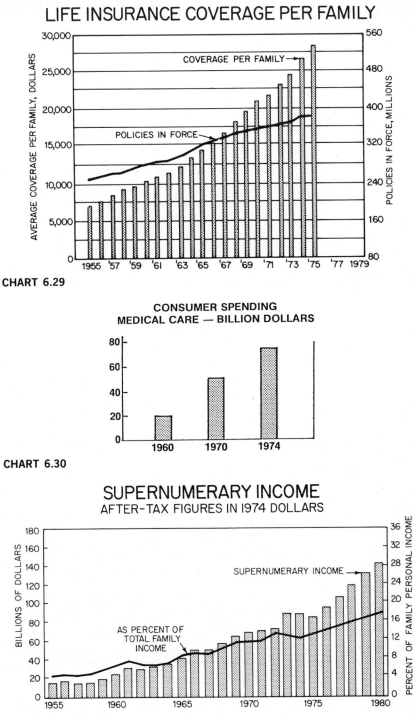

# LIFE INSURANCE COVERAGE PER FAMILY

AVERAGE COVERAGE PER FAMILY, DOLLARS

POLICIES IN FORCE, MILLIONS

COVERAGE PER FAMILY →

POLICIES IN FORCE

1955 '57 '59 '61 '63 '65 '67 '69 '71 '73 '75 '77 1979

**CHART 6.29**

### CONSUMER SPENDING
### MEDICAL CARE — BILLION DOLLARS

1960   1970   1974

**CHART 6.30**

## SUPERNUMERARY INCOME
### AFTER-TAX FIGURES IN 1974 DOLLARS

BILLIONS OF DOLLARS

PERCENT OF FAMILY PERSONAL INCOME

SUPERNUMERARY INCOME →

AS PERCENT OF
TOTAL FAMILY
INCOME

1955   1960   1965   1970   1975   1980

**CHART 6.31**

for ordinary needs such as food, clothing, housing, medical expense, and the like. This is sometimes called *discretionary income*. It can be used for traveling or other recreation, or the money can be invested so as to improve future income. As a percent of total income, it has increased from about 5 percent in 1955 to about 15 percent today. It is expected to be around 18 percent in 1980.

**Summary**   The preceding charts seem to confirm the premise that the key to improved living standards is in the wise use of energy to run machines and to improve productivity. It appears that the American economic system is working, that it does result in higher income and the well-being of the people. To repeat, the system is not perfect. Wise leaders in government and in management should constantly aim for improvement. In doing so, care should be exercised to weigh all factors before any drastic changes are made in a system that is working.

## THE IMPORTANCE OF UNDERSTANDING ECONOMIC FACTS AND MANAGEMENT'S ROLE

Over the past 35 or 40 years, a great deal of work has been done in the field of analyzing human behavior patterns. With the highly developed process of sampling and analyzing public opinion, the activity has been raised almost to the level of a science. It is no longer necessary to remain in the dark as to what people think and why they think as they do.

Many thoughtful people have expressed concern regarding the direction that Americans may be traveling with regard to socioeconomic systems. Some say that the federal government is getting too large, with more and more people calling upon government to do for them those things which, in a free society, people ordinarily do for themselves. Some say America is drifting toward a government planned economy with the resultant loss of freedom. Wise management asks itself: "Is this so? If so, why? Is there a relationship between this trend and knowledge of facts? What can I do to present those facts, the ignorance of which may have a bearing on attitude? What is my responsibility?"

Management wants to know how many of the employees are dissatisfied and why, as well as the extent to which their behavior is based upon lack of knowledge of facts. As management has the facts, or easy access to them, management has some responsibility in keeping employees informed.

## ANALYSIS OF PROBLEM

In a free society, the government is responsive to the people. The problem has to do with people—their attitudes, emotions, fears, hopes, desires, and beliefs—and with their knowledge or lack of knowledge of business affairs and public affairs.

Scientific research has disclosed a strong correlation between knowledge of facts and attitudes and desires. The research indicates that facts can be communicated to people at all levels and walks of life—and further, that people's attitudes change in proportion to their knowledge of the facts. When people know and understand economic facts, they tend to believe more of those principles which are the foundation of individual freedom.

The analysis shows:

1.  People want more things. This is the strongest motivating factor having a bearing on the problem.

2.  Most people believe that workers are not getting a proper share of the fruits of their labor. They think profits are too high. They believe most of the corporate income goes to owners and top management, not the workers.

3.  They believe that rich people get too large a portion of the total income and thus deprive others of a fair share.

4.  People who believe these things seek to remedy what they see as inequities by turning to government. They call on government to control profits, to get them a bigger share of the national income, to give them money, services, and help. They believe government can tax corporations for this purpose, not realizing such taxes are generally included in the prices the consumer must pay.

Until the facts about the economic system are widely known and believed, it is likely that there will be continued demands for high government spending, for spending beyond income, and more demands for wage increases that cannot be met without raising prices—all of which tend to create greater inflation. And, further, without proper education it is probable that there will continue to be demands for increasing government operation of the economy.

A research project some years ago sought to determine how well the American people understood their economic system. The questions dealt chiefly with work and jobs and division of income among groups of people. It should be kept in mind that about 80 percent of all workers

# EMPLOYEE OPINION

20%

**People think
manufacturers make**

10%

**People think
is fair**

CHART 6.32

(excluding workers in government) are employees of companies or corporations.

**Questions of Fact—Profits**   This question was asked of a balanced cross section of all workers: "What is the average profit made by industry in peacetime?" The average answer was "20 percent." The next question asked was "What percentage do you think is fair?" The answer: "10 percent" (Chart 6.32). The fact is, however, that profits run consistently about 3.5 to 4 percent on the average (Chart 6.33).

**Question of Fact—Division of Income**   Then this question was asked: "Let us take all the money of a corporation that goes to people—all the money left over after paying for material, supplies, rent, heat and the like. This is the money that is shared by owners and employees. In your judgment, what is the percentage of this money going to employees and the percentage going to owners?"

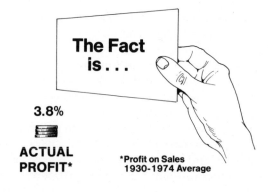

**The Fact
is . . .**

3.8%

**ACTUAL
PROFIT***

*Profit on Sales
1930-1974 Average

CHART 6.33

The average answer indicated that people believed 75 cents out of every dollar of this divisible income went to owners and only 25 cents out of every dollar to employees. The fact is that 88 percent of this divisible income goes to employees and only 12 percent to owners (Charts 6.34 and 6.35).

**Correlations**   The next step in the study was to analyze the relationship between ignorance of economic facts and a person's ideology. To do this, the survey cards were divided into two groups. The cards of all the people who were informed as to the facts were put in one group, and those of people who were uninformed were put in another group. Then both sets of cards were run through a computer to determine how the questions respecting basic philosophy were answered.

The results were enlightening. In the group of people who were well informed as to facts, only 25 percent tended to favor the government planned economy ideology while 75 percent tended to support the free society. Among the group uninformed on facts, only 17 percent favored the free society. An overwhelming 83 percent of them favored a government planned economy (see Chart 6.36).

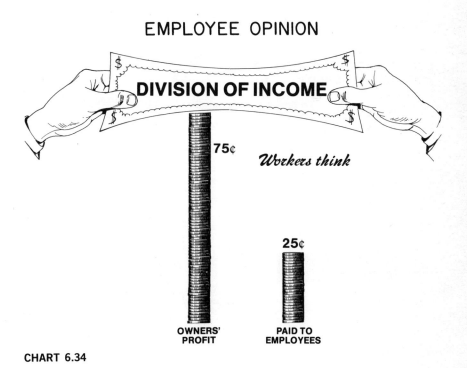

## EMPLOYEE OPINION

DIVISION OF INCOME

75¢

*Workers think*

25¢

OWNERS'
PROFIT

PAID TO
EMPLOYEES

**CHART 6.34**

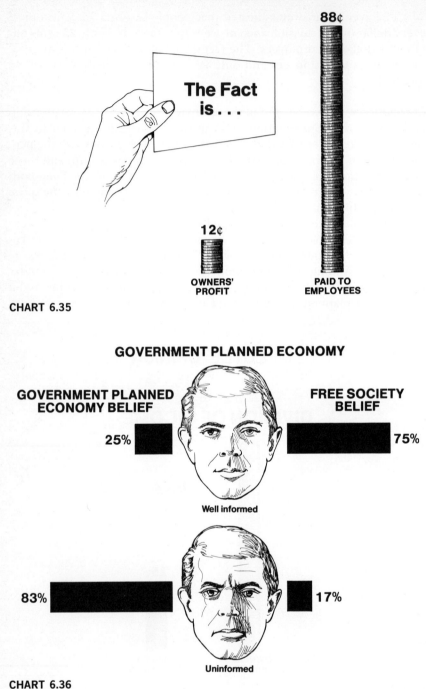

**The Fact is . . .**

88¢

12¢

OWNERS'
PROFIT

PAID TO
EMPLOYEES

CHART 6.35

**GOVERNMENT PLANNED ECONOMY**

GOVERNMENT PLANNED
ECONOMY BELIEF

FREE SOCIETY
BELIEF

25%

75%

Well informed

83%

17%

Uninformed

CHART 6.36

**Management's Role**   Management should give serious consideration to the significance of these findings:

1. People want more things (better living standards).

2. Most people work for a company under business management. They are employees.

3. They think profits are 20 percent (few distinguish between profits on sales and profits on investment).

4. They think 10 percent is a fair profit. Owners are making double the fair profit. Management works for owners. This makes employees angry.

5. On top of that, people think owners get 75 percent and workers 25 percent of the money available for division between owners and workers, while actually workers get 88 percent. Believing as they do, is there any wonder they are upset?

These are a few of some 17 facts about the economic system, the ignorance of which has a bearing on attitude. These are the kinds of facts that wise management wants to communicate to employees.

**Communication**   To accomplish this goal, there should be courses for employees on company time, providing facts backed by evidence in which the employees have confidence.
The course should cover:

1. The principles of the American system

2. How machines run by energy result in increased production and higher living standards

3. How to acquire more machinery

4. Evidence that the system works in raising living standards

5. Profits, wages, and jobs

6. The division of income between owners and workers

7. The division of income between rich and nonrich

8. The place of government and its limitations in "giving to people"

For effective communication (see Chapter 10), each fact must be:

1. Heard or seen or both

2. Understood

3. Believed

4. Retained

There are proven ways to accomplish and measure this communication.

**Sample Communication**   An employee course may involve the following:

- 1½ h a month on company time

- Groups of about 40 employees

- Trained conference leaders who are near to the level of employees in the course but who have been through a 2-week training course

- Charts, color slides, props, and full discussion

- Books of backup material acceptable to employees on the table

**Profits**   The profit of a company or a corporation is the money remaining after expenses, including depreciation, taxes, bond interest, and the like. This is the money available to owners or stockholders. There is always risk in investment, as all corporations do not make a profit every year.

Chart 6.37 shows all corporations in the United States from 1930 through 1975 [16]. Even during normal years, the percentage of corporations having losses runs about 35 percent. During the years of the Great Depression, in the early 1930s, the percentage was 60 to 70 percent.

The actual percent profit is presented first in Chart 6.38 [17]. The leader refers to the reference material, and the first reaction of employees in the course is disbelief.

The percent profit ranges from 5.3 percent during the highest years to zero or a loss in other years. The average of all the years where a profit is made is about 4 percent. People think that profit is 20 percent, which is exorbitant and more than double what the people think is fair. The actual percent profit is half of what people think is fair.

The leader should not leave the subject until all have had full discussion. In particular it should be noted that in the public mind there is little distinction between the average corporation and a utility. In Chapter 4 it

# PERCENT OF CORPORATIONS
# MAKING PROFITS OR LOSSES

## ALL ACTIVE CORPORATIONS IN THE U.S.

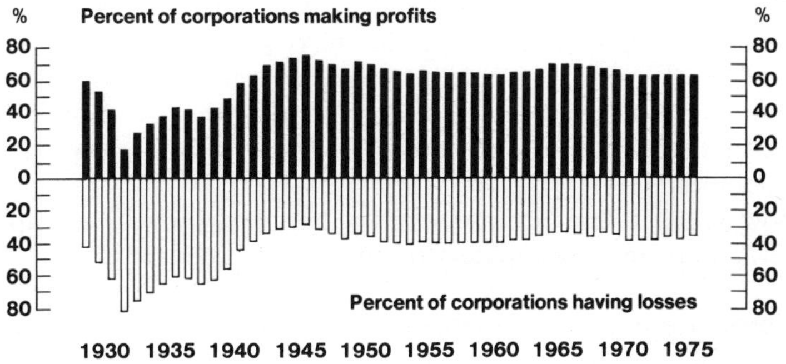

CHART 6.37

was pointed out that the average corporation has a capital investment of some 50 cents for each $1 of annual gross revenue. The electric utility has an investment of somewhere around $4 for each $1 of revenue. Therefore, the economics of a utility is unlike the economics of the ordinary business. This is why the utility is spoken of as a natural monopoly.

## CORPORATIONS EARNED ONLY 3½ PERCENT PROFIT
## ON ALL SALES FROM 1930-1973

| YEAR | SALES* | NET PROFITS* | % PROFIT ON SALES | YEAR | SALES* | NET PROFITS* | % PROFIT ON SALES |
|------|--------|--------------|-------------------|------|--------|--------------|-------------------|
| 1930 | $118,294 | 2,480 | 2.1% | 1952 | 499,454 | 17,232 | 3.4% |
| 1931 | 92,365 | 1,278 | Loss | 1953 | 523,307 | 18,089 | 3.5% |
| 1932 | 69,185 | 3,402 | Loss | 1954 | 516,502 | 16,841 | 3.3% |
| 1933 | 73,027 | 370 | Loss | 1955 | 599,390 | 23,035 | 3.8% |
| 1934 | 89,553 | 972 | 1.1% | 1957 | 671,801 | 22,286 | 3.3% |
| 1935 | 101,953 | 2,194 | 2.2% | 1958 | 658,153 | 18,764 | 2.9% |
| 1936 | 119,462 | 4,331 | 3.6% | 1959 | 739,427 | 24,469 | 3.3% |
| 1937 | 128,884 | 4,733 | 3.7% | 1960 | 761,212 | 22,989 | 3.0% |
| 1938 | 108,551 | 2,271 | 2.1% | 1961 | 783,123 | 23,302 | 3.0% |
| 1939 | 120,789 | 4,962 | 4.1% | 1962 | 849,102 | 31,229 | 3.7% |
| 1940 | 135,248 | 6,486 | 4.8% | 1963 | 892,629 | 33,077 | 3.7% |
| 1941 | 176,181 | 9,372 | 5.3% | 1964 | 963,910 | 38,667 | 4.0% |
| 1942 | 202,777 | 9,467 | 4.7% | 1965 | 1,056,762 | 44,493 | 4.2% |
| 1943 | 233,435 | 10,480 | 4.5% | 1966 | 1,158,265 | 49,943 | 4.3% |
| 1944 | 246,737 | 10,371 | 4.2% | 1967 | 1,212,642 | 46,638 | 3.9% |
| 1945 | 239,512 | 8,288 | 3.5% | 1968 | 1,313,851 | 48,180 | 3.7% |
| 1946 | 270,898 | 13,440 | 5.0% | 1969 | 1,420,060 | 48,542 | 3.4% |
| 1947 | 347,801 | 18,242 | 5.2% | 1970 | 1,502,197 | 40,151 | 2.7% |
| 1948 | 388,744 | 20,517 | 5.3% | 1971 | 1,614,790 | 45,902 | 2.8% |
| 1949 | 370,079 | 15,995 | 4.3% | 1972 | 1,850,506 | 57,653 | 3.1% |
| 1950 | 431,857 | 22,763 | 5.3% | 1973 | 2,181,085 | 72,914 | 3.3% |
| 1951 | 488,446 | 19,706 | 4.0% | *Sales and net profits in millions. | | | |

CHART 6.38

The duplication of the expensive plant would be against public interest. This is why government regulation is a substitute for competition. When a course of this kind is given to electric utility employees, this distinction should be made.

**Questions of Fact—How Income Is Shared**  First, where does national income originate? See Charts 6.1 and 6.39 [18]. Seventy percent of national income comes from work, while thirty percent is from income on property. Workers also share the property income to the extent that they receive interest on investments, savings, rent, or the like. Note in Chart 6.40 that most of the national income is created in business, largely in corporations [19].

How is all national income shared? Chart 6.41 shows that employees receive 77 percent. Stockholders receive 2½ percent [20].

**How Corporate Income Is Shared**  As most of the income originates in corporations where most employees work, and as that is where there is lack of understanding (see Chart 6.34), this subject needs careful presentation. Chart 6.42 shows how the income of manufacturing corporations was shared from 1929 through 1975 [21]. Notice the percent of the gross revenue required for outside expense and replacement. This has been running consistently in the neighborhood of 70 percent and higher. The percent of the gross revenue that is paid to employees is in the range of 20 to 30 percent. A relatively small percentage goes to the owners. This is the part that has been averaging around 4 percent of sales. Now consider only those dollars going to employees and to owners. Of this amount, the employees think that the owners get 75 percent and employees get 25 percent.

**NATIONAL INCOME**

**30% from PROPERTY**
**70% from WORK**

**CHART 6.39**

## WHERE DOES NATIONAL INCOME ORIGINATE

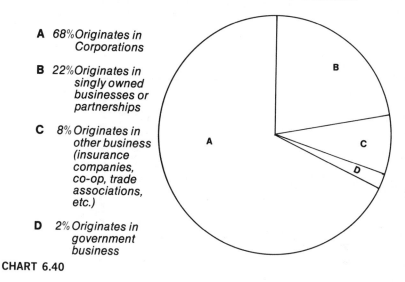

**A** 68% Originates in Corporations

**B** 22% Originates in singly owned businesses or partnerships

**C** 8% Originates in other business (insurance companies, co-op, trade associations, etc.)

**D** 2% Originates in government business

**CHART 6.40**

In a classroom of employees, the first reaction to this chart, and to the chart showing profit, is disbelief. This is why it is so important that there be appropriate reference books on the table for employees to examine during their discussion.

For some time, a company used statistics from annual reports. Some

## HOW NATIONAL INCOME IS SHARED

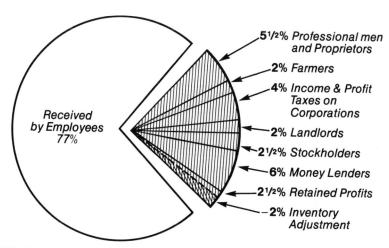

Received by Employees 77%

5½% Professional men and Proprietors

2% Farmers

4% Income & Profit Taxes on Corporations

2% Landlords

2½% Stockholders

6% Money Lenders

2½% Retained Profits

−2% Inventory Adjustment

**CHART 6.41**

## BREAKDOWN OF INCOME OF
## ALL MANUFACTURING CORPORATIONS

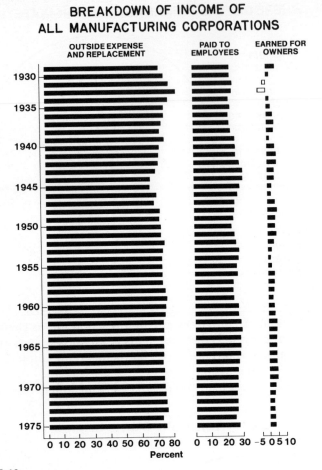

CHART 6.42

of the employees said that they did not believe company figures. The company then used statistics put out by the U.S. Department of Commerce. They were used until one employee said, "Why do you use the Department of Commerce—why not use our Department of Labor?" That was a happy suggestion. At each meeting that followed, the reference books included U.S. Department of Labor statistics. So contrary to general beliefs are the facts shown here that only after extensive discussion are the facts understood, believed, and retained.

Chart 6.43 gives a summary of all these figures for the whole period 1929–1975 [22]. Notice that about 70 percent of the revenue goes to neither employees nor owners; about 26 percent goes to the employees, and about 3.5 to 4 percent goes to the owners. This is further illustrated in

## EMPLOYEE WAGES, SALARIES, AND BENEFITS
## COMPARED WITH OWNER PROFIT
### 1929-1975

|  | $BILLION | % |
|---|---|---|
| TOTAL MANUFACTURING CORP. SALES | 15,057 | 100.0 |
| OUTSIDE EXPENSE & REPLACEMENT COST | 10,641 | 70.5 |
| REMAINDER FOR EMPLOYEES & OWNERS | 4,416 | |
| EMPLOYEE WAGES, SALARIES & BENEFITS | 3,885 | 26.0 |
| REMAINDER FOR OWNERS | 531 | 3.5 |

*FOR EVERY $100 DIVIDED BETWEEN EMPLOYEES & OWNERS*

**EMPLOYEE WAGES, SALARIES & BENEFITS**                **$88**

**OWNER PROFIT**                                        **$12**

CHART 6.43

Chart 6.44 [23]. There is a total of $4416 billion (3885 + 531) going to people: owners and employees. Let the $4416 billion represent $100. Then, employees get $88 (3885 ÷ 4416 × $100). Owners get $12 (531 ÷ 4416 × $100).

Chart 6.45 shows that 52 percent of the profit (or about half of the $12) is paid out in dividends [24]. About 48 percent (the other half of the $12) is kept in the business for reserves and for the purchase of improved machinery so that the company can do better next year.

Chart 6.46 shows the breakdown between employees and owners for a number of different years. The size of each circle represents the total dollars available to both employees and owners. The size of the pie became much smaller in 1933, the Depression year, compared with 1929. By 1939, the economy was back up to the level of 1929. There was a big increase in the size of the pie between 1939 and 1958, and then a bigger increase between 1958 and 1975. But the division between owners and employees remained about the same under all the circumstances. In 1933, the size of the slice going to owners seems to have shrunk more than the size of the piece going to employees. It can be assumed, however, that the ratio is usually about $88 to employees and $12 to owners.

**The Silver-Dollar Illustration**    The amount available to employees and owners was reduced to $100 because the conference leader may want to use 100 silver dollars to illustrate the division. (Three-dimen-

# 4 PERCENT IS THE AVERAGE PROFIT

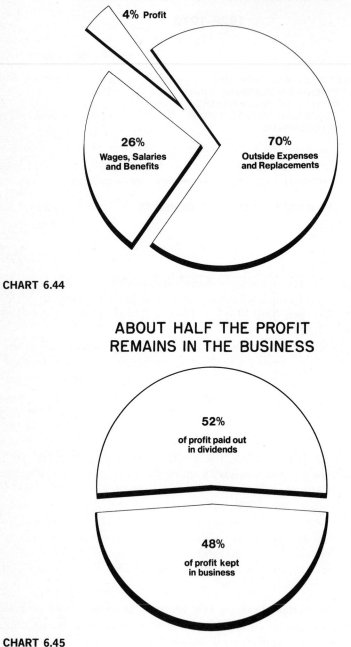

**4%** Profit

**26%**
Wages, Salaries
and Benefits

**70%**
Outside Expenses
and Replacements

**CHART 6.44**

# ABOUT HALF THE PROFIT
# REMAINS IN THE BUSINESS

**52%**
of profit paid out
in dividends

**48%**
of profit kept
in business

**CHART 6.45**

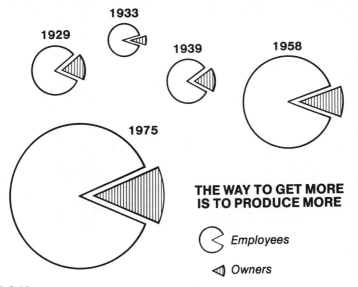

**1929**

**1933**

**1939**

**1958**

**1975**

**THE WAY TO GET MORE
IS TO PRODUCE MORE**

*Employees*

*Owners*

**CHART 6.46**

sional props are valuable tools in communication.) At this point, then, the conference leader brings out 100 silver dollars, first breaking them into two piles, one of 25 and the other of 75, and pointing out that this is the way many employees think the money is divided between the employees and the owners. Then, dividing the dollars properly, the leader asks five course participants to come forward—four to represent employees and one to represent owners. The four employees are given 88 silver dollars, with 12 going to the owner, who is told that only $6 will be received as dividends because the other $6 remains in the company for reserves (Chart 6.47). Now one person has $6 a year income and the others have $88 income.

Assume that the four representing employees want a 10 percent increase in salaries, or $8.80. The owner has only $6. Without any increase in productivity and production, the only way that the owner can pay such excessive demands for higher wages is by raising prices, which may result in other manufacturers doing the same thing. This is one of the primary causes of inflation, which, among other things, produces a situation in which employees do not get in real terms the increases they thought they were going to receive.

The conference leader emphasizes the point that the only way to get more real income is to produce more, and the only way to produce more is to increase the output of goods and services per worker-hour, i.e., to increase productivity. This is done by using efficient machinery and

## DIVISION OF INCOME BETWEEN EMPLOYEES AND OWNERS

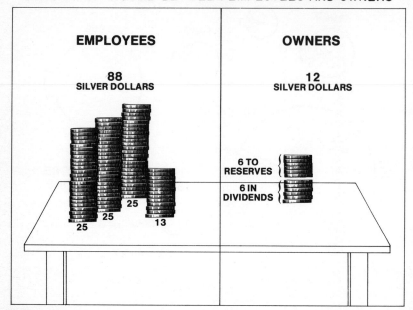

**CHART 6.47**

more of it. Thus, the employees have a direct interest in seeing that the owner receives some return on investment so that others will be encouraged to save and invest.

Consider now the employees of 1939, who might have thought that they could have obtained more income by taking the income of the owner (Chart 6.46). Then look at the much bigger increase in income employees received in 1958 by making a bigger pie. To produce more means that the society must encourage people to save so as to invest their savings in more machinery and thus increase the productivity of the employees. People are encouraged to save with some kind of reward as incentive, such as interest or a return on their savings. The return to the owners and the wages to employees go hand in hand.

**What Percent Increase May Employees Expect?**  Naturally, there is no general answer. Many factors affect this, and individual cases vary. The question is how much is available without contributing to inflation?

A few guidelines might help:

1.  For the period under study, 1929–1975, the average annual increase in gross revenue of all manufacturing corporations was 5.92 percent [25].

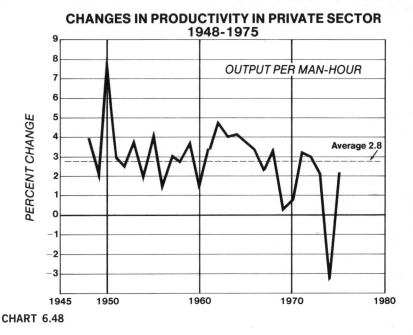

**CHANGES IN PRODUCTIVITY IN PRIVATE SECTOR
1948-1975**

CHART 6.48

2. The average employee's share of total manufacturing gross dollars was 26 percent.

3. The employee's share increased at the rate of 6.2 percent per year.

4. The productivity index increased an average of 2.8 percent per year, as shown in Chart 6.48 [26].

5. There should be sufficient return to investors to encourage them to save and to invest, so as to improve and add machinery in order to increase productivity.

**Rich and Poor**   At this point in the course someone may say (or think) something like this: "Okay buddy, you made a point, but the rich get more than their share. We, the ordinary employees, do not get our share of the $88." So the conference leader suggests that the question be examined. Chart 6.49 indicates the number of income tax returns going to people making less than $50,000 per year and to those making more than $50,000 per year [27]. Of all of the income-tax dollars 91 percent comes from people making less than $50,000 per year, while 9 percent is from those making more. Chart 6.50 shows that there are 81 million tax returns from people making less than $50,000 per year and only

## DIVISION OF INCOME BETWEEN
## RICH AND NONRICH IN FIGURES

| BILLIONS OF DOLLARS | | | | | | |
|---|---|---|---|---|---|---|
| 1975 | INCOME TAX | % | REPORTED INCOME | % | INCOME AFTER TAX | % |
| PEOPLE WITH INCOMES OF LESS THAN $50,000 | 864 | 91 | 99 | 79 | 765 | 93 |
| PEOPLE WITH INCOMES OF MORE THAN $50,000 | 84 | 9 | 26 | 21 | 58 | 7 |
| **TOTAL** | 948 | 100 | 125 | 100 | 823 | 100 |

CHART 6.49

960,000 from people making more [28]. If all the earnings of the people making $50,000 a year and above were added to the existing total earnings of all the other people, the annual income of the majority would be increased by only 9.7 percent per year, or $2.84 per day. People cannot become wealthy by taking from those who are.

**"We'll Get from Government"**  Most surveys of public opinion indicate that a large percentage of the people feel that they can turn to their government for the solution of their problems. They do not realize that government has no income except that which it first takes from people. As shown in Chart 6.51, of every $100 of income to the federal government, $65 is from individuals, either in the form of income taxes or as individual contributions toward pension funds or social security [29]. Of the $100, $30 is from corporations. People should know that substantially all the taxes paid by a corporation are passed on to the consumer in the price of the commodity being sold. In the final analysis, it is the consumer who pays practically all the taxes of a corporation. Part of this $30 is the amount the corporation contributes for the individual's social security. In other words, about 95 percent of all the income to the federal government comes directly and indirectly from individuals.

# DIVISION OF INCOME BETWEEN
# RICH AND NONRICH IN GRAPHS

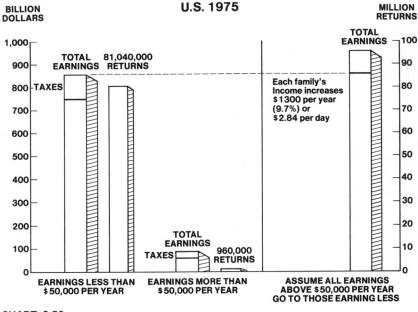

CHART 6.50

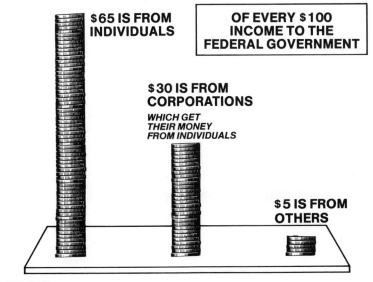

CHART 6.51

Nevertheless, a large number of the American people continue to call for more and more government spending. As shown in Chart 4.3, in most recent years government spent more than it took in. Government will continue to spend in this fashion as long as it is the will of the people, or as long as the people allow it. The budget will be balanced when the majority of the people demand it.

**Does Employee Education Work?**  Individual Examples    It has been found that, as a result of an 8-h discussion program held over a 6-month period, the areas of ignorance on economic facts and principles can be cut in half, which causes a corresponding reduction in support for the ideology of the government planned economy. Chart 6.52 shows some "before and after" results of one such program of conference discussion.

Large-Scale Example   During the 1940s and 1950s, many community leaders were expressing concern over the trend toward a government

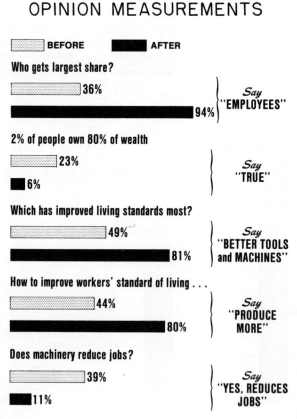

OPINION MEASUREMENTS

BEFORE        AFTER

Who gets largest share?
36%
94%  Say "EMPLOYEES"

2% of people own 80% of wealth
23%
6%  Say "TRUE"

Which has improved living standards most?
49%
81%  Say "BETTER TOOLS and MACHINES"

How to improve workers' standard of living . . .
44%
80%  Say "PRODUCE MORE"

Does machinery reduce jobs?
39%
11%  Say "YES, REDUCES JOBS"

CHART 6.52

planned economy just as they are today. Opinion Research Corporation and others carried on extensive research to determine:

1.  The percentage of people leaning toward the government planned economy

2.  Why they so leaned

3.  Knowledge of economic facts

4.  Correlation between attitudes and knowledge

Over a number of years, during which hundreds of test questions were asked, one question developed as the one that most nearly reflected the trend toward a government planned economy: "Do you expect more or less government intervention in business?"

A number of national organizations were formed, including the Foundation for Economic Education, the American Economic Foundation, the Freedom Foundation, and others. Many courses in economic education were designed by such groups and by individual companies. The author took an active part in them.

Chart 6.53 shows the national results of this work. Since 1952, however, the number of companies conducting courses of this kind has

## RELATIONSHIP BETWEEN ECONOMIC EDUCATION AND EMPLOYEE ATTITUDE

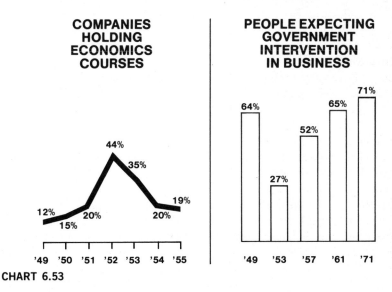

**CHART 6.53**

**Percentage Change of Electricity and Employment**

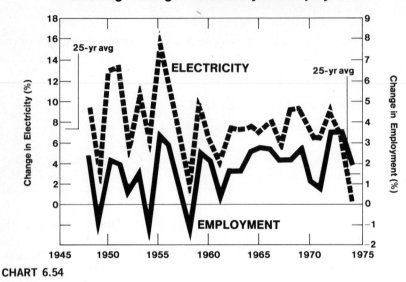

CHART 6.54

dropped sharply. By 1971 the percentage of people expecting governmental intervention in business had risen to 71 percent. Perhaps even more ominous, by 1977 less than 50 percent of those surveyed in a national opinion poll could say they had a lot of confidence in the free enterprise system, and almost half favored greater government intervention in the economy in the form of wage and price controls [30].

The following three charts give a pertinent summary of some information in this chapter. Chart 6.54 shows how the pattern of jobs follows the pattern of electric energy use. Chart 6.55 was prepared from the exhaustive work of Fremont Felix, which dramatically illustrated the relationship between income per capita and energy per capita [31]. In making this study, he examined statistics of 150 nations. Not all the nations could be shown. By extending the line, it is possible to estimate the U.S. position in the year 2000 (provided this country continues to adhere to the principles being described).

A similar relationship is shown in Chart 6.56 [32] for the electric energy used in the United States. From this chart one can also see the time element in the increase in living standards. The chart begins about 1930. Notice that at that time the per capita personal income in the United States was not unlike that of a number of developing countries today. The forecast here is based upon the electric energy forecast that will be described in the following chapter.

# ENERGY and INCOME / *WORLD 1961*

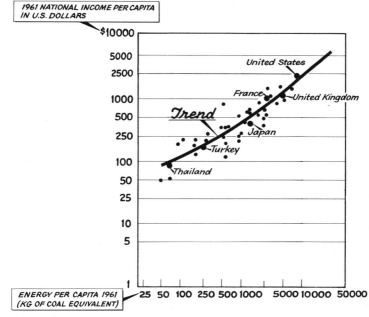

CHART 6.55

## KILOWATT-HOURS AND INCOME
### *USA · 1930 — 2000*

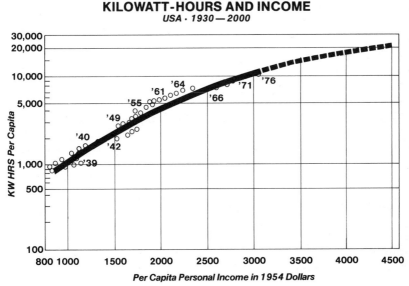

CHART 6.56

## Summary: 17 Key Facts about the American Economic System

1. Each person has the right to own property.

2. Each person has the right to buy and sell that property.

3. Each person is free to enter a trade or business and to compete with anyone else.

4. People are paid for mental and physical labor according to what they can produce.

5. Government acts as a regulator to maintain order and fair play, not as the operator of business.

6. The way to improve the workers' standard of living is for all workers to produce more.

7. This system produces more goods and services than any system the world has ever known.

8. There are good results in terms of human values as well as material goods.

9. Machines increase production, decrease prices, make jobs, cut work time, bring more leisure, and increase buying power.

10. The money to buy tools and machines comes from savings and profits.

11. The hope of profit produces the incentive to save.

12. On the average, corporation profits on sales are about 4 percent.

13. Of every $1 to be divided among employees and owners of a company, the owners get 12 cents (about half of which remains in the business as reserve) and the employees get 88 cents.

14. Employees receive about two-thirds of every $1 of national income.

15. Government has no income except that which it takes from the people.

16. Most of government's tax income comes from people with average incomes.

17. When a business is taxed, the tax is passed on to the consumer in the price of the product.

# SOURCES FOR CHARTS

1. U.S. Department of Commerce, *Statistical Abstract of the United States, 1976*, Tables 628 and 629, p. 393, Table 630, p. 394, Table 638, p. 398, Table 641, p. 400, Table 645, p. 403. *Survey of Current Business*, March 1977, pp. 5–6, 8–9, S-1, S-3.

2. U.S. Department of Commerce, *Historical Statistics of the United States, 1789–1945*, Table G-192, p. 157, and Table D-69, p. 65, for data prior to 1946. *I Want to Know about the Electric Utility Industry*, 1962–1963 ed., p. 20, and prior editions for 1946–1960 data. *Questions and Answers about the Electric Utility Industry*, p. 24, for 1961–1969 data. *31 Answers to 32 Questions about the Electric Utility Industry*, 1976–1977 ed., p. 9, for 1970–1974 data.

3. J. Frederick Dewhurst and Associates, *America's Needs and Resources* (New York: Twentieth Century Fund, 1955), p. 908.

4. U.S. Department of Commerce, *Statistical Abstract of the United States, 1976*, Table 1445, p. 867, for population data. U.S. Department of State, Agency for International Development, in *The Associated Press Almanac, 1975*, p. 175, for GNP data.

5. U.S. Department of Labor, *Employment and Earnings*, April 1977, Table B-2, p. 68.

6. U.S. Department of Commerce, *Historical Statistics of the United States: Colonial Times to 1970*, Table A1-5, p. 8, for 1870 population data. U.S. Department of Commerce, *Statistical Abstract of the United States, 1976*, Table 2, p. 5, for 1975 population data. *Historical Statistics of the United States*, Chapter D, Series 11–25 and 152–166 for 1870 occupational data. *Statistical Abstract of the United States, 1976*, Table 569, p. 355, Table 591, p. 365, Table 594, p. 366, for 1975 occupational data.

7. J. Frederick Dewhurst and Associates, *America's Needs and Resources* (New York: The Twentieth Century Fund, 1955), pp. 1116 and 1121, for 1850 –1930 horsepower-hours per man-hour worked. Continuation of the series 1940–1968 derived from historical relationship to Total BTU Energy Consumption Series of Bureau of Mines for previous years. Continuation of the series 1969–1974 derived from historical relationship to commercial energy consumption data from United Nations, *World Energy Supplies, 1950–1974* (New York: 1976), Table 1, p. 5. Private national income per man-hour in 1958 prices recalculation over the period 1950–1968 following Dewhurst method of calculation for data in 1950 prices shown in Table 14, p. 40. Private national income per man-hour in 1958 prices for period 1969–1974 derived from annual increase in output per man-hour series of U.S. Department of Commerce, *Statistical Abstract of the United States, 1976*, Table 598, p. 371.

8. U.S. Department of Commerce, *Historical Abstract of the United States*, *Colonial Times to 1957*, Chapter D, Series 37-38, p. 72, for 1820–1930 data. *Statistical Abstract of the United States*, *1962*, Table 280, p. 215, for 1940–1960 data. *Statistical Abstract of the United States*, *1976*, Table 601, p. 373, for 1960–1975 data.

9. *Statistical Abstract of the United States*, *1976*, Table 1065, p. 632.

10. *Mill and Factor Magazine*, July 1946, for 1850–1910 data. National Industrial Conference Board, *The Economic Almanac*, *1962*, pp. 54–55, for 1914–1960 data. U.S. Department of Labor, Bureau of Labor Statistics, *Employment and Earnings*, September 1971, Table C-1, p. 85, for 1961–1970 data. *Employment and Earnings*, *United States*, *1909–75*, p. 759, for 1970–1974 data. *Employment and Earnings*, July 1976, Table C-4, p. 104, for 1975 data.

    (Note: Decline in real buying power of factory wages, 1973–1975, was due to high rates of inflation. Price index utilized was from Conference Board, *A Guide to Consumer Markets*, *1976–1977*, p. 241.)

11. J. Frederick Dewhurst and Associates, *America's Needs and Resources* (New York: The Twentieth Century Fund, 1955), Appendix 20-4, Column 5, p. 1073, for 1850–1960 data. U.S. Department of Labor, Bureau of Labor Statistics, *Employment and Earnings*, August 1971, p. 79, for 1961–1970 data. *Employment and Earnings*, April 1977, Table C-1, p. 89, for 1971–1976 data.

12. Wilford I. King, *The Keys to Prosperity* (New York: Constitution and Free Enterprise Foundation, 1948), p. 78, for 1850–1939 data, and the Conference Board for 1940–1972 data.

13. U.S. Department of Commerce, *Statistical Abstract of the United States*, *1976*, Table 1306, p. 762, for manufacturing data. Edison Electric Institute, *31 Answers to 32 Questions about the Electric Utility Industry*, 1976–1977 ed., p. 23, for electric utility data.

    (Note: The figures from the 1976 *Abstract*, which were derived from Conference Board data, and the figures from the 1976–77 ed. of *Questions about the Electric Utility Industry* are for 1972. This is the last year for which comparable data are published. The most recent Conference Board data use a new method of calculation, one which excludes financial capital in computing productive investment per employee. These most recent figures are therefore substantially lower than the figures for 1972.)

14. The electric industry figure is for 1972, as that is the latest year available for the other classifications in this chart. The figure for the electric utility industry for 1976 was $425,000 per employee, as shown in Chapter 3.

15. Charts 6.19 through 6.31 are from data supplied by the Conference Board.

16. U.S. Treasury Department, Internal Revenue Service, *Statistics of Income for 1949, Part 2,* Table 2, pp. 336–337, for 1930–1944 data. *Statistics of Income for 1951, Part 2,* Table 15, p. 176, for 1945–1950 data. *Statistics of Income, 1950–1960,* Table 41, p. 231, for 1951–1960 data. *Statistics of Income, 1967,* Table 21, p. 169, for 1961–1967 data. *Statistics of Income, 1972,* Table 29, p. 187, for 1968–1972 data. *Preliminary Statistics for Income, 1974,* Table A, p. 2, for 1973–1974 data.

17. U.S. Department of Commerce, *National Income,* 1954 edition, Table 20, pp. 188–189, for 1930–1945 data. *U.S. Income and Output,* November 1958, Table VI-7, p. 205, Table VI-17, p. 215, for 1946–1955 data. *Survey of Current Business,* July 1962, Table 49, p. 27, Table 58, p. 31, for 1956–1961 data. *Survey of Current Business,* July 1966, Table 6.15, p. 33, Table 6.19, p. 35, for 1962–1965 data. *Survey of Current Business,* July 1970, Table 6.15, p. 42, Table 6.19, p. 44, for 1966–1969 data. *Survey of Current Business,* July 1972, Table 6.15, p. 42, Table 6.19, p. 44, for 1970–1971 data. *Survey of Current Business,* July 1974, Table 6.15, p. 39, Table 6.19, p. 41, for 1972–1973 data.

(Note: After 1973, figures are available for corporate net profits, but figures for corporate sales, all industries, are not. Figures for "business sales" are given past 1973, but these include manufacturing and trade only.)

18. U.S. Department of Commerce, *Survey of Current Business,* July 1974, Table 1.13, p. 17.

19. U.S. Department of Commerce, *Survey of Current Business,* July 1974, Table 1.3, p. 17.

20. U.S. Department of Commerce, *Survey of Current Business,* March 1977, Table 7, p. 7.

21. U.S. Department of Commerce, *The National Income and Product Accounts of the United States, 1929–1965,* Table 6.19, pp. 142–145, for 1929–1965 sales data. *Survey of Current Business,* July 1970, Table 6.19, p. 44, for 1966–1969 sales data. *The National Income and Product Accounts of the United States, 1929–1974,* Table 6.5 p. 194, Table 6.21, p. 248, for 1929–1974 profits and compensation data. *Survey of Current Business,* July 1974, Table 6.19, p. 41, for 1970–1973 sales data. *Survey of Current Business,* July 1976, Table 6.5, p. 51, Table 6.21, p. 56, for 1974–1975 profits and compensation data. *Survey of Current Business,* December 1976, p. S-5, for 1974–1975 sales data.

22. Calculated from data for Chart 6.42.

23. This chart summarizes data from Chart 6.43.

24. U.S. Department of Commerce, *The National Income and Product Accounts, 1929–1974,* Table 6.22, pp. 252–255, Table 6.23, pp. 256–259, for 1929–

1974 data. *Survey of Current Business*, March 1977, Table 7, p. 7, for 1975 data.

25. U.S. Department of Commerce, *The National Income and Product Accounts of the United States, 1929–1965*, Table 6.19, p. 142, and *The National Income and Product Accounts of the United States, 1929–1974*, Table 6.5, p. 194, for 1929 data. *Survey of Current Business*, July 1976, Table 6.5, p. 51, and *Survey of Current Business*, December 1976, p. S-5, for 1975 data.

26. Council of Economic Advisors, *Economic Report of the President, 1977*, Table B-35, p. 228, Table B-36, p. 229.

   (Note: For gas and electric utilities, average annual gains were much higher, about 6.5 percent for the period 1947–1974. See National Commission on Productivity and Work Quality, *Fourth Annual Report, 1975*, pp. 9–13.)

27. U.S. Department of the Treasury, Internal Revenue Service, *Preliminary Statistics of Income, 1975: Individual Income Tax Returns*, Table 2, p. 11.

28. Data computed from source 27.

29. Office of Management and Budget, *The United States Budget in Brief: Fiscal Year 1976*, p. 12.

30. *The Gallup Opinion Index*, January 1977, Report No. 138, p. 26, and March 1977, Report No. 140, p. 16.

31. Fremont Felix, *World Markets of Tomorrow* (New York: Harper & Row, 1972).

32. Edison Electric Institute, *Year Book of the Electric Utility Industry for 1975* (New York: Edison Electric Institute, 1976), Table 7-S, p. 14. U.S. Department of Commerce, *Statistical Abstract of the United States, 1976*, Table 641, p. 400, for personal income data. Conference Board, *A Guide to Consumer Markets, 1976–1977*, for price index.

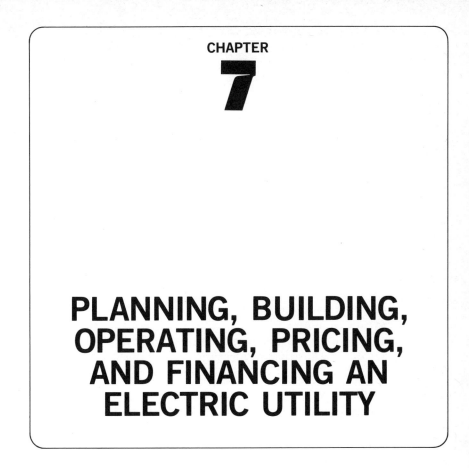

# CHAPTER

# 7

# PLANNING, BUILDING, OPERATING, PRICING, AND FINANCING AN ELECTRIC UTILITY

## HOW MANAGEMENT PLANS

In Chapter 5 there was a discussion of the management profession, especially as it applies to an electric utility. The various objectives of electric utility management were listed there, and before reading further, it might be advisable to review them and the skills, tools, and methods that management uses in carrying out those objectives.

Now will be discussed, in brief form, how top management designs an overall plan for the whole company for a period of, say, 5 years in the future. It is obvious that physical plant expenses must be planned 5 to 10 years in advance (and longer), as decisions and commitments must be made that far ahead. This is the time required to complete a plant. But planning does not stop there. There is long-range planning for every phase of the business.

At the moment, management is especially preoccupied with plans to build the percent return on investment to a full fair return and keep it there. This is not easy, considering that the customers and the regulatory bodies have become accustomed to witnessing declining prices, whereas rising prices are the pattern now and will remain so in the immediate future. It is precisely because of this severe change that management wants a long-range plan on financing, pricing, load management, expense control, employee and customer information, and all the other phases of the business. All departments contribute to the plan, and when the final plan is adopted, each department and subdepartment knows the overall company program and how each department fits into that program.

Some may argue that because conditions change so fast, the plan will always be changing. Of course, the plan is changed from time to time as necessary to meet new conditions. But this is no reason for having no plan.

Every company has a 1-year budget. It is changed during the year if warranted. The 5-year plan is somewhat like a 5-year budget. The company puts on paper the projected balance sheet and income statement by years for 5 years, in figures which management considers desirable and attainable. Added to these is the plan or program under which management expects to (or will strive to) meet the desired objectives.

## THE CORPORATE MODEL

The terms financial model and corporate model have different meanings to different people. As used herein, the term *corporate model* means the overall company plan, or objective, by years for, say, 5 years, as envisioned by all departments. In the preliminary study, various plans are reviewed. Various computer models are used in making the studies. Finally, one is decided upon. That plan, that program, is then put into a computer. This is the corporate model.

For example, the program may call for earnings of 8 to 9 percent return, or whatever may be required, by years for 5 years. If someone from within or outside the company proposes a change, management can promptly determine the future effect of that change on the percent return. Also, management can easily and quickly consider various alternative courses of action.

**Changes in the Plan**   About once each year, the plan should be reviewed and changed as desired. The fact that it needs changes does not lessen its value. Having a plan promotes coordination of staff functions, helps in lowering costs, and promotes efficiency. Each department and subdepartment has a plan that fits into the whole. Where appropriate, the separate departments have models, as the engineers have had their models for years.

**General Format of this Presentation** In effect, a case study will be presented here—not in detail, but in general form. The hypothetical company to be used is Edison Power Company, as described in Chapter 3. To repeat, for emphasis, this is no particular company. Conditions vary. Plans will vary to meet local conditions. There are about 200 members of the EEI comprising the investor-owned electric utilities. The top 100 serve most of the customers. Dividing the total investor-owned industry by 100 results in a typical investor-owned electric utility company.

This company is used because:

- Yearly statistics are readily available.

- Its economic characteristics are typical of most electric utilities.

- Forecasting, planning, and analyses have been applied to this company by the author for over 30 years. There has been time to note the reasons for errors and to make corrections, so as to do a better job next time. Also, new tools are continually being developed.

- This is the company used in the author's book *The Electric Power Business*,"[1] first published in 1962 and revised in 1970.

The aim is to develop a program and a plan for Edison Power Company that will earn the full cost of capital by years for each of the next 5 to 10 years. The study and presentation will be in two major parts. The first part deals with forecasting every element in the company for the next 5 to 10 years:

- Kilowatthours
- Kilowatts
- Plant investment
- Gross revenue
- All expenses
- Depreciation
- Taxes
- Net for return
- Percent return
- Earnings per share

[1] McGraw-Hill, New York.

Until recently the percent return was considerably below the cost of money, and it is still much too low. This reflects the new economic climate described in Chapter 4. In making this first forecast, it will be assumed that there will be no change in company policies or practices. The forecast will be followed by an analysis of all factors that have changed the economic climate and caused the percent return on investment to drop.

The second part of the study is a presentation dealing with the planning, designing, building, operating, pricing, and financing of the company in a way that will earn the full cost of capital and make it financially and economically sound. In this second part, all the operations of the company will be examined under the following major classifications:

1. System planning

2. Cost control (operating expenses)

3. Marketing (load management, or customer services)

4. Pricing

This is the process which results in the building of what is termed a corporate model. Once the overall program and 5-year plan have been established, the computer programmer will put all the figures into a computer. The model is then programmed to bring out the results of any proposed changes. Once management has this tool at its command, it can then simulate on the model any changes that come to mind or come to the attention of the company manager. Thus the manager will be able to see in advance the results of alternative methods of reaching the objective.

## FORECASTING

Forecasting is essential in the energy business and is one of the principal skills of management. It includes the skill of planning.

This portion of the study and the forecast of all energy will be based on figures for the total electric utility industry in the United States. At other times, figures for the Edison Power Company will be used.

Of all the items in the forecast, experience has shown that the item that lends itself to the highest degree of accuracy is the forecast of kilowatthours. Once this has been established, the other factors can then be related to kilowatthours in terms of various ratios.

## STUDIES LEADING TO THE
## KILOWATTHOUR FORECAST

In 1957 the 45 power company executives who comprised the board of directors of the EEI authorized a comprehensive forecast of all the kilowatthour needs and requirements of people in the United States through the year 2000. Forecasts had been made in the past by many organizations and individuals. The passing of time showed that most of these had varying degrees of error. Most of them tended to level off too soon in the long range.

The studies leading to the forecast were prepared by the staff of the Institute, and required almost 2 years.

1. All the approximately 200 member companies of the Institute were sent guidelines for making local forecasts. These forecasts were made by towns, by classes of service, by districts, and for each company as a whole. Studies were made of the saturation of appliances. Numbers of customers were trended. Consideration was given to new industries and to industries leaving the area. Judgments were formed on things to come.

2. An examination was made of the world patterns of energy growth, both in total energy and in electric energy. Studies were made of the mathematical relationships between energy use, population trends, GNP, and living standards.

3. One part of the study dealt with the economics of each of the competitive fuels, such as oil and gas, which were being used by customers as an end product.

4. Another study dealt with the mix of primary fuels used in past years and how much of them were converted to electric energy. A judgment was formed as to what the mix might be through the year 2000.

5. Another study dealt with the existing literature relating to reserves of all primary fuels. From trends, a judgment was made as to when the reserves might be exhausted.

Parenthetically, these studies, made about 19 years ago, disclosed that oil and gas would be the first of the fuels to reach "short supply" and that there was a good chance that the effect might be seen during the 1970s. This was one of the factors that prompted electric utilities to press urgently for research to make nuclear energy feasible.

These studies were also discussed with authorities in Washington with the hope that a national energy policy would be established that would lessen, to some extent, the dependence on oil, which was becoming scarce, and give the country more energy from coal, which was in long supply. The country was then and still is obtaining about 80 percent of its primary energy needs from gas and oil, the fuels that are in short supply, and only 20 percent from coal, which is in long supply. If a more realistic policy had been established then, the country might have avoided the energy crisis that now plagues it.

6. All the new fuels and new possible conversion methods were studied, such as:

    a. The nuclear light-water reactors

    b. The breeder

    c. Solar energy

    d. Fuel cells

    e. Magnetohydrodynamics (MHD)

    f. Geothermal energy

    g. Fusion

7. Computer analyses were made of all the principal correlations. Sometimes mistakes are made in forecasting by putting in more parameters than are justfied. Because a computer can handle great quantities of figures is no reason for putting into the study correlations that have little bearing on the results. Sometimes this can lead to errors. It is found there are definite correlations between:

    a. Total energy use and total living standards

    b. Electric energy use and total living standards as expressed in personal income per capita

    c. Total kilowatthours and gross national product

    d. Residential kilowatthours and personal income per capita

    e. Industrial kilowatthours and the Federal Reserve Board index of industrial production

    f. "Service" personal consumption expenditures and kilowatthour sales to commercial customers

8. Records showed that about 15 new appliances had been developed over the previous 10 years. Would there continue to be 1½ new appliances invented per year? Some "phantom" new appliances were assumed to come into being.

9. On the average, people double their knowledge about every 10 years. What new uses would this new knowledge develop? This fact was given consideration.

10.   There have been constant inventions to improve the efficiency of conversion and the efficiency of power systems. With increased knowledge would this continue? It was assumed that it would.

## PATTERNS THAT HELP IN FORECASTING

In Chapter 6 reference was made to the patterns developed by Fremont Felix, who made a study of these matters concerning all 155 nations on earth.

Here are two more patterns based upon Felix's work. Chart 7.1 shows the correlation between the percent increase in consumption of total energy and total energy consumed per capita. The same pattern holds true in the use of electric energy. Countries with low electric energy use have a higher rate of increase than countries, like the United States, which have a relatively high level of use (Chart 7.2). There are other patterns which help in understanding the basic nature of the electric business, and the way it will be evolving in the years to come. Chart 7.3 shows the relationship for the United States alone, in energy use *per capita*.

It was noted in Chart 4.10 that the pattern of growth, or index of growth, of electric energy has been consistently about 2½ times the rate of growth of total energy. This observation was made in 1958. The question of the forecaster was: Will this trend continue to 2000?

To answer the question, the causes of the trend had to be known. Then a judgment could be made as to whether the causes would continue. It was found that the customer's choice is most influenced by:

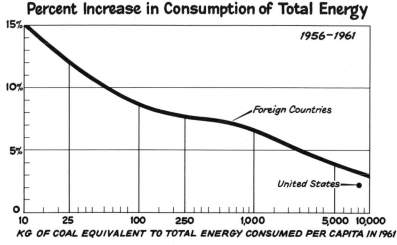

**Percent Increase in Consumption of Total Energy**

*1956–1961*

*Foreign Countries*

*United States*

KG OF COAL EQUIVALENT TO TOTAL ENERGY CONSUMED PER CAPITA IN 1961

**CHART 7.1**

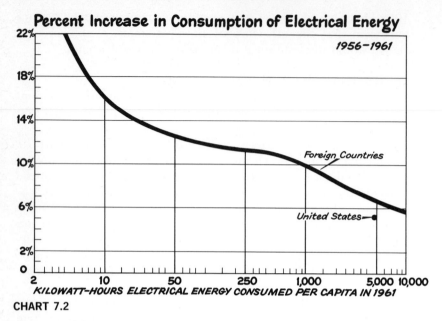

CHART 7.2

- Price
- Reliability
- Convenience

It was known that the price of electricity declined where there was an inflation rate of 2 to 3 percent. The questions then became: Would the price *relationship* between electric energy and total energy continue, and what caused electricity prices to decline?

The causes of the price decline might be simply mentioned here:

- Increased volume of electricity
- Larger and larger generating units
- Higher and higher transmission voltages
- More and more interconnections
- Higher and higher load factors
- Constant improvements in thermal efficiency
- More and more diversity

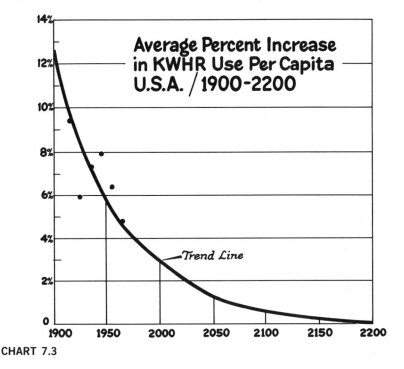

CHART 7.3

- Less and less self-generation

- Constant availability of electric energy

In 1959 the forecasters concluded that the end of the road had not been reached in any of these causes. The improvements would continue—but to a lesser degree. Therefore, the price *relationship* between electricity and primary fuels (such as coal, gas, or oil) would continue, even with the increased inflation anticipated. (Admittedly, the forecasters did not foresee the magnitude of the increase in the rate of inflation, but apparently this has not affected the pattern in energy growth).

Chart 4.10 shows that the pattern did continue from 1959 through 1976. In a moment, mention will be made as to the reasons the forecasters concluded that this pattern would continue to 2000, and probably beyond.

**Effect on the Price of Electricity**  For all the reasons mentioned above and others, the price trend between 1930 and 1968 was as shown in Chart 7.4. The left-hand side of the chart shows the rate of inflation

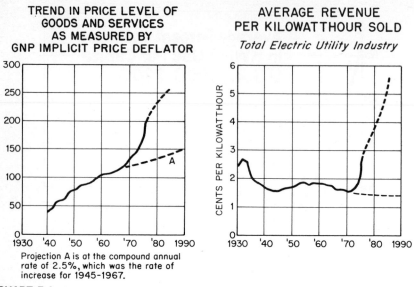

TREND IN PRICE LEVEL OF
GOODS AND SERVICES
AS MEASURED BY
GNP IMPLICIT PRICE DEFLATOR

AVERAGE REVENUE
PER KILOWATTHOUR SOLD
*Total Electric Utility Industry*

Projection A is at the compound annual
rate of 2.5%, which was the rate of
increase for 1945–1967.

**CHART 7.4**

averaging about 2½ percent per year from 1930 to 1968. During this period the average price of electricity dropped from about 2.5 to 1.5 cents per kilowatthour, or about 40 percent.

Calculations show that if the rate of inflation had continued at 2½ to 3 percent per year as shown by the dotted line in the left-hand portion of the price-level chart, the price trend would have followed the dotted line on the right-hand side. But inflation increased as indicated by the solid line. The price of electricity changed as shown in the chart to the right. All the primary fuels also increased in price.

In 1959, the forecasters concluded:

- The price benefits of electricity as compared to primary fuels would continue.

- The reliability and convenience of electricity would continue.

- The pattern of Chart 4.10 would continue.

- By 2000, half the primary fuels used would be converted to electricity (Chart 4.19).

**Other Factors Considered in the Kilowatthour Forecast**   Here are a few of the many other factors that need to be considered in forecasting kilowatthour usage:

CHART 7.5

1. GNP (Chart 7.5)
2. Disposable personal income (Chart 7.6)
3. Correlations:
   a. Total energy consumption and GNP (Chart 7.7)
   b. Industrial kilowatthours and Federal Reserve Board index of industrial production (Chart 7.8)
   c. Disposable personal income and residential kilowatthours (Chart 7.9)

CHART 7.6

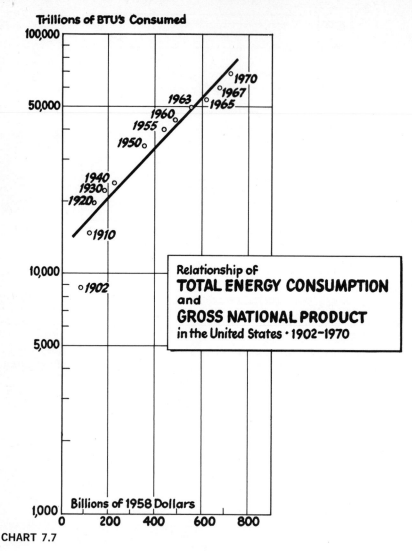

**CHART 7.7**

## THE FORECAST OF KILOWATTHOURS

The result of a thorough consideration of all these factors was a forecast as shown in Chart 7.10, first published in 1959. At that time, it was considerably higher than other forecasts. It caused some consternation.

Most other forecasts leveled off much sooner in the long term. (Notice that the vertical scale is logarithmic. Any commodity having a constant percent increase per year is represented by a straight line on this kind of scale.)

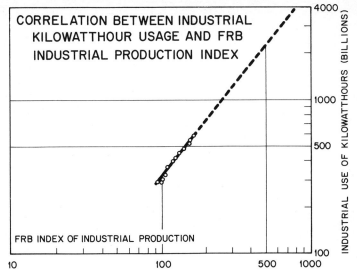

**CHART 7.8**

Correlation Between Disposable Personal Income
(1954 Dollars) and Residential KWHR Sales

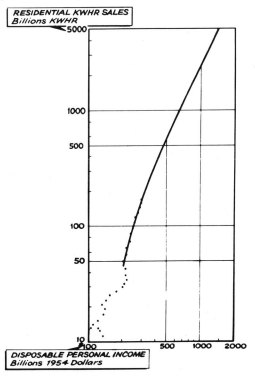

**CHART 7.9**

193

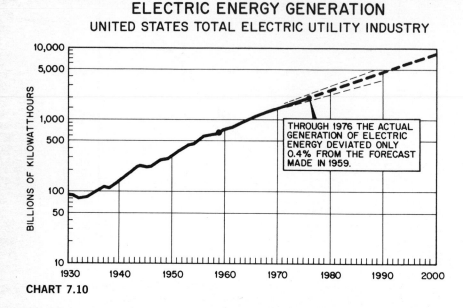

## ELECTRIC ENERGY GENERATION
### UNITED STATES TOTAL ELECTRIC UTILITY INDUSTRY

THROUGH 1976 THE ACTUAL GENERATION OF ELECTRIC ENERGY DEVIATED ONLY 0.4% FROM THE FORECAST MADE IN 1959.

**CHART 7.10**

The forecast has been checked every year. It has not been changed in 19 years. The *cumulative* error always remained in the range of 1 percent. On a cumulative basis, the difference in 1976 between the actual and the forecast was 0.4 percent (despite the low years of 1974 and 1975).

Naturally, there were some individual years when the use was above and some when it was below the EEI long-term forecast.

This is shown in Chart 7.11 and might be called the "wobble effect." This effect is given some consideration in planning capacity, but care must be exercised in the weight given to it. If capacity is planned 6 years hence based on a 9 percent growth each year and the actual is only 5 percent, there will be considerable excess capacity. Similarly, shortages could result if plans were made on the basis of a 5 percent annual growth rate and the actual growth rate was 9 percent.

Over the long term, it appears that even the large fluctuations in the annual growth rate of generation that occurred during the Great Depression and during World War II were only long-range wobbles in the overall trend.

In the past, two major schools of thought have prevailed in the making of long-term forecasts. One school was of the opinion that the long-term trend in the use of electric energy was at a fixed compound rate per year, which would result in a straight line on semilog paper. The other school assumed that saturations of major appliances and other energy

## ACTUAL KILOWATTHOUR PERCENT INCREASE
## AS COMPARED WITH THE FORECAST INCREASE

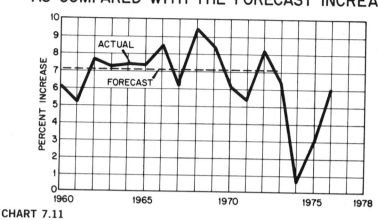

**CHART 7.11**

applications would occur, resulting in a leveling off of energy use at a much earlier date than has been the case.

It now appears that neither of these is correct. The curve appears to be one of a very gradual decline in the *rate of increase* per year.

**Patterns to 2000**  With a cumulative error of only 0.4 percent (sometimes above and sometimes below the forecast) after 19 years, it is probable that the error will be small over the next 25 years to 2000.

Chart 7.12 shows the 100-year pattern. The curve is actual from

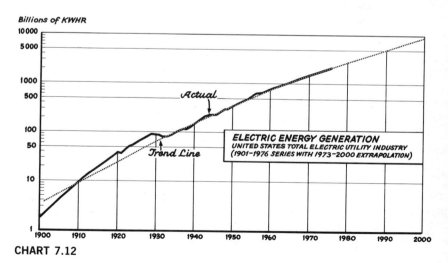

**CHART 7.12**

1900 to 1976. The dotted line is a computer-developed trend, based upon actual data for the period 1930–1959. This was one of the guides used in developing the forecast in Chart 7.10. Notice that it is a slightly curved line on semilog paper. The rate of growth from 1902 to 1907 was about 18 percent per year. The rate of growth now is about 7 percent per year. The rate of growth in the year 2000 will be about 6 percent per year.

Note the "close fit" between the actual rate and the dotted line (mathematical equation) for the years 1932 through 1976. This forecast, published in 1959, has a high degree of accuracy for the period 1959–1976, suggesting some basis for confidence in the 1976–2000 forecast.

There are two rather distinct patterns in the actual consumption: (1) the period 1900–1930 and (2) the period 1930–1976. Neither is a straight line. Both have a very slight curve, indicating a slight decline in the rate of increase for all increases in energy use per capita per year.

The two patterns probably occur for this reason: Prior to 1930, a large contributing factor to the increase was the number of new customers who had not been receiving electric service prior to this period. By 1930, practically all the small towns had service. Rural connections were only about 10 percent complete, but rural use of energy was only a small percent of the total at that time.

Therefore, prior to 1930, the increase was caused by three major factors: (1) addition of new customers previously without service, (2) population (family formation) increase, and (3) increased use per capita. After 1930, the second and third factors played the principal role.

## USE OF THE STUDY

Although the study is on a national basis for one country, it may have value to individual companies in the United States or abroad, in making their forecasts. These thoughts occur:

1.  Because of the large base of the EEI study, it probably has a more stable pattern than that of a smaller entity. The degree of accuracy may be higher than a single company or smaller country may expect.

2.  If the forecasting company or country has not yet reached all the prospective customers that need service, the pattern of the years 1900–1930 should apply.

3.  If all prospective customers have service, a new pattern, similar to the pattern of 1930–1976, will apply.

4.  If the subject company has a use per capita above the U.S. av-

erage and a rate of increase *higher* than the U.S. rate, the chances are it will decline in *rate of increase* more steeply than the U.S. average.

5. If the subject company has a use per capita lower than the U.S. average, the chances are it will tend to approach the U.S. average and experience a higher rate of increase.

6. These world and U.S. patterns should be used *only* as "backdrops." Each situation is different. Each situation should be thoroughly analyzed, and decisions based upon *local* conditions.

**Conservation** In the United States and in many parts of the world, conservation of energy is receiving much attention. Obviously, conservation and avoidance of waste are worthy objectives. This is discussed more fully in Chapters 12 and 13.

From the standpoint of the forecast being studied here, there are these considerations:

1. While waste is ever present and should constantly be lowered, the waste in the United States is not as great as most people think (Chapter 13).
2. While the conservation of energy is good:
   a. There is no shortage now, nor is one in prospect in the immediate future, in coal and nuclear fuels. Oil and gas are causing a problem.
   b. Care should be exercised to avoid curtailing the use of *useful* (as distinguished from *wasteful*) energy if it is desirable to maintain and improve living standards and provide job opportunities (Chapter 6).
   c. The benefits should always be weighed against the costs.
   d. The "no growth" or "forced slow-growth" philosophies should be considered in light of the relationship between energy use and living standards.[2]

---

[2] An article by Leonard Silk in *The New York Times* of April 12, 1976, is of interest. Here are excerpts:

The Club of Rome, which aroused intense controversy three years ago by a report it commissioned on "The Limits to Growth," now recognized that further global growth is essential if the problems of world poverty and threats to world peace are to be solved.

At an international conference of scholars and businessmen here at the University of Pennsylvania, Aurelio Peccei, founder of the Club of Rome and former managing director of Olivetti, stated that the limits-to-growth report had served its purpose of "getting the world's attention" focused on the ecological dangers of unplanned and uncontrolled population and industrial expansion. The report sold more than two million copies world-wide.

The Club of Rome is a group of scholars and businessmen from many countries

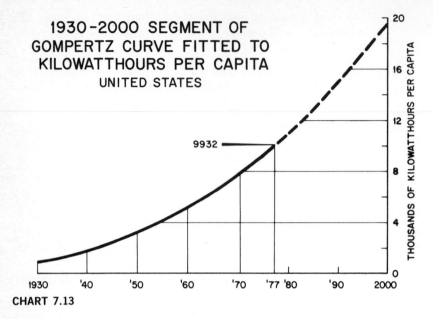

## 1930-2000 SEGMENT OF GOMPERTZ CURVE FITTED TO KILOWATTHOURS PER CAPITA
### UNITED STATES

9932

THOUSANDS OF KILOWATTHOURS PER CAPITA

1930   '40   '50   '60   '70   '77 '80   '90   2000

**CHART 7.13**

e.   The dip in energy growth in 1974 and 1975 can be traced largely to the dip in the GNP, not energy conservation.

**Energy per Capita**   Notice in Chart 7.13 the smooth pattern that prevails when the actual data to 1977 and the forecast figures to 2000 are plotted on arithmetic graph paper. A few of the forecasters in the 1959 study recognized the similarity between the long-term shape of the energy curve being developed and the Gompertz curve (named after an English mathematician and actuary of the eighteenth century).

Chart 7.14 shows the Gompertz curve fitted to the kilowatthours per capita, actual and forecast, to 2000. If the predicted kilowatthour curve is reasonably accurate (and all signs indicate that it probably is), then this curve also fits the Gompertz curve.

*Mathematically,* the curve takes the shape of an elongated "S." The total use per capita is 20,000 kWh in the year 2000. The rise continues

who are concerned with global problems affecting the environment, food, energy and other resources as well as the question of human survival.

The original study, based on a computerized model developed at the Massachusetts Institute of Technology, warned of a disaster to humankind within a century if present growth trends continue.

## GOMPERTZ CURVE FITTED
## TO KILOWATTHOURS PER CAPITA
### UNITED STATES

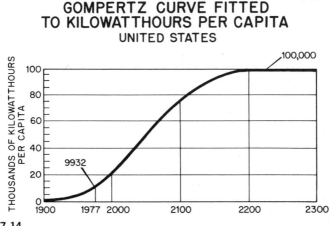

**CHART 7.14**

and does not level off until 100,000 kWh per capita in the year 2300. (This is not a forecast beyond 2000, but is the mathematical result of applying 100 years of kilowatthour data to a Gompertz curve.)

Mathematically, the percent increase per year continues to decline, reaching zero increase per year at 100,000 kWh per capita in about the year 2300. Similarly, the percent of primary fuel to electricity continues to rise, mathematically, reaching an all-electric economy in the year 2300.

## FORECAST OF OTHER FACTORS

The study so far has dealt with the forecasting of all factors from kilowatthours to percent return for 5 years, with no change in policy or procedure. So far, there has been developed the forecast of kilowatthours, this being the factor that lends itself to the greatest degree of accuracy. This is the first step in forecasting all elements in the power business over the next 10 years. Later, in the second part of this chapter an examination will be made of all the factors under control of management, and a program will be designed that will earn the full cost of money for each year for the next 10 years.

Following are the elements that are now to be forecast for the next 5 or 10 years without any change in policy and without any rate changes:

- Kilowatts
- Plant investment
- Depreciation
- Taxes

- Gross revenue     - Percent return

- All expenses      - Earnings per share (where applicable)

- Net for return

The kilowatts are derived by applying the appropriate load factors to the kilowatthours. Using the appropriate percent reserves, the plant capacity can be determined.

In forecasting kilowatts, it is advisable to go back 5 or 10 years and trend the summer peak as compared with the winter peak. As a general rule, summer peaks tend to grow more rapidly with air conditioning.

Using 65 percent load factor, Chart 7.15 shows the kilowatts needed for the entire U.S. electric utility industry. About 80 percent of the industry is investor-owned. Dividing the investor-owned portion by 100 gives the approximate capacity of Edison Power Company. Chart 7.16 shows the corresponding investment. Other factors, such as revenue, expenses, etc., will be discussed in the second part of the chapter.

**Percent Return**   Of particular interest is Chart 7.17 showing the percent return for the entire investor-owned electric industry and for Edison Power Company. From 1967 through 1974, there was a problem because of the low rate of return on investment. This situation was corrected to some extent in 1975 although the return on investment is still not adequate.

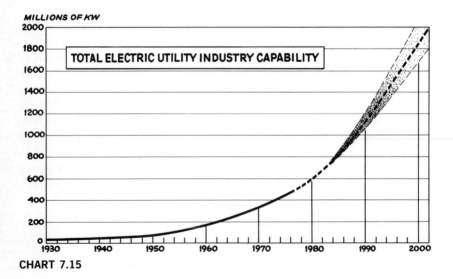

CHART 7.15

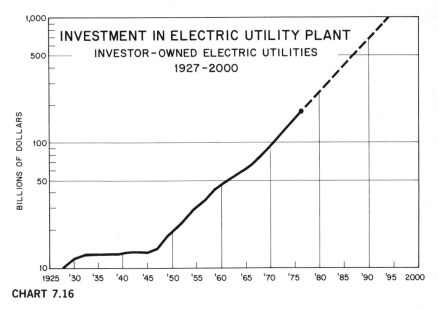

**CHART 7.16**

This chart shows the actual percent return of the industry and for Edison Power Company from 1967 to 1977. The dotted lines shows the actual cost of money for each of the years. The results have been:

1.  From 1967 to 1974 earnings were depressed.

2.  The ratings on bonds were lowered.

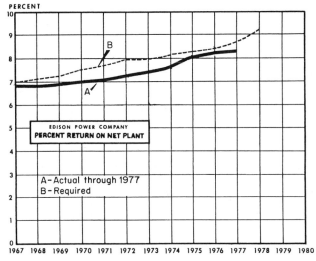

**CHART 7.17**

3. Common stocks sold for less than book value (most have since improved).

4. Many companies experienced difficulty in financing.

5. Capital was costing more than it would have if the companies had been allowed a reasonable return.

6. Many power plants were cancelled or postponed, partly because the increase in load leveled off somewhat in 1974 and part of 1975 and partly because of financial difficulties.

What happened? Basically, the trouble stemmed from the fact that, traditionally, the average price of electricity was declining. The increased use of service was healthy. Efficiencies were increasing, enabling a lowering of price of electricity when the inflation rate was 2 to 3 percent per year. For reasons outlined in Chapter 4, there was a complete reversal around 1968, causing the unit cost (price) to rise.

Regulating an industry is easier, economically and politically, when prices are going down, but becomes difficult when prices must be increased. The complicated economics for the increase must be understood. Some public dissatisfaction is inevitable.

It takes considerable time and expense to prepare and present a rate case. The commission and its staff must study it. By the time an order is issued, the new rates are obsolete. Higher rates are already needed. Time lag is the culprit.

Chart 7.18 gives a clue. This is the same as Chart 7.17 except that there is a dot for each year, beginning in 1971. This dot is the arithmetic average of all the percent returns allowed by the state regulatory bodies for that year. From 1971 through 1974 the dots are almost on the line representing the cost of money.[3] The commissions have been allowing almost the cost of money, but on a rate base that is obsolete by the time the rates are effective. When the rates go into effect, investments and operating costs have increased to such a degree that the resultant percent return realized is much less than that ordered.

---

[3] There will be more on fair return and cost of money in Chapter 9. It is sufficient to mention here that "cost of money" is not necessarily the fair return. During a period when expanding plant is needed to serve increased loads, it appears that the bare cost of money is the *least* return the regulatory body can allow if the regulatory body is to fulfill its obligation of assuring good service for the future, which carries with it the obligation to allow (indeed require) the company to earn not less than the cost of money. Without that bare cost, the company cannot carry out its part of the obligation imposed by law.

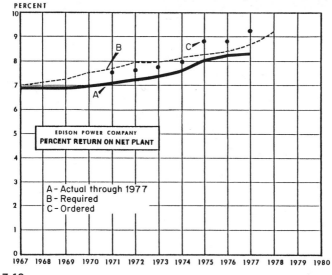

CHART 7.18

However, a commission can remedy this situation by using the rate base and operating costs of the year or years for which the rates are designed. This means using a forward rate base, which has already been adopted by some regulatory bodies.

This means basing rates on a forecast. As this study shows, the electric utility business, by its economic nature, is subject to a higher degree of accuracy of forecasting than is generally realized. The plant is committed. Its costs are well known. Fixed charges on plant take nearly half of gross revenue. Labor costs are fairly well known and are fairly stable. Most of the remaining costs are for fuel, and those costs are covered (or should be) by a fuel clause.

In recent months, the situation has improved somewhat:

- Commissions have shortened the time lag.

- Some interim rate increases have been allowed.

- Some examination is made of the year or years for which the rates apply.

**A Problem that Needs a Solution**    The problem has been stated and analyzed. The percent return must move up. Now there is need for designing a number of 5-year programs, each of which will earn the full cost of money. Managers will select the best one. This will be the company's overall plan. It will be the corporate model.

In the next phase (the second part of this chapter) an analysis will be made of the primary factors under the control of management, in the design of various alternative programs each of which will earn the complete cost of capital. Through computer models, management selects that alternative which, in its judgment, is the best.

## PLANNING A PROGRAM TO EARN
## THE COST OF MONEY

The purpose now is to design and plan a number of alternative programs that will earn the required percent return for each of the next 5 or 10 years. The study will indicate the percent return that will be required and the alternative that management selects as being the best for Edison Power Company.

In this presentation, there is a description of the steps that management takes in planning and building a program. The case study applies to no particular company. Each company has different quantities and different figures. Described here are some of the various skills management needs in building a program of this kind and some of the tools that are available in reaching excellence.

The forecast of kilowatthours is now known from the first part of this chapter. From this the kilowatts of demand and capital investment can be determined. Now the task will be to examine four of the five primary areas under the control of management in order to determine what improvements are possible. The fifth area, public and employee relations, is analyzed separately. This examination is made under the following headings:

- System planning

- Cost control (operating expenses)

- Marketing (load management, or customer services)

- Pricing

The purpose of this arrangement is as follows: The aim of management is to design the most efficient power system possible before setting the price of the service. The goal is to plan the most economical kind of operation possible consistent with good service.

Next management must determine how much *net* revenue it can obtain through the marketing (or load management) process; from this is determined how much the percent return can be improved through marketing. Any improvement here lessens, by that amount, the total funds that need to be raised through increased price.

All three of these factors (system planning, cost control, and marketing) are then forecast in the best plan that management can design. With the unusual economic climate now being experienced, managements in most parts of the world are finding that, without increased prices, the best plans that can be devised are not sufficient to produce the required net revenue to enable the payment of the full cost of money.

All required net revenue that cannot be obtained through improvements in system planning, cost control, and marketing must be obtained through the pricing process.

Here are the steps which most American companies follow:

1. Estimate the cost of money by years for 5 years.

2. Using this cost of money and the forecast of plant investment, calculate the required net revenue (see Chart 7.17).

3. Compare this figure with the net revenue expected from the forecast with no change in policy. (The comparison will probably show a need for increased net revenue. If not, go through the procedure described here and plan for lower rates.)

4. Calculate the amount of increased net revenue that can be expected by reexamining (a) system planning, (b) cost control, and (c) marketing.

5. With the improvements in the above, calculate the increase in net revenue to earn the full return.

6. Add the new net revenue to expenses to arrive at the required gross revenue.

7. Design rates to produce that gross revenue year by year for the next 5 years.

In Chapter 4 it was noted that:

- The wholesale price index, contrary to previous patterns, continued rising after World War II and took a sudden further steep rise around 1968.

- Inflation rose sharply.

- The index of plant construction rose sharply.

■ About 1968 or 1969, unit electricity cost (price) reversed its long-term trend and began to rise.

How much will the rise be through 1985?

## SYSTEM PLANNING

System planning is a long, complicated process, frequently performed by employing specialized engineering firms. Probably the existing plan is fairly well fixed for the next few years. Nevertheless, most companies periodically review the long-term plan and change it as necessary. Preliminary checks are made by selecting appropriate ratios for comparison with other companies operating under somewhat similar conditions. Care should be exercised not to draw conclusions from these comparisons. Skilled technicians analyze each "flag" to see whether or not improvements can or should be made.

The companies in the United States have the advantage in being members of the EEI, their national trade association. There is a committee for each major subject, some 70 in all. The committee members are individuals from member companies. Thus, there is a continuing exchange of ideas, information, and studies.

**The Process of System Planning**   Here is an outline of the steps taken in system planning.

The engineers put on paper various alternative system plans that may be practical and feasible for generation and transmission. There may be four or five or possibly more trial system plans. Plan A may call for a combination of conventional steam, hydro, pumped-storage, and nuclear plants, whereas alternative B may contain another mix of plants with another mix in the size of units. Each of the alternative plans includes appropriate transmission and substation additions designed to fit the particular plan. Each of these plans is tested on computer models to be sure that it remains stable under various assumed forced outages. The engineers may decide upon what they consider the proper degree of reliability. Obviously, the engineers will plan for scheduled outages for maintenance and other purposes. Then they will assume, for example, that the largest generating unit fails during some peak hour 3 or 4 years hence. The computer model can tell whether or not the system is stable (see Chapter 3) under those criteria. If not, the engineers will change the design so that it is stable.

In other words, a number of alternative plans are programmed into the computer, and each one is tested in advance for its reliability and stability.

In turn, each of the system plans goes through a process of economic loading. This is also tested on a computer model. Each generating unit may have a different efficiency. Each may be using fuel of a different type. The plants all have different distances from loading centers, and the power is carried over lines of different voltages.

For each hour of the day there is a combination of loads on each of the generating units that results in the greatest overall economy for the system. There may be a different mix for each hour. Once this is established, the computer can print out the total operating costs for plan A for the entire period of 5 or 10 years. To these operating costs are added the fixed charges on the investment of plan A. Thus the planner knows the total costs of financing and operating plan A. The same process is followed for alternatives B, C, D, and E.

All these alternative plans are discussed with management, and management decides which plan it will adopt. Thus, the company has exercised all its managerial and engineering skills in planning the system in the most economical fashion consistent with the reliability of service.

Chart 7.19 shows the system plan finally adopted by management. Note the economic characteristics. The total investment per kilowatt remained steady at about $350 per kilowatt from 1955 through 1968, when inflation was running about 2½ to 3 percent. The fixed charges on plant constitute a large item in the total cost. With this cost holding steady, and operating cost per kilowatthour declining, the total cost (or price) per kilowatthour could decline.

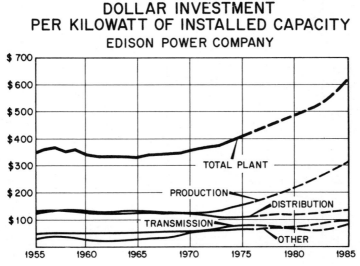

### DOLLAR INVESTMENT
### PER KILOWATT OF INSTALLED CAPACITY
#### EDISON POWER COMPANY

**CHART 7.19**

Then notice the sharp rise beginning about 1968. (See Chart 4.6 for the increased cost of new plant.) The unit average embedded cost has already risen to about $450. It will be about $600 in 1985. Fixed charges have also risen with higher cost of money. Add to this the large increase in fuel prices, and one can see where the price of electricity will have to be in 1985.

In Chart 7.20 is shown the adopted plan in total kilowatts of capacity through 1985 with the proper reserves. The investment in plant was shown in Chart 7.16.

## COST CONTROL

Reducing costs is a never-ending preoccupation of management. As a member of the EEI, Edison Power Company has the assistance of the Institute. The company has 38 employees serving on 38 EEI committees. The committees meet two or three times a year. The members exchange ideas and information, one with the other, in an endeavor to determine the most efficient tools and methods to perform each function. Group studies are made. Group research is carried on. The company receives continuing reports on industry matters at home and abroad and has direct access to new and improved methods developed by EEI member

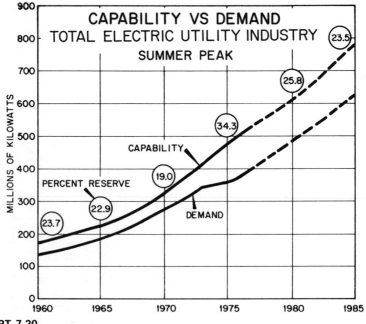

**CHART 7.20**

companies. Special computer committees periodically develop new methods, which are available to all companies.

Nevertheless, periodically (possibly every 5 years) the company makes an in-depth study, or audit, of its current position on expense control. Frequently, an outside consulting firm is employed, as it is wholly objective. However, to be most effective, the outside firm should work in close cooperation with the company staff.

For such an audit the management wants a reexamination of every function, such as meter reading, billing, planning, gathering statistics, disseminating information, storing information, purchasing, construction, etc., in order to be sure it is performing in the most efficient fashion. Frequently, the analysis deals with the number of people required to carry out each task. The question naturally arises: Is the task now being done really necessary? It may have been started some time ago and served its purpose. The aim here is to discontinue any routine work that is not needed. This pertains to the many forms that a company fills out, all reports that might be eliminated, all duplication of forms and reports. Then a detailed study is made of the whole organization to find out which people are really not necessary in carrying out the function of the company economically.

In the United States and throughout Europe, where all companies have ready access to cost comparisons with other companies, use is made of the comparison process in analyzing the economy of operation. For example, the following procedure is followed in the United States. It has been found quite effective. It can be used only where there are ample statistics available to compare the subject company with others operating under similar conditions.

This kind of study is carried on as follows:

- Select appropriate ratios.

- Compare with similar companies.

- Analyze high ratios.

There are approximately 30 to 40 pertinent ratios that, when analyzed, help to compare the efficiency of one company to another. These include such ratios as total power-plant investment, cost per kilowatt, total utility plant investment per kilowatt, transmission investment per mile of line, administrative and general expense per dollar of revenue, customer accounting and collecting expense per customer, production cost per kilowatthour, fuel cost per kilowatthour, transmission and distribution expense per customer and per mile of line.

All these statistics are available in the United States either through

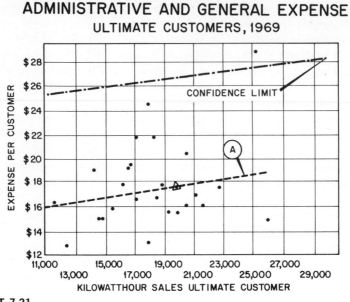

# ADMINISTRATIVE AND GENERAL EXPENSE
## ULTIMATE CUSTOMERS, 1969

**CHART 7.21**

the Federal Energy Regulatory Commission or through the EEI. For a study of a particular company, for example company A, a selection is made of 10 or 15 companies similar to company A in size and characteristics and serving similar areas. This is called the *test group*. All the ratios for the test group are placed into a computer model. Usually, these figures and ratios are given by years for 5 years. The computer prints out the average figures for all these test companies and computes a trend line by years. Thus the subject company A can be compared with the average of the group or to an individual company in the group.

Chart 7.21 is a plot of computer data for one of the ratios with the average of 30 companies being compared with one company. This ratio is the administrative and general expense per dollar of revenue.

In using this ratio method, *care must be exercised to avoid reaching a conclusion based on the ratio alone*. There may be good reason for the ratio being as it is. But this is a flag, suggesting an examination. The expert examines this expense by going down the line and finding all the components that are included in this expense. It is this detailed analysis that shows the reason why the expense of this company is high compared with other companies. Furthermore, it shows whether or not the expense can be reduced, and if it can, by how much.

If this process means the elimination of people (and it usually does), the company will card-index these people and place the cards in a computer file along with their qualifications. Management will hire these people, where qualified, before hiring others. In this process some retraining may be required.

Usually outside experts are called in to make this study, as experience has shown that it is difficult to amputate one's own arm. A completely detached analyst is not encumbered by these matters. Nevertheless, in making the final decision, management gives the appropriate weight to the opinions of local company people who have a greater knowledge and experience with local conditions.

For this particular company some 30 to 40 ratios of this kind were analyzed. In each case a closer examination was necessary to determine the cause. In some cases it was a justifiable cause; in others it was not.

Chart 7.22 shows the total operating expense under the final alternative operating plan adopted for Edison Power Company. Notice from this chart the sharp change in the trend that occurred in 1968, reflecting the factors described in Chapter 4.

## MARKETING

The term "marketing" has different meanings for people in different kinds of business. To most people and for most businesses, "marketing" refers to the methods used for increasing the volume of sales. This is not necessarily so in the electric utility business. Volume increase, when taken alone, does not fit any of the eight purposes of electric utility management. A primary index of performance of management in the electric utility business is the percent return on total investment. Increased

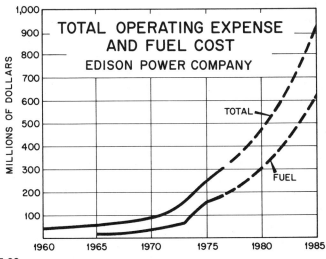

**CHART 7.22**

volume may occur without due regard for net revenue. Even an increase in net revenue is not the criterion for excellence. Net revenue may be rising, but if plant investment is rising faster, percent return may be declining.

The term "marketing," as used by electric utilities, is broad in scope. Some choose to call it "customer service," as it is a function which directly concerns the customers' interest. Also, since a primary function of management is to keep the price as low as practical and since the price can be lower than otherwise at higher load factors, a primary purpose of marketing is to build load factor. Some call this "load management." But this is only a part of the picture.

As energy conservation is an important concern, some choose to call the "marketing" function "conservation." As always, this is important, but conservation also is but a part of the broad marketing concept. Conservation of electric energy should be practiced to avoid any waste or inefficiency. Oil and gas should be conserved because these declining resources must be stretched out for as long as possible. If more energy is needed to improve living standards, coal and nuclear fuels should be used. And better living embraces environmental improvement. This requires more, not less, electric energy. Better living requires increased production, which in turn requires more useful energy.

**Goals of Marketing**   Possibly the term "marketing" as it is used herein, can be clarified by a statement of the areas in which electric utility management has an interest.

1.   Customers should be informed on all pertinent matters regarding their electric service.

2.   Conservation means avoidance of waste and improvements in efficiency. Customers should be informed and advised on how to accomplish it. Conservation pertains largely to the fuels in short supply—oil and gas. This means more electric heat as a means of shifting to coal and nuclear fuels. A major cause of waste is inadequate or poor insulation. This requires education and promotion, i.e., marketing.

3.   Electric utilities work closely on all matters of safety, in committees of the industry and with national safety organizations. This is done not only in the interest of safety as it applies to utility equipment but also as it applies to uses by the customer. The utility wants to work closely with manufacturers, contractors, and dealers so as to encourage safety features and practices.

4.   Management wants to assist the customers in the efficient use of electricity. This requires adequate wiring, proper standards, and efficient appliances and motors. Following these matters and keeping the customers informed is part of marketing.

5.  Part of the "good life" desired by most customers depends on the use of such home appliances as the iron, vacuum cleaner, washer, dryer, refrigerator, and freezer. In the past, as each of these became available, management assisted in providing the information to customers. The information included the types available for each purpose, the method of use, the servicing, the first cost and the operating cost. In short, customers were assisted in making a choice, to use or not to use. Further, appliances are being improved from time to time. The first refrigerators had no freezing compartment. Then, there were available those with a small freezing compartment in the top. Later there were models with a larger freezer in the bottom or at the side. Then came the separate freezer. All these improvements required more energy. Some customers wanted them, some did not.

6.  All things have not yet been invented. As new developments become available, customers are entitled to know about them.

7.  With respect to lighting, exhaustive research and testing have demonstrated that:

- In industry, adequate and proper lighting improves productivity, safety, and worker satisfaction.

- In commercial establishments good lighting improves sales, increases efficiency and reduces eyestrain.

- In the home proper lighting reduces eyestrain, helps children's grades in school, and improves efficiency in the performance of home tasks.

- On the highway and in the streets, good lighting improves safety and reduces crime.

These are facts, known and encouraged by management. The value is in telling others about them. This is marketing.

8.  Customers need a place in the company's office where all complaints can be made and questions answered. New customers need information. This is marketing.

9.  Customers want to be assured of adequate service on all appliances. Management works toward encouraging and promoting such service among all contractors and dealers.

10.  As all improvements in load factor help keep the price low and earnings adequate, management works through many avenues to bring this about. More will be said on this. It involves load management, pricing policy, improved diversity, the encouraging of off-peak service daily, monthly, and yearly, and the encouraging of a balanced load among

classes of customers. All this involves customer information programs which are part of marketing.

11.  Keeping customers informed requires communication. Management strives to carry out this function in the most effective and efficient manner. It may involve personal calls, such as among industrial and large commercial customers. Opinion leaders such as those in the news media need personal calls. At times, bill inserts or personal letters are best. Then there is the whole array of media involved in mass communication. The aim is to have the customer:

- Hear or see the message

- Understand it

- Believe it

- Retain it

There are ways to measure the efficiency of various methods in terms of "cost per recall."

12.  Customers are interested in having their community grow and prosper. They want to be assured of jobs and adequate wages. They want a good, clean community. This requires what some utilities call "area development" or "industrial development." Wise utility management takes an interest in this. There is a department to encourage it, to determine what type of industry would be fitting, and to seek it. It works closely with chambers of commerce and community leaders.

13.  Personal calls are made on industrial customers to help in the efficient use of energy and to encourage increased load factor. In a two-part rate system (see Chapter 8) this lowers the price of electricity.

All this work in the customers' interest takes manpower and money. In the standard classification of accounts it is listed under "sales expense." Management tries to operate the department in the most efficient fashion by ensuring that the customer benefits exceed the cost.

**Marketing to Build Load Factor**   As it is the purpose of this chapter to develop a 5-year program that will earn the cost of capital with the lowest practical price to the customer, there will now be a discussion of some of the steps taken by management to build load factor. Any improvement here lessens the amount that will be needed through price increases.

It is not the purpose in this overview to give a detailed discussion of marketing. This analysis will simply touch upon the manner in which management uses this tool as a means of keeping the price low. There will be demonstrated here the kind of economic analysis that marketing people use so as to assure management that every dollar spent in the marketing field results in an improvement in the overall percent return of a company. This means that the marketing effort is concentrated upon those uses of energy that occur during the off-peak periods and, therefore, require no additions to plant. A classic example of this is now taking place in the United States, where companies have their annual system peak in the summer because of the growth of the air-condition-ing load. This air-conditioning load comes on without marketing. People demand it. Electric house heating is promoted as a means of balancing the summer air-conditioning peak. The heat pump is encouraged be-cause it cools in the summer and heats in the winter.

In short, this study shows the way in which a company may plan a marketing program to increase load factor and result in an increase in percent return, after allowing for sales expense.

For convenience, a three-dimensional annual load chart of Edison Power Company is repeated (Chart 7.23). This model was built by pre-paring on cardboard a separate load curve for each day of the year. When all 365 cardboard charts are stacked together, they form the three-dimensional load-factor chart shown here. There are many hours during the evening and nighttime when there is capacity in the power plant not being used. Also, note the seasonal valley. The total volume of the overall cube represents the capacity of the power company. Notice that the peak load never reaches the top of the box. The difference is the reserve. Examination of the chart, from the standpoint of the annual

# ANNUAL LOAD
## EDISON POWER COMPANY

**CHART 7.23**

peaks, discloses that the highest peak load occurs only a few hours out of the year. The aim of the marketing manager is twofold: (1) hold down the peak as far as practical and (2) fill the valleys.

The encouraging of energy use during off-peak periods does not constitute a waste as long as the energy serves a useful purpose. Remember, the primary fuel being imported is oil. Only 18.3 percent of the fuel used to generate electricity is oil.

Capital is a resource that also needs consideration from the standpoint of conservation. The utilization of off-peak energy requires little or no increased capital and results in improved percent return, thus enabling the financing of needed new capacity, resulting in a lower overall price than would otherwise be possible.

In brief, the public interest calls for increasing load factor (utilizing investments) by all practical means.

Repeated for convenience is Chart 7.24, showing the relationship between load factor and percent return. Every change of 10 percentage points in the load factor (as from 50 to 60 percent) is equivalent to a change of 1 whole percentage point in percent return, i.e., 6 to 7 percent (see Table 7.1).

**The Competitive Markets**   Before discussing how Edison Power Company conducts its marketing program, it will be helpful to have a general discussion of the entire field of energy markets.

The electric utility business is one that converts and upgrades primary energy into electric energy. By and large, there are four principal

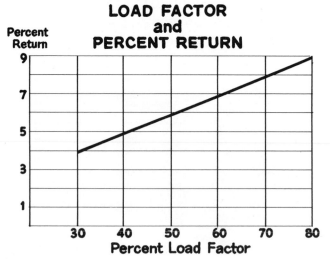

CHART 7.24

**TABLE 7.1**  RELATIONSHIP OF LOAD FACTOR
AND RETURN*—1 KW

| | |
|---|---:|
| (1) Plant investment | $ 425.00 |
| (2) Gross revenue: 5344† @ 3 cents | 160.00 |
| (3) Return 8 percent of (1) | 34.00 |
| (4) Increased kWh @ 10 percentage points increase in load factor = 10 percent of 8760 = 876   say   900 kWh | |
| (5) Increased net revenue | |
|     Increased gross revenue   2.5 cents (not 3 cents) | |
|     Increased expense   <u>1.0</u> cents (mostly fuel) | |
|     Increased net   1.5 cents × 900 kWh | $ 13.50 |
| (6) Federal income tax: 50 percent | 6.75 |
| (7) Increased net revenue | $ 6.75 |
| (8) New net revenue (3) + (7) | $ 40.75 |
| (9) New percent return (8) ÷ (1) | 9.6 percent |

* Given an output of 1 kW at 61 percent load factor, a 10 percentage point increase in load factor equals about 1 whole percentage point increase in percent return.

† 1 kW × 8760 (hours in year) × 0.61 load factor = 5344 kWh.

markets for the use of energy, namely: (1) lighting; (2) stationary motors in homes, business, and factories; (3) heating; and (4) transportation. For reasons previously mentioned, it has been possible, as a result of research and innovation, to improve the efficiency of converting primary fuels to electric energy and to bring about a decrease in the average price (until 1968) of electricity sold to the consumer. Those who provide primary fuels are also constantly carrying on research to improve their efficiency in removing fuel from the ground and delivering it to the consumer. However, the opportunities for improvement in the efficiency of these operations are not as great as the opportunities for improvement in the efficiency of converting primary fuel to electric energy. For these reasons, and because of the increase in electric utilization equipment, the growth of electric energy in this country has been about 2½ times the rate of the growth of primary energy (Chart 4.10). This relationship is expected to continue.

By the early part of the century most of the lighting in the nation was changed from oil lamps and gas to electricity. Today the lighting market is almost universally electric. Beginning in the 1920s, manufacturers and industry began to realize they could change the gas, oil, and steam engines in their factories to electric motors and buy electric energy from one of the interconnected power systems, thus obtaining energy more reliably and more economically. As a consequence, there has been a gradual and steady changeover to electric energy in the stationary motor market.

The third major market for energy is heating. Electricity began making inroads in this market through heating appliances such as the toaster, the coffee percolator, the hot plate, the electric range, the water heater, and the dryer. In commercial establishments, electric fry kettles, hot plates, and electric ranges appeared. In industry, heating applications took the form of large electric furnaces.

Electric space heating began to be competitive with gas and oil in the last decade. Only a few years ago few homes were heated electrically. In 1976 there were about 10 million, representing 13 percent of all dwellings. It is estimated that by 1985 there will be about 27 million. By the year 2000 about 70 million homes in the country will be heated electrically.

In commercial establishments and in industry, electric heating has become competitive in many areas. There are numerous all-electric shopping centers, office buildings, schools, churches, and hotels. One innovative technique is to use the heat from lights to heat the building. A multistory office building, for example, can be heated entirely by the heat from the lighting required to give adequate lighting levels in the offices. In the summer, heat from the lighting is removed through ducts and air conditioning. As mentioned in Chapter 13, if national policy calls for the lowering of oil imports, then it will be necessary to replace oil with coal and nuclear fuels for heating. As this must be through the electric process, electric heating could grow more rapidly than indicated here.

Electric energy has been used in transportation for some time. Electric streetcars, commuter trains, and railroads are not new. In some countries, such as Switzerland, France, and Japan, railway electrification is more advanced than it is in the United States.

As mentioned in Chapter 13, another large use of petroleum products is in transportation. A national effort to conserve oil and gasoline could add a stimulus to electric transportation, including the electric automobile.

**Edison Power Company Marketing Program**   Lighting was the first popular use for electricity in the home. The next was the electric iron, followed in a few years by such applications as the refrigerator, the electric carpet sweeper, and the clothes washer. Each had to be demonstrated to convince customers that the particular appliance would actually perform its designated task through the use of electricity. The acceptance of appliances as well as lighting, resulted in improved load factor, which enabled the lowering of the price of electricity.

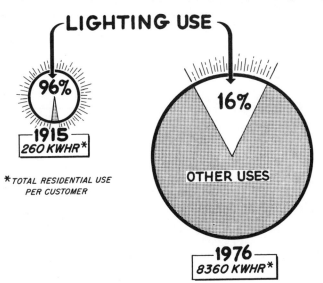

LIGHTING USE

96%
1915
260 KWHR*

*TOTAL RESIDENTIAL USE
PER CUSTOMER

16%

OTHER USES

1976
8360 KWHR*

**CHART 7.25**

The first electric ranges were rather crude devices, not nearly as efficient as modern ranges. People had to be convinced that this method of cooking was practical and convenient. To help do this, Edison Power Company gradually built up a home service department which sent representatives to visit customers' homes to help them get the most out of their electric ranges.

Lighting  Lighting is a good example of the nature of the growing market for electricity. Chart 7.25 shows the lighting use in the home in 1915 and 1976. Although electric lighting has been on the market since the first days of the industry, customers want and need more for convenience and efficiency in performing home tasks.

In the book *Seeing* by Mathew Luckiesh and Frank K. Moss,[4] some interesting observations are made on lighting. Primitive people were accustomed to seeing distant objects under the light of thousands of footcandles produced by the sun. [The *footcandle* (fc) is a measure of light intensity. It is equal to the intensity of the light of a candle held at a distance of 1 ft.] Today, people tend to do most of their seeing under a few footcandles while doing closeup work such as reading or operating a machine.

Chart 7.26 shows the footcandles of lighting in the noonday sun on the beach as compared with footcandles in the daytime under the shade

[4] Williams and Wilkins, Baltimore, 1931.

**CHART 7.26**

of a tree. Compare these with the 20 to 150 fc that might be available for reading purposes in the home or office and for work in the factory. In recent years, people have become aware that good lighting can help improve efficiency and increase production. Careful experiments show that when there is proper lighting, less energy is wasted on useless work.

A few years back, a simple research project was devised to determine how many footcandles of light people wanted for ease in reading. The researchers rigged up a chair with a light which could be made dimmer or brighter by turning a knob. The customer was asked to sit in the chair, read a newspaper, and turn the knob to the level of light which seemed most comfortable for reading. A dial attached to the controls showed the amount of light selected. Invariably the person would turn to a lighting intensity five or even ten times as great as that in the home.

**Refrigeration**   From the standpoint of load factor, only the completely off-peak electric energy uses can match the refrigerator-freezer. They are "on" roughly one third of the time. These appliances have a diversified load factor of 85 percent.

Utility marketing in widely accepted appliances such as the refrigerator is devoted to encouraging safety, efficiency, good insulation, good appliance servicing, and standards.

People are not so well informed on such matters as:

- The benefits of electric heating.
- The need for proper insulation.

- The value of the heat pump in energy conservation.
- The differences in the efficiency of various room coolers.
- The value of good lighting.
- The value to the customer of high company-wide load factor.

Where such knowledge is needed and useful, the utility has the responsibility to supply it. (Table 7.2 shows the national saturation of key appliances.)

**Air Conditioning**   The air-conditioning load has had a marked effect on load characteristics, first in the South and Southwest, then through most parts of the country. There has been a gradual change from a winter to a summer peak. The developing countries can benefit from this record by early preparation for the shift.

The summer peak has become higher than the winter peak had been, thus causing a decline in load factor. For a while, the electric-house-heating load was growing fast enough to offset the growth in load from air conditioning. However, in recent years, some regulatory bodies and others have discouraged all marketing by utilities, with the result that national load factor has dropped from 65 percent in 1968 to about 61 percent in 1976. For the same percent return, the required price at 61 percent load factor is higher than it would be at 65 percent load factor. Some companies have continued marketing and have prevented load factor from declining. A few companies maintain 70 percent load factor. Their price can be lower than average for the same percent return.

Lately, there has been recognition of the need to improve load factor, with the result that utilities are urged to engage in "load management," which is a part of marketing under a different name.

**Electric Heating**   Edison Power Company has done considerable work in electric space heating. There are many electrically heated homes in the company's service area. Some have radiant panel heating, others the heat pump. Each has its place, based on the particular local conditions.

The heat pump is simply an ordinary refrigeration device which both heats and cools. The refrigerator in the home pumps heat out of the refrigerator into the kitchen. Warm air can be felt coming from the back or bottom of a refrigerator. This is the heat that was taken out of the inside of the refrigerator to make it cold. The air conditioner is a refrigeration device which pumps heat from inside the house to the outside. If the unit is turned around so that the part normally facing inward is made to face outward, it will pump heat from the outside into the house. The heat pump accomplishes both of these functions by pumping

**TABLE 7.2**  SATURATION INDEX FOR KEY PRODUCTS
AS OF DECEMBER 31, 1976

| Appliance | Percent of Wired Homes* |
|---|---|
| Air conditioners, room | 54.4 |
| Bed coverings, electric | 58.6 |
| Blenders | 47.6 |
| Calculators, electronic | 83.0 |
| Can openers | 56.4 |
| Coffeemakers | 99.4 |
| Cornpoppers | 40.7 |
| Digital watches, electronic | 16.9 |
| Dishwashers | 39.6 |
| Disposers, food waste | 40.7 |
| Dryers, clothes, electric and gas | 58.6 |
| Fondues, electric | 4.5 |
| Freezers, home | 44.4 |
| Frypans and skillets | 63.6 |
| Hair dryers, hand-held styling | 27.3 |
| Hair setters | 41.3 |
| Hotplates and buffet ranges | 26.6 |
| Irons, total | 99.9 |
| Irons, steam and steam-spray | 97.7 |
| Knives, slicing | 41.6 |
| Makeup mirrors | 17.9 |
| Microwave ovens | 5.1 |
| Mixers, food | 91.1 |
| Portable steam pressers | 3.4 |
| Radios | 99.9 |
| Ranges, electric free-standing | 50.4 |
| Ranges, electric built-in | 19.7 |
| Refrigerators | 99.8 |
| Slow cookers | 25.5 |
| Smooth-top ranges | 1.4 |
| Styling combs | 26.2 |
| Television, black and white | 99.9 |
| Television, color | 77.7 |
| Toasters | 99.4 |
| Trash compactors | 2.0 |
| Vacuum cleaners | 99.5 |
| Washers, clothes | 72.5 |
| Water heaters, electric | 41.7 |
| Water-pulsating units | 14.4 |

*Source: Merchandising,* 1977 Statistical and Marketing Report.

* All figures based on 74,084,000 domestic and farm electric customers.

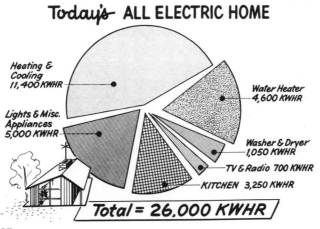

*Today's* ALL ELECTRIC HOME

Heating & Cooling 11,400 KWHR

Lights & Misc. Appliances 5,000 KWHR

Water Heater 4,600 KWHR

Washer & Dryer 1,050 KWHR

TV & Radio 700 KWHR

KITCHEN 3,250 KWHR

Total = 26,000 KWHR

**CHART 7.27**

heat from within the house to the outside during the summer months and pumping heat from the outside to the inside when heat is needed. Even when it is quite cold outside, there is heat in the air which can be pumped indoors.

As shown in Chart 7.27 the heating and cooling market is about as large as the combined market for all other uses in the home. The all-electric home uses about 26,000 kWh/year.

**The Farm Market**  The farm market has shown a substantial growth. The number of farms has been decreasing as shown in Chart 7.28, but total electricity use on the farm has been growing steadily. In 1940 the

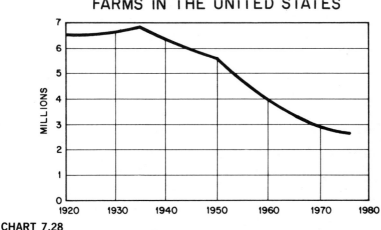

FARMS IN THE UNITED STATES

MILLIONS

**CHART 7.28**

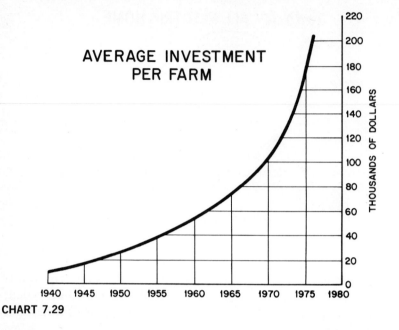

## AVERAGE INVESTMENT PER FARM

THOUSANDS OF DOLLARS

1940 1945 1950 1955 1960 1965 1970 1975 1980

**CHART 7.29**

average investment per farm was only about $8000 (Chart 7.29). By 1950, capital requirements had nearly tripled, to about $23,000 per farm. In 1976 the average investment was about $212,000. The average size of farms is increasing, and more machinery is being employed. Much of this machinery is in the form of electrically power-driven tools. The progressive farmer is using more and more electrically operated equipment to help get work done. The growth in energy use per farm is the key to the large U.S. farm production, feeding all Americans plus millions of people abroad.

Commercial Sales  The power company usually maintains a staff of people who regularly call upon commercial customers to acquaint them with good practices in lighting, air conditioning, cooking, and heating. Trained in the efficient use of energy, these representatives advise customers on conservation and the avoidance of waste. Many of their customers are billed under a rate schedule reflecting both kilowatthours and kilowatts of demand. The customer is advised on load management so as to improve load factor and thus lower the average price of energy.

The Industrial Class  Industrial customers are only 0.5 percent of total customers, but they use about 40 percent of all the kilowatthours sold by the company and provide 28 percent of the company's revenue.

There are several types of power available to industrial concerns.

They can use oil, gas, or steam engines. These engines can be applied directly in the industrial process or they can be used to generate electricity to be used in the plant. However, most industries have found that it is cheaper and more reliable to purchase electric energy from the electric company supplying the area. Usually the cost is less than the cost of other forms of energy. The electric utility company is able to produce the energy cheaper because of the larger, more efficient equipment used by the power company, because of diversity, and because of the many interconnections.

The electrification of industry has resulted in substantial improvements in the environment, from the standpoint of air pollution. Also, this results in energy conservation. Inefficient engines with uncontrolled stack and exhaust gases are replaced by a few efficient central generating stations with extensive pollution control.

The cost of electricity is one of the smallest items in the total cost of finished manufactured products, as illustrated in Table 7.3.

However, this does not mean that the electric utility company should not maintain power engineers to call upon the industrial customers. On the contrary, the electric utility company considers these contacts quite important. The power engineer's job is to help the customer realize the maximum benefits from the use of electric service. The power engineer may suggest ways to reduce losses within the plant, which may include further mechanization and electrification to reduce costs. Advice may be given as to how load factor may be improved in order that the customer may earn a lower rate. The power engineer is constantly checking to be sure that the service is adequate and the customer is satisfied.

When new uses of electricity are developed, the electric utility company's power engineers acquaint their customers with these new developments.

Chart 7.30 shows the percent of electric motor use in relation to total mechanical horsepower in manufacturing.

**Off-Peak Service**   Any practice or policy that gives the customer the incentive to use less energy on peak and more off peak helps improve load factor. When a practical and efficient storage battery becomes available (see Chapter 11), increasing load factor will be given a big boost. Until then, here are a few ways companies are bringing about desired results.

1. *Storage-Water Heater.*   Many companies have established special rates for off-peak water-heating service. These reductions may be achieved through a separate rate, or a special feature in the existing schedule. If desirable, there may be a time clock on the customer's prem-

**TABLE 7.3** ELECTRICITY IN U.S. MANUFACTURING
INDUSTRIES (1975)

*Cost of Purchased Power as a Percent of Product Value\**

| Industry | Electricity Consumption (Millions of Kilowatthours) | | Cost of Purchased Power (Percent of Product Value) | |
|---|---|---|---|---|
| | 1939 | 1975 | 1939 | 1975 |
| Primary metal industries ⎫ | 18,191 | 136,653⎫ | 1.8 | 2.3 |
| Fabricated metal products ⎬ | | 24,441⎭ | | 0.8 |
| Chemicals and allied products | 9,811 | 142,983 | 2.0 | 2.3 |
| Paper and allied products | 9,394 | 63,269 | 3.9 | 2.5 |
| Food and kindred products | 6,388 | 40,951 | 1.1 | 0.5 |
| Petroleum and coal products | 3,440 | 30,684 | 1.2 | 0.7 |
| Transportation equipment | 2,949 | 27,557 | 0.7 | 0.5 |
| Stone, clay, and glass products | 4,852 | 28,331 | 3.2 | 2.0 |
| Textile mill products | 6,805 | 26,996 | 2.3 | 1.7 |
| Electrical machinery | 1,432 | 23,756 | 1.1 | 0.8 |
| Machinery, except electrical | 1,985 | 27,837 | 0.9 | 0.6 |
| Rubber and plastic products | 1,584 | 19,277 | 1.6 | 1.5 |
| Lumber and wood products | 1,238 | 14,812 | 1.8 | 1.0 |
| Printing and publishing | 859 | 9,977 | 0.8 | 0.6 |
| Apparel and related products | 353 | 6,819 | 0.4 | 0.5 |
| Furniture and fixtures | 605 | 3,925 | 1.0 | 0.7 |
| Instruments and related products | — | 5,157 | — | 0.5 |
| Leather and leather products | 402 | 1,732 | 0.6 | 0.6 |
| Tobacco manufactures | 115 | 1,071 | 0.2 | 0.3 |
| Miscellaneous | 466 | 4,061 | 0.9 | 0.7 |
| Total manufacturing | 70,869 | 656,245† | 1.41 | 1.10 |

*Source:* U.S. Census of Manufactures, 1939, and Annual Survey of Manufacturers, 1975.

\* Use of electricity by manufacturing industries in 1975 (latest data available) was more than nine times that of 1939, but during this time cost of purchased power decreased from 1.41 to 1.10 percent of product value.

Purchasing power of the manufacturer's dollar has decreased less since 1939 in buying electricity than in buying other items. The 1976 dollar purchased only 12 cents of its 1939 value in labor, 10 cents in construction, and 21 cents in raw materials, while in industrial electricity, the 1976 dollar purchased 57 cents of its 1939 value.

† Total includes an additional amount that could not be allocated to particular industries.

ises. At one time these were quite popular. But as they are expensive to maintain and keep on time, most companies have discontinued them.

Or the company may connect all water heaters to the load dispatch center through a carrier signal system in which a signal is sent over the electric line. In this way, the load dispatcher may disconnect all water heaters during the peak hours.

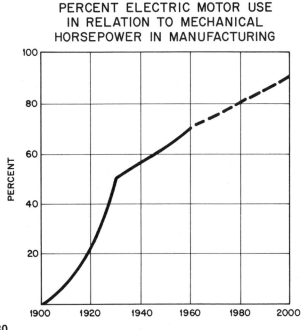

PERCENT ELECTRIC MOTOR USE
IN RELATION TO MECHANICAL
HORSEPOWER IN MANUFACTURING

**CHART 7.30**

2.  *The British Plan*.  Electric house heating offers an excellent opportunity for off-peak services. In England, there is the so-called white
meter, one part of which measures the energy used from 7 A.M. to 10 P.M.
while the other part measures the energy used from 10 P.M. to 7 A.M. The
price of the off-peak energy is something like half the on-peak cost. This
system is being studied in the United States.

In England the manufacturers make a storage space heater. It is
about the size of an ordinary radiator, has about 3 in of insulation and
contains cast-iron blocks. The iron blocks are heated to high temperatures at night. The heat is taken off at a reasonable temperature. If the
customer desires, the storage water heater can have two elements, the
main one connected to the off-peak circuit, and an emergency element
connected to the on-peak service. (Some U.S. water heaters also have two
elements for such purposes.)

Naturally, in England customers may choose to do other chores,
such as clothes washing and drying, dishwashing, and the like during
the off-peak hours. There is unusually high diversity in the daytime operation.

3.  *Electric Blanket*.  This is an article much desired by customers
because of its convenience and comfort and the resultant savings in bedcovers. It operates entirely off peak.

4. *The Electric Vehicle.* The electric automobile is coming and should be encouraged. It results in petroleum conservation, lower air pollution, and more economical operation, and it can be charged off peak.

5. *Off-Peak Rates.* Pricing policy can encourage off-peak operation. Winter-summer rates are popular where there is a distinct summer peak, with little chance that the winter peak will equal that of summer in the immediate future. These rates provide for lower intermediate and bottom steps in the rate for winter use.

6. *Demand-Type Rates* (see Chapter 8). The two-part rate schedule automatically encourages higher load factor, including off-peak operation. Some commercial and industrial customers, especially those in which the cost of electricity is a significant part of their total operating cost, can shift part of the operation to off-peak hours. Any such shift automatically lowers the average price.

7. *Time-of-Day Metering* (see Chapter 8). The new method called time-of-day metering is still in the research stage to determine whether the benefits justify the cost. If practical, it offers opportunity.

8. *Interruptible Service.* The *load-duration curve* (Chart 7.31) shows the number of hours in the year that the peak load is experienced. Notice that the sharp peak is experienced for only a very few hours, yet this peak determines the plant capacity and necessitates the high investment on which fixed charges are paid the year round. Anything that can be done to keep customers off this sharp peak is beneficial.

Some very large industrial customers are prospects for interruptible service. Typically, these are customers whose cost of electricity is significant and whose product can be stored. Here is how the process works:

1. The customer is offered a discount of, say, 30 to 40 percent in the cost of electricity, provided the service may be interrupted a total of, say, 600 h in any 1 year. Notice from the chart how many kilowatts the 600 h represents on the power system.

2. The company has the right to choose the hours of interruption. As total load varies, the utility cannot predict far in advance when the interruptible hours will occur. The company tries to give the customer reasonable advance notice.

3. In practice, the utility may carry the entire load over the peak, by operating reserve capacity. If an unscheduled failure occurs, the interruptible service may be dropped. Or the utility may choose to cut off the interruptible service over the peak.

4. The service for the full allowed hours in any year should be

# TYPICAL LOAD-DURATION CURVE
## SUMMER-PEAKING
## REGION

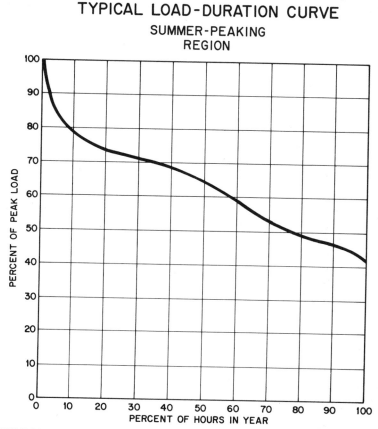

**CHART 7.31**

interrupted even in cases where it is not needed. To do otherwise would be contrary to the rate schedule and might cause the customer to assume that service will never be curtailed.

All these, and other avenues for off-peak service, require that the customers be informed. This means communication. This is marketing.

**The Specific Design of a Program**  Following this preliminary discussion of the objectives of marketing, now there will be an outline of the steps that Edison Power Company takes in designing a specific 5- or 10-year marketing program, as part of the overall company program.
This program will be designed to:

1.   Keep the customer properly informed on relevant matters.

2. Increase load factor so as to increase percent return and thus keep the price from rising more than otherwise.

3. Control marketing expense to ensure sufficient increase in *net* revenue to enable increased percent return. Increased net revenue is not enough. There must be sufficient increased net revenue to pay full return on any new capital required, plus an additional amount to enable an improvement in the company's overall percent return. As one marketing man put it: "We spend $1 to bring back $2 in net revenue."

**Load Research**  To design a program of this kind requires knowledge and use of the load curves and characteristics of each appliance and each class of service, with their relative diversities. For more than 30 years, research in this field has been conducted by the Load Research Committee of the Association of Edison Illuminating Companies, one of the industry's national associations. Thousands of recording demand meters have been installed on a balanced cross section of ranges, water heaters, refrigerators, dryers, washers, and other major appliances. Similarly, recordings have been made of residential lighting, various types of commercial and industrial customers, and various classes of services.

Samples of these curves, singly and in groups, are shown and discussed in Chapter 8. Such research is needed both in rate design and in marketing. It is suggested that the reader review that part of Chapter 8 before proceeding further with marketing, as reference will be made to those charts. To design a marketing program to build, not gross energy use, not gross revenue, not even net revenue, but to build *increased* percent return on capital, requires that marketing and pricing go hand in hand.

Parenthetically, the patterns of these load curves fit almost all parts of the United States. Comparisons have been made with companies abroad. By and large, they fit there also. Before any company uses the U.S. national charts, there should be spot checks made locally. If there is a reasonable fit, the local company can save considerable expense by using these curves. Naturally, each company should make local checks on the characteristics of those energy uses which are temperature oriented, such as heating and cooling.

**The Steps in Design**  The following steps are basic to the design process:

1. By the sampling process, a survey is made of all the major appliances in the home. An accurate survey can be made by telephone. It is found that a 2 or 3 percent sample is sufficient for the purpose.

2. From utility, manufacturer, and dealer records, estimate the saturation of each appliance by years for each of the past 5 years. Trend

these over the next 10 years to estimate the saturation of each at the end of the period.

3.   From the diversified load curves of each major appliance, determine the demand of each on the day of system peak. With the knowledge of the number of each of the appliances now installed, and the hourly diversified demand, prepare the residential load curve, illustrated hypothetically in Chart 7.32. With these data, build a computer model. Program the model so that it can print out the residential load curve for any combination in saturation of appliances.

4.   From Chart 7.32 calculate the load factor of residential service. Check with any residential load curves the company may have prepared from other data.

5.   Using the same process, prepare the residential load curve for the year 10 years hence, assuming no change in the company's current marketing program. Calculate the load factor.

6.   Using (a) incremental costs and (b) average embedded costs, calculate the percent return to the company when each new appliance is added.

7.   From the study in (6) above and other studies, calculate the amount of money the company can afford to spend to market the concept of improved load factor, or in modern language, how much to spend in load management.

In this concept, an amount representing the *estimated annual revenue* (EAR) is calculated for each major appliance and each kilowatt of added commercial and industrial service. Then a calculation can be made of the approximate amount per dollar of EAR the company can afford to spend for each class of service. The *added* cost of serving each additional kilowatt should take into account the added investment and the added operating cost. The revenue to be expected in each class will be in accordance with the rates, soon to be designed, which will earn the full cost of capital year by year in the next 5 years. The *marketing* aim should be to bring about earnings *above* those required for bare cost of money, as earning the *bare* cost on *added* business will not contribute to the necessary increase in the *average embedded* percent return.

It is this extra benefit that needs to be evaluated. How much can the company afford to spend, as a percent of EAR, to get this *extra* net revenue, the amount above the bare cost of money? In making the study, remember, the expense dollar is spent once. The EAR, obtained from that effort, may go on indefinitely.

To be conservative, assume that the increased net revenue carries on for 10 years. Say the *added* net revenue (above all costs) is $100 per year for 10 years. The problem then is: How much can be spent this year to get $100 a year for 10 years?

8.   The general procedure described for residential service can be

# DAILY RESIDENTIAL LOAD CURVE
## EDISON POWER COMPANY
### DECEMBER 1976

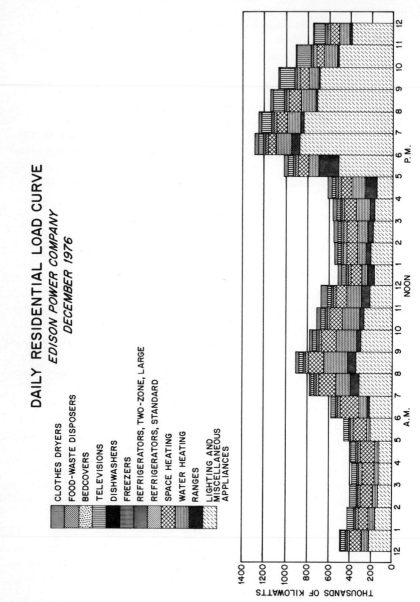

CHART 7.32

232

applied to commercial service and industrial service. The commercial service is largely lighting, cooking, space heating, and cooling. The industrial service requires a study of individual types of industry and, where the customers are large, individual customers. As a matter of practice, recording demand meters are frequently installed on large customers.

9.   Now the analyst is ready to design various new alternative marketing programs to build percent return over the next 10 years. Remember this is the third part of a four-part analysis of system planning, expense control, marketing, and pricing to plan a program to earn the full cost of capital, by years, for the next 5 years. Any improvement in net revenue to be obtained through marketing will lessen, by that amount, the added net revenue to be obtained through price increases.

The analyst, with the marketing and rate people, puts on paper a marketing program, trial A, possibly unlike the existing program, designed to increase load factor and percent return. This may or may not call for increased marketing expense. If it does, that added expense is taken into account. Trial A is programmed into the computer model with all its changed revenues and expenses and added capital (if called for) to determine the company's overall percent return 10 years hence. Then try another, trial B, a different grouping of appliances, commercial and industrial activities. Get its percent return from the model. Run through seven or eight different trials. Each trial is checked with the marketing people for its expectation of accomplishment, with the budget expenditures allowed.

Management reviews all the trial marketing programs and selects the one which it thinks best. Chart 7.33 shows how the residential load might look about 10 years hence, with the trial selected.

10.   Local conditions govern the design for the best program. However, here are a few general hints.

- If the company has a summer peak, house heating will increase load factor and percent return.

- If the company has a summer peak, lighting will be largely off peak and will increase percent return.

- Any controlled or off-peak water heating is good.

- The storage space-heating system as used in England has merit.

- The summer air-conditioning peak can be lowered by increasing the efficiency of room coolers and central cooling equipment.

- Because of their high annual load factor, the refrigerator and freezer improve company annual load factor.

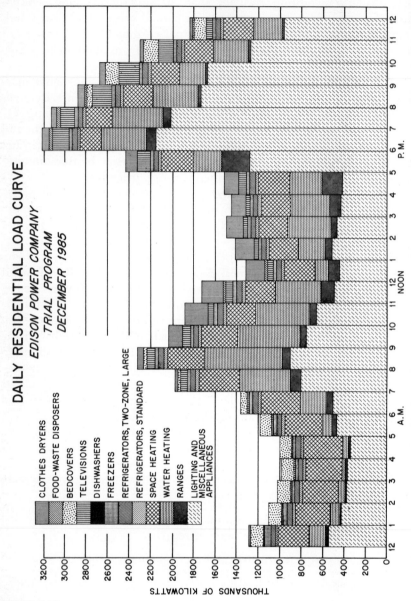

DAILY RESIDENTIAL LOAD CURVE
*EDISON POWER COMPANY*
*TRIAL PROGRAM*
*DECEMBER 1985*

CLOTHES DRYERS
FOOD-WASTE DISPOSERS
BEDCOVERS
TELEVISIONS
DISHWASHERS
FREEZERS
REFRIGERATORS, TWO-ZONE, LARGE
REFRIGERATORS, STANDARD
SPACE HEATING
WATER HEATING
RANGES
LIGHTING AND MISCELLANEOUS APPLIANCES

THOUSANDS OF KILOWATTS

**CHART 7.33**

234

- Electric bedcovers—blankets, sheets, and mattress covers—are good.

- If the company has a summer peak (and sees no change in the next 8 or 10 years), summer-winter rates should be tried.

- Extend the demand feature of rates as far as practical and so long as the benefits justify the cost.

- Encourage better insulation through all avenues. Try to get the regulatory body to approve a feature in the electric space heating schedule that limits its availability to customers having a minimum specified insulation.

- Keep customers properly informed in the most efficient fashion. Use mass communication with residential and small commercial customers. Bill inserts are excellent and inexpensive. Use personal calls on large commercial and industrial customers.

11.  Once management decides upon the marketing program, calculate the annual improvement in net revenue that can be expected. Account for this before calculating the increased rates required.

## PRICING

In designing a plan to earn the cost of money each year for the next 5 or 10 years, there has been a discussion of three of the primary areas under the control of management and required in planning, building, and operating an electric energy system: system planning, cost control, and marketing. All the additional money that is now required to increase net revenue and gross revenue sufficiently to pay the cost of money must be obtained from the pricing.

To determine the necessary level of rate schedules, by years for the next 5 or 10 years, the study goes through several steps, as illustrated by the following charts.

Chart 7.34, in some respects, is a duplication of Chart 7.19 and represents the most important factor in the power business affecting the cost of furnishing service. This chart is expressed in a little different fashion from the previous one, though it still represents Edison Power Company (the whole industry divided by 100). The solid line represents the total plant investment (production, transmission, distribution, and other plant) per kilowatt. Notice how this remained steady at about $350 per kilowatt until about 1966 and 1967 and then began to rise. The dotted line shows the incremental cost (meaning the added or marginal cost) per kilowatt of plant capacity. Notice that prior to 1966 the cost per unit of the plant was less than the average embedded cost. Under these

# TOTAL PLANT INVESTMENT
## PER KILOWATT OF CAPACITY
### EDISON POWER COMPANY

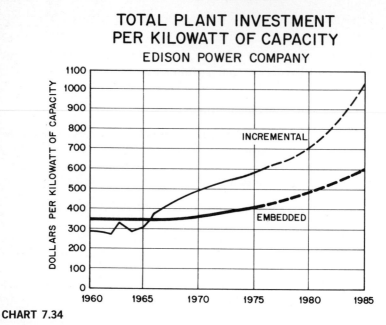

**CHART 7.34**

circumstances (which prevailed practically from the beginning of the industry) the total unit embedded cost per kilowatt could decline. With this factor holding steady or declining and with operating efficiency improving, the companies could lower their rates over the years.

Notice that in 1970 the incremental investment cost per kilowatt was $500, whereas the average embedded cost was less than $400. By 1973, the added cost of plant was $550 per kilowatt. In 1976 it was $600 per kilowatt. New plant is now being committed for 1985. The cost is expected to be about $1000 per kilowatt. This means the average embedded cost, upon which rates are based, will continue to rise through 1985.

The next step is to reduce this plant investment per kilowatt to the fixed cost per kilowatthour. Remember that rate design is based upon both the unit investment cost per kilowatt and the unit operating cost per kilowatthour. This step is to determine the unit fixed cost per kilowatthour. For this study, estimates based upon experience were made to determine the cost of money, depreciation, and taxes by years over the years 1977 through 1985. After the study is computerized, it can easily be revised should this become necessary.

Prior to 1968 the running overall cost of money for electric utilities was about 6 percent. The cost of bond interest rates has been unusually high lately—over 10 percent in some of the years under study. These high costs of bonds are averaged into all the other embedded bonds in the company to get the prevailing average cost of bond money for the particular period under study. Even if bond money drops, the average embedded bond interest will keep rising because much of the bond interest now in plant is at or near 4.5 and 5 percent. For this study, the load factor remains in the range of 60 to 61 percent. Notice, from the previous load factor chart, that this has been dropping slightly over the past 3 or 4 years. This has probably been caused by the slowing of the marketing activity. When the "no growth"philosophy subsides, as it will, companies may resume marketing in the public interest, so as to better ensure service and keep the price lower than otherwise.

Chart 7.35 was obtained by using the capital investment figures of Chart 7.34 for both the embedded unit cost and the incremental unit cost. These investment forecasts enabled a determination of the fixed costs for each year. These were divided by the forecast of kilowatthours for each year.

The result is the fixed charge per kilowatthour. This factor is shown for embedded costs and incremental costs. Notice that this figure remained nearly constant at 8 mills per kilowatthour until about 1967, when the big rise began. In 1976, this figure was about 1.13 cent. This is a rise of about 43 percent in this part of the price since 1967.

Now look at the incremental cost per kilowatthour in the fixed charges. This was below the average until about 1966. At one time this was down to about 6 mills per kilowatthour for the fixed charge. Now it

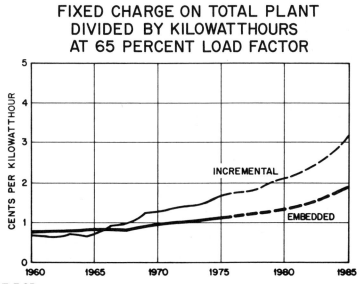

FIXED CHARGE ON TOTAL PLANT
DIVIDED BY KILOWATTHOURS
AT 65 PERCENT LOAD FACTOR

**CHART 7.35**

has crossed and is rising rapidly to correspond to the rapid rise of the incremental investment cost per kilowatt. Notice that for 1976 this figure was about 1.7 cent per kilowatthour. This is 114 percent higher than the average embedded cost of 8 mills in 1966.

The next step is to find the total operating cost per kilowatthour. This is obtained by dividing the total operating cost, as shown previously in Chart 7.22 by the kilowatthours generated that year. The result is shown in Chart 7.36. Notice that the unit operating cost dropped from something like 7.5 mills in 1961 to close to 6 mills, the low point, in 1968. Then came the sudden dramatic rise in operating cost. This was caused by the unusually large increase in unit labor costs, the overall inflation factor affecting all operating expenses, and in addition came the dramatic increase in the cost of fuel.

Chart 7.37 shows the total combined *cost* per kilowatthour (Chart 7.35 plus Chart 7.36) for electric service in the United States from 1961 through 1976, actual and that forecast from this study to 1985. This cost is based upon the minimum required return. The return allowed by the regulatory bodies has been in this range. However, because of the time lag, the actual return realized has been much less (see Chart 7.18).

The purpose of this study is to find the price that will equal the cost, for each of the years.

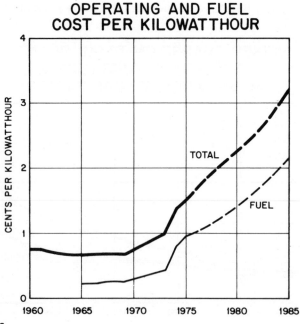

CHART 7.36

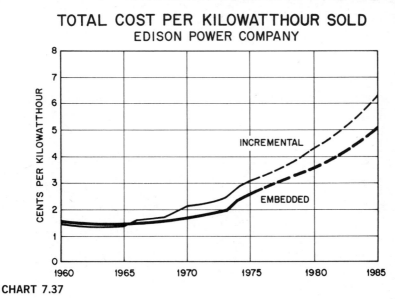

## TOTAL COST PER KILOWATTHOUR SOLD
### EDISON POWER COMPANY

CHART 7.37

Notice from Chart 7.37 that the total average price or cost of electricity was declining gradually from 1961 through about 1968 despite the fact that the rate of inflation was around 3 percent. Note too that up until 1966 the incremental unit cost was less than average and then went above average. The total average cost had risen 90 percent by 1976. In 1985 the cost may be about 5 cents, unless, of course, there is a reduction in the price of fuel.

How accurate are these forecasts and studies? Probably, more accurate than generally realized. A similar study was made in 1970, with a forecast to 1976. If it is assumed that the fuel clause was operative, the forecast total cost per kilowatthour for 1976 came within 5 percent of the actual cost required. How does this compare with what may be expected by a single company? Probably, the larger data base in the national study enables greater stability in the kilowatthours. On the other hand, the local company is more familiar with local conditions.

**Revenue Requirements**    Here is one way to calculate revenue requirements.

1. Forecast cost of money.

2. Forecast depreciation and taxes.

3. Forecast operating expenses after making all practical improvements in efficiency.

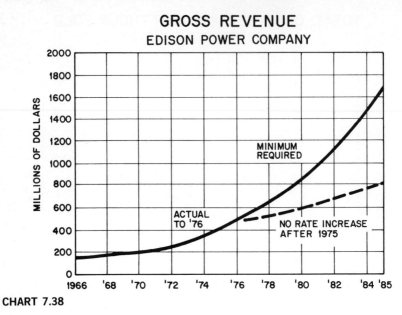

## GROSS REVENUE
### EDISON POWER COMPANY

**CHART 7.38**

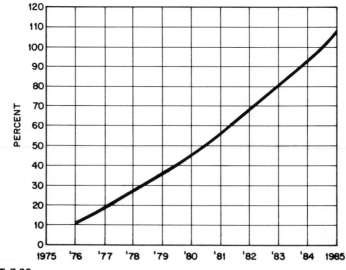

## CUMULATIVE REQUIRED RATE INCREASE
### EDISON POWER COMPANY

**CHART 7.39**

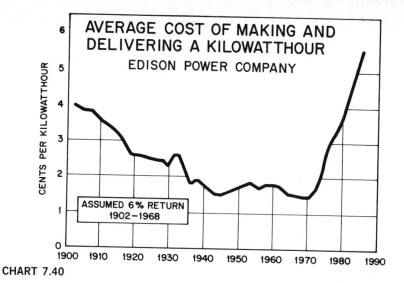

CHART 7.40

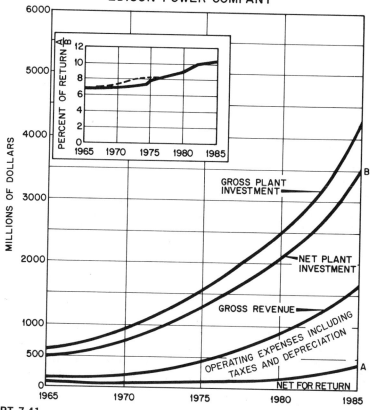

INVESTMENT, REVENUE, EXPENSES, AND RETURN
EDISON POWER COMPANY

CHART 7.41

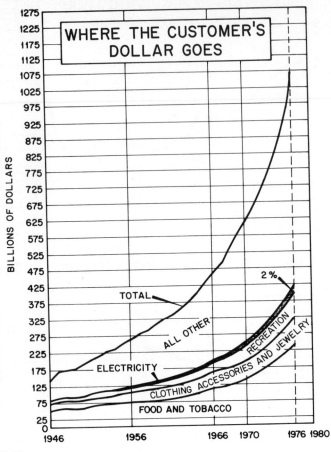

**CHART 7.42**

4.  Forecast all estimated improvements in net revenue expected through marketing.

5.  The gross revenue required as (1) + (2) + (3) − (4) and taxes are adjusted for (4).

Chart 7.38 shows the revenue requirements of Edison Power Company.

Chart 7.39 shows the percent increase in rates required by years for each of the years.

Chart 7.40 is a repetition of Chart 4.5 and extends through 1985. (Again review Charts 4.8 and 4.9 for comparisons with increases of other commodities).

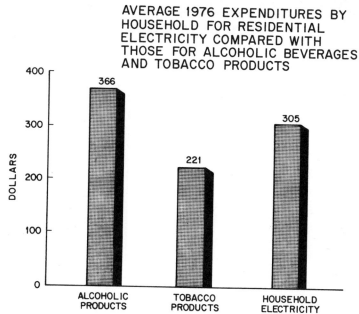

AVERAGE 1976 EXPENDITURES BY
HOUSEHOLD FOR RESIDENTIAL
ELECTRICITY COMPARED WITH
THOSE FOR ALCOHOLIC BEVERAGES
AND TOBACCO PRODUCTS

**CHART 7.43**

Chart 7.41 shows the corporate model in graph form and the over-all results of the planning study described in Chapter 7.

Chart 7.42 shows where the customer's dollar went, from 1946 through 1976. The electric bill in the home was 2 percent of the family budget. See Table 7.3 for the electricity cost to industry as a percent of cost of the finished product.

Chart 7.43 shows the cost of household electricity compared with the average household cost of alcoholic and tobacco products.

Chart 7.44 shows some rate comparisons with Germany, France, Italy, Netherlands, Belgium, and the United Kingdom.

Management now instructs the rate department to prepare various alternative sets of rates to produce the forecast of required revenue. These are discussed in Chapter 8.

## SUMMARY

The analysts have now designed a number of alternative overall programs to earn the full cost of capital, after adopting the best practical programs in system planning, cost control, and marketing. Management decides upon the one the company will adopt.

## AVERAGE REVENUE PER KILOWATTHOUR
### LOW VOLTAGE SALES, 1975

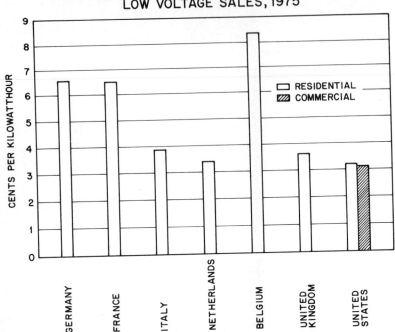

**CHART 7.44** (Prices: Statistical Office of the European Communities, Energy Statistics Yearbook 1970–1975, p. 269; Exchange Rates: U.S. Department of Commerce, Statistical Abstract of the United States, 1976, table 1471, p. 896.)

Once management decides upon its 5-year program, the entire plan can be programmed into a computer and becomes the corporate model. The model is designed to print out any information management wants. If a department proposes a significant change, or if some conditions change, the model can be adjusted to show the results of such changes on the company's percent return. Top management should be presented various alternative plans with the results over a period of years known before the decision is made. Various departments may have subsidiary models, tying into the overall model.

Notice, finally, that the plan itself is designed by utility experts—with the help of various computer experts and models.

# CHAPTER

# 8

# COST ANALYSES, COST ALLOCATION, AND RATE DESIGN

Before reading this chapter, the reader may want to review Chapter 6.

The principles of the free society, as outlined in Chapter 6 are based upon private ownership of property, wage and profit incentives, and encouragement of savings. Competition regulates the profit in nonutility companies. Government regulates the profit of utilities. The word "profit" for a utility really means a reasonable return on the invested capital. The profit in the case of a nonutility and a fair return in the case of a utility are determined after all other costs have been paid.

The point is that a fundamental principle of the free society is that the prices of goods and services are strongly related to the costs of furnishing those goods and services. In the case of utilities, the whole body of law in the legislatures and the courts is based upon these concepts of relating the price of service to the cost of service. This does not mean that the price to each customer is determined precisely according to the cost of furnishing service to each customer. Rate making is largely a matter of judgment. Nevertheless, that judgment is guided by cost studies.

This chapter aims to briefly outline the manner in which electric utilities determine the costs of furnishing electric service for all classes of customers over the whole range of use.

## COST ANALYSES

### LOAD CHARACTERISTICS

All appliances and equipment that use electricity have certain characteristics in the way they consume electricity. Taken as a whole, these characteristics form the basis for rate making, for design of electrical systems to meet possible demands, and for the design of marketing programs. Because of their importance, committees of the industry are constantly studying these characteristics. (See Chapter 7 for mention of the Load Research Committee.) To make the studies, instruments to record the demand at intervals during the day are installed on typical appliances used by typical customers.

**The Range**   Chart 8.1 shows what a recording demand meter might show for a typical day, when installed on a range. For this particular day the range had a peak demand of a little less than 3 kW.

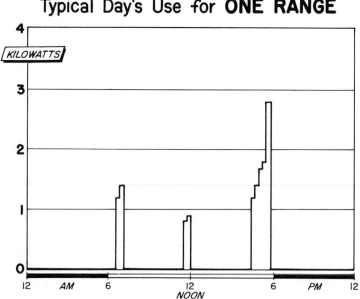

CHART 8.1

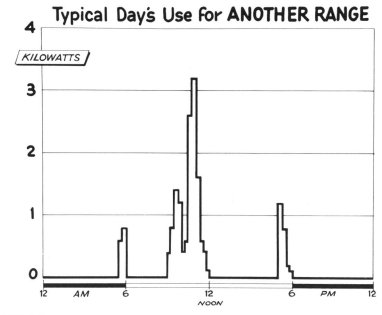

**CHART 8.2**

Chart 8.2 shows another typical load curve of a range in a home where a customer's big meal occurs at a different hour of the day. The peak demand of this range occurs at a different time from that of the other customer. If the curves of these two ranges are put together, the combined load curve will be as shown in Chart 8.3. Notice that the maximum peak demand is the same for the two ranges as for the second range. This illustrates the diversity among range users.

Chart 8.4 shows an average load curve for 100 ranges. On the average, the electric range has an annual group load factor of 20 percent. The maximum composite demand of these ranges is 65 kW.

Ranges may have a connected load as high as 10 or even 20 kW, but the diversified demand on the power plant is only about 0.75 kW per range.

**The Water Heater**    Chart 8.5 shows a daily load curve of a typical water heater which has no controls to limit the use during any hours of the day. The water heater is off and on depending on the habits of the customer. Chart 8.6 shows a composite demand of 100 water heaters which are uncontrolled.

Chart 8.7 shows the load curve of a typical range with the load curve of a controlled water heater superimposed on it.

## Load Curve of TWO RANGES COMBINED

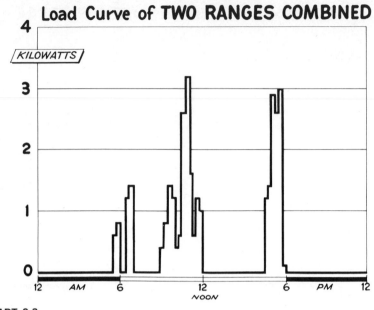

CHART 8.3

## Average Daily Load Curve for 100 RANGES

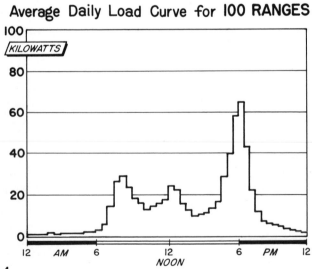

CHART 8.4

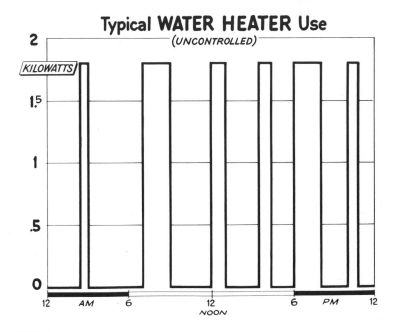

CHART 8.5

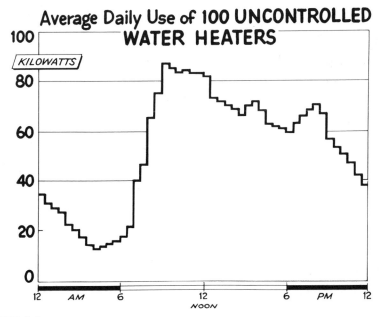

CHART 8.6

# Typical Daily Load Curve of a RANGE and CONTROLLED WATER HEATER Combined

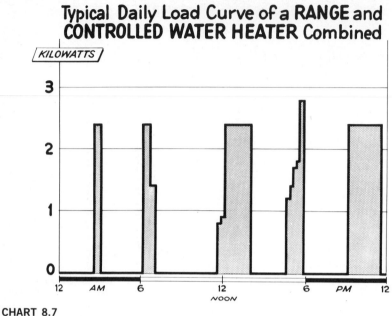

**CHART 8.7**

**The Refrigerator-Freezer**   Chart 8.8 shows a typical load curve of a single refrigerator-freezer. Note that it is operating off and on, actually running about 30 percent of the total time. Chart 8.9 shows a composite load curve of 100 refrigerator-freezers.

**Lighting and Small Appliances**   Chart 8.10 shows typical load characteristics of lighting and small appliances in a home, and chart 8.11 shows a composite load curve of small appliances and lighting in 100 homes.

**The Room Air Conditioner**   Chart 8.12 shows a typical daily load curve of a room air conditioner during a hot day, and chart 8.13 shows a daily load curve of 100 room air conditioners.

**Commercial Load Characteristics**   Among the commercial customers lighting is more prominent than in the residential load. Nevertheless commercial customers use air conditioning, some power devices, and some heating devices. Chart 8.14 shows a few daily load curves of typical commercial uses.

**Industrial Load Curves**   Committees of the industry constantly study the load characteristics of industrial and other large light and power cus-

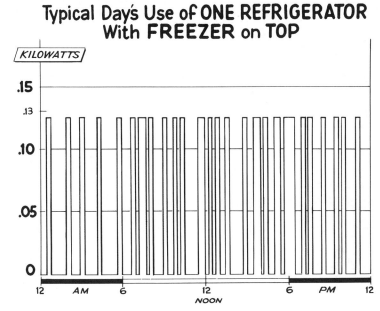

# Typical Day's Use of ONE REFRIGERATOR With FREEZER on TOP

CHART 8.8

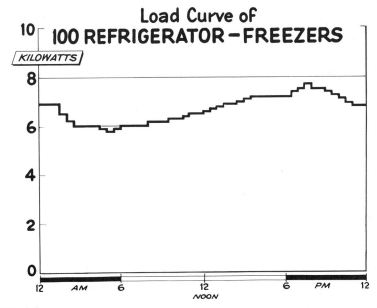

# Load Curve of 100 REFRIGERATOR-FREEZERS

CHART 8.9

# Typical Day's Use of LIGHTING and SMALL APPLIANCES - One Home

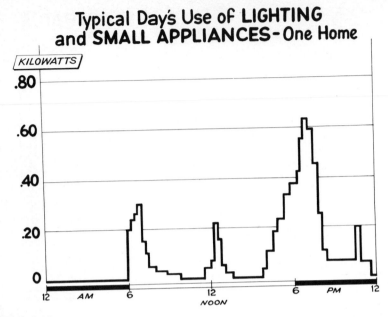

CHART 8.10

# Load Curve of LIGHTING and SMALL APPLIANCES - 100 Homes

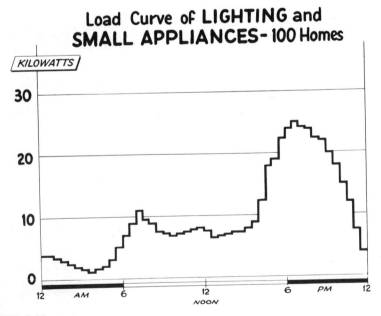

CHART 8.11

## Daily Load Curve of
# ONE ROOM AIR CONDITIONER

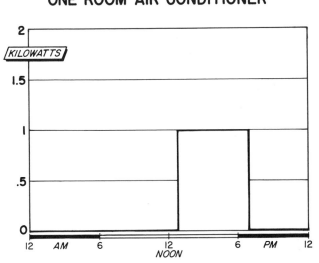

**CHART 8.12**

tomers. Frequently, in the case of large industrial customers the company installs recording demand meters as a matter of practice.

Chart 8.15 shows daily load curves of a few typical large light and power customers, and Chart 8.16 shows the composite industrial load curve for the Edison Power Company.

## Daily Load Curve of **100 ROOM AIR CONDITIONERS**

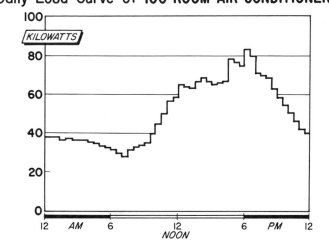

**CHART 8.13**

# Typical COMMERCIAL Uses

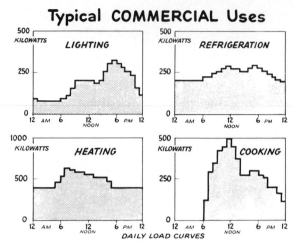

**CHART 8.14**

**Summary of Load Curves**  If the load curves of all customers are combined, the result is a composite system load curve. Shown in Chart 8.17 is the composite load curve of Edison Power Company for a typical winter day. Chart 8.18 is for a typical summer day.

The top line on Chart 8.19 shows the annual load curve for Edison Power Company for 1976. The maximum hourly load of each month is plotted to produce this curve.

# Typical LARGE LIGHT and POWER CUSTOMERS

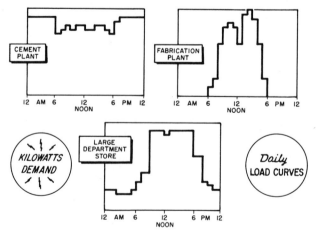

**CHART 8.15**

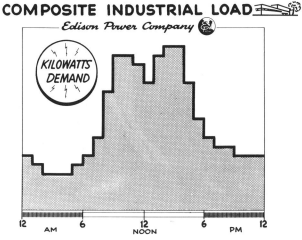

CHART 8.16

## FACTORS IN COST STUDIES

There are many factors having a bearing on the price of electricity. The cost of furnishing the service is one of the principal ones, which varies mainly in relation to three factors, namely, the quantity used, the load factor at which it is used, and the customer density. The aim in rate making is to set a pricing policy that has a reasonable relationship to cost over the whole range of use and the whole range of load factors. In the inter-

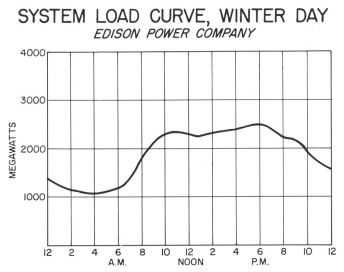

CHART 8.17

# SYSTEM LOAD CURVE, SUMMER DAY
### *EDISON POWER COMPANY*

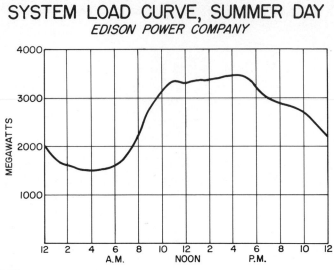

**CHART 8.18**

# MONTHLY PEAK LOAD
### *EDISON POWER COMPANY, 1966-1976*

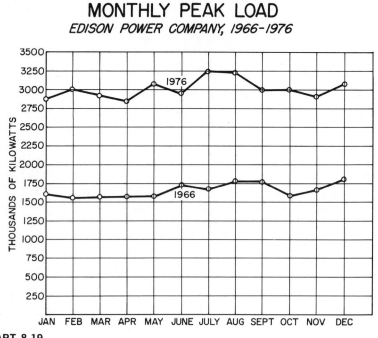

**CHART 8.19**

est of simplicity and practicability certain compromises are necessary and advisable. (The recent innovations in pricing are discussed later in this chapter.)

**Fixed Costs and Variable Costs**    An electric utility company must keep a keen eye on the various elements of cost so that it can arrive at sound rates. Certain costs are *fixed costs*; that is, they remain the same whether the business is idle or running at full capacity. Fixed costs can be demonstrated using the cost of an automobile as an example. For example, one pays, say, $5000 for an automobile, and the installment payments toward this cost are the same each month, regardless of whether the automobile travels 100 or 1000 mi. These fixed costs may be $150 a month.

Other costs are *variable*; that is, they go up or down according to how much or how little of a product is made. They can be compared to the cost of gasoline to run an automobile. If an automobile uses 1 gal of gasoline to go 15 mi and gasoline is 60 cents a gallon, it will cost 4 cents a mile for the gasoline. The total cost per mile of running the automobile will be as shown in Table 8.1.

The cost of any article is made up of both fixed and variable costs. Different economic principles apply to situations where, on the one hand, variable costs make up most of the total cost and where, on the other hand, fixed costs are greater. Charts 8.20 and 8.21 illustrate the effect of these two conditions on the price of the product sold.

Chart 8.20 shows costs which are made up mostly of fixed costs. The variable costs are fairly small. This might be the case in a gravity-flow water system where water is free and about the only costs are for the dam and water mains.

Chart 8.21 depicts costs made up mainly of variable costs.

In each of these cases the area showing variable costs is the same width at every point on the chart. The area representing fixed costs diminishes as more units are sold. The downward slope of the top line of the chart is due entirely to the decline in the fixed cost element. Where the fixed costs are relatively high, as in Chart 8.20, the decline in total cost per unit is rapid. Where fixed costs are relatively small, there is a smaller decrease in total cost with increased production.

Where the capital investment is high, fixed costs are high. Increas-

**TABLE 8.1**    COST OF OPERATING AN AUTOMOBILE

|  | *100 mi* | *1000 mi* |
|---|---|---|
| Fixed cost per mile | $1.50 | $0.15 |
| Operating (variable) cost per mile | .04 | .04 |
| Total cost per mile | $1.54 | $0.19 |

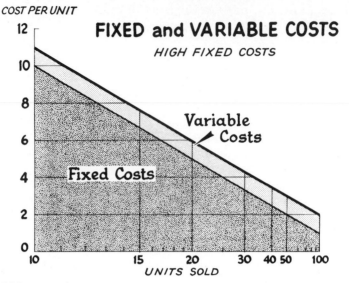

CHART 8.20

ing the amount produced from this investment results in lowering the cost of each item substantially. This is the case in the electric utility industry. A large portion of the income of an electric utility is required to "service" the investment, that is, to pay interest on the money invested, to renew and repair the property, and to pay taxes and insurance on it.

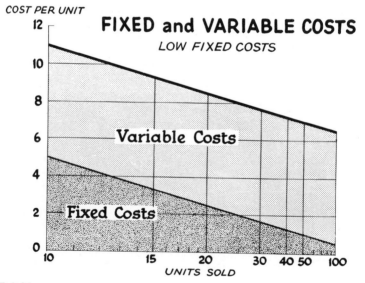

CHART 8.21

These are called *fixed capital costs*. They remain the same whether the plant operates at 100 percent capacity or is closed down completely.

The *variable costs* of providing electric service consist, in the main, of fuel, materials, and some maintenance and wages. These vary more or less directly with the plant output and do not decrease on a unit basis with increased output in the same manner as do the fixed costs.

**Electric Service Cost Mainly Fixed Cost**    Assuming labor as fixed expense, about 65 percent of the total costs of providing electric service are fixed costs; the remaining 35 percent are variable costs. This is significant from an economic point of view in that it places the industry in the category of diminishing unit costs within the limits of an existing plant.

Chart 8.22 illustrates the effect of diminishing costs with increased output. It shows Edison Power Company's cost per kilowatthour sold in relation to total sales. Note that its actual sales in 1976 were almost 17 billion kWh and that its cost as shown on the chart is 3 cents per kilowatthour. The chart indicates that if the company could sell more energy without installing more generating equipment, the cost would go even lower.

**Effect on Cost of Electricity**    Chart 8.23 shows the share of fixed and variable costs making up the cost to serve a residential customer. This ratio also holds true for commercial and industrial service, although the cost of furnishing service tends to level off for the larger users. For

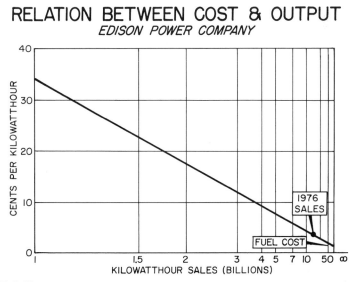

# RELATION BETWEEN COST & OUTPUT
## *EDISON POWER COMPANY*

**CHART 8.22**

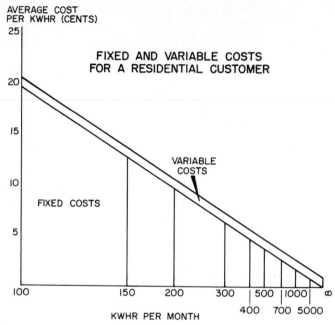

**CHART 8.23**

them, there is not such a great decrease in cost with increased use of service.

The investment needed to serve a customer depends on kilowatts of demand. If demand does not increase, the fixed charges on the plant used to serve that customer do not increase. If more kilowatthours are used at the same demand, then the fixed charges for each kilowatthour will go down. In other words the higher the load factor, the less it costs per kilowatthour to furnish service to the customer. Chart 8.24 shows how fixed charges drop with increased load factor.

Power companies take this into account, whenever it is practical to do so, in setting the price for electric service. Rates for large users provide, in effect, a discount for customers with high load factor. This encourages the customer to operate at a higher load factor. Also this shows why power companies strive to balance their load to get the maximum overall load factor.

**Value and Limitations of Cost Analyses**   As it is not practical to make a separate rate for each customer, customers are grouped into classes. All those customers having somewhat similar characteristics are grouped into a class for rate-making purposes. These classes may vary slightly by states or by countries, but generally speaking they are as follows:

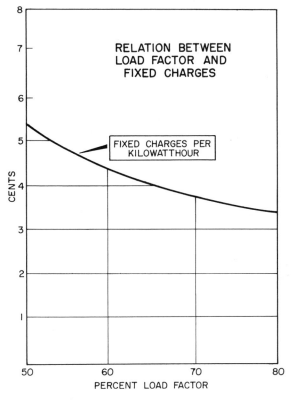

CHART 8.24

**Residential Service**   All dwellings used for home living are in this category. It includes individual dwellings, apartments, and mobile homes.

**Farm Service**   These are all the customers living in rural areas and usually operating farms. Some companies classify them the same as the city dwellers.

**Commercial Service**   These are the shops, restaurants, stores, and the like—places of business, but not factories. Sometimes churches and schools are included here.

**Small Power Service**   Some companies have a rate schedule for the small power users with characteristics somewhat different from commercial service. Most companies have been dropping this classification.

**Large Lighting and Power Service**   These are the large commercial and power customers. They include the hotels, department stores, food stores, shopping centers, large filling stations, and any other large users.

Some companies also use this classification to serve the large industrial customers.

**Industrial Service**   Many companies have a separate rate schedule for large industries and manufacturing establishments. In the past, there were separate schedules for the separate classes of industries, such as cement plants, oil refineries, cotton gins, pipeline pumping, textile mills, pulp and paper mills, and the like. The later tendency is toward simplification and standardization. Most industries are served under one rate schedule.

**Miscellaneous Services**   These may include such services as street lighting, municipal water pumping, temporary service, and the like.

Basically, there are four major classes:

- Residential
- Commercial
- Large lighting and power
- Industrial

A separate rate schedule is designed for each class of service. But company records are not kept by classes. This would be an added unnecessary expense. Therefore, from time to time, the company prepares cost analyses and cost allocations. Management and regulatory bodies want to know the approximate cost of furnishing service for each class, and over the whole range of use of energy in each class. Furthermore, to avoid unjust discrimination, the aim is to maintain a *reasonable* balance in the percent return for each class. The word "reasonable" is used advisedly, as it is neither possible nor practical to have each class always earn the same percent return. Also, there are other factors in rate making besides cost, although cost is the primary factor.

There is a further complicating factor in electric utility cost allocation and rate making. Previous charts have demonstrated that the fixed costs to the electric utility are higher than the variable costs. The fixed costs are controlled by kilowatts of demand, whereas the variable costs are controlled by kilowatthours.

Traditionally, the bulk of the residential and commercial customers have been serviced under a price schedule based upon kilowatthours only—the element reflecting variable costs. Such a schedule is simple and easy to understand. Various attempts have been made in the past (as discussed in more detail later) to introduce a kilowatt, or load factor, principle into these rates. The public and the regulatory bodies, however, would not allow them to remain.

The telephone utilities have avoided this difficulty. Generally, the customer pays so much per month for one phone, whether the phone is used much or not at all. If two phones are wanted, the monthly charge is higher, and so on for 100 phones or 1000 phones. This is something like a kilowatt charge to cover fixed charges. Usually, the telephone company charges an additional amount per call.

With the handicap of the rate schedule based upon kilowatthours only, even further approximations must be made. For these and other reasons, the cost allocation is an approximation. It is a useful tool and a good guide, but management judgment must be exercised in using it.

Any cost analysis requires that the various components of cost items be sorted out or allocated. It is desirable to know what share of a cost item should be charged to a single customer or to one of the classes of service, for example. The cost analysis may be used to estimate the cost of serving one community or a particular large customer. The analysis might be used to approximate the cost of furnishing service in each of the divisions of a company, all of which obtain their power supply from one common pool. Cost allocations can also be used to determine the way costs of different companies in a pool are to be shared.

Naturally, the cost analysis is an estimate and in some measure reflects the judgment of the person making it. The costs allocated to a particular customer or class will vary from time to time as circumstances change with changes of load patterns, changes in characteristics of customers, and changes brought about by growth.

Sometimes the cost analysis is used to calculate the percent return being earned by a company for each of the classes of service. Cost analyses can serve as a guide to assist in maintaining a reasonable balance of earnings by classes of customers. However, the analyses are guides only, as the earnings of the company as a whole are taken into account in determining fair return on the property.

## COST ALLOCATION

In a cost allocation, all costs necessary to provide the service to the consumer are divided between customers in as equitable a manner as possible. The cost to serve a customer may be apportioned between three elements, namely, customer cost, demand cost, and energy cost.

The customer cost is made up of those costs that vary with the number of customers. They include such costs as the reading of the meter, record keeping, billing, and other costs of a similar nature including some distribution costs.

Demand costs are those incurred in providing for the system load. They include such costs as interest, depreciation, and taxes on generat-

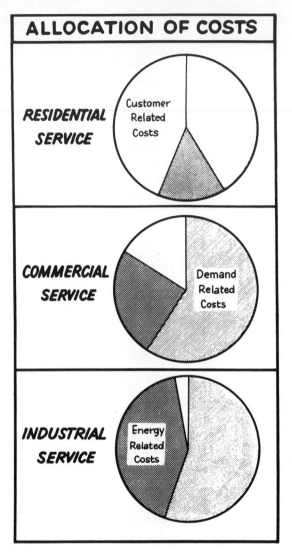

**ALLOCATION OF COSTS**

RESIDENTIAL SERVICE — Customer Related Costs

COMMERCIAL SERVICE — Demand Related Costs

INDUSTRIAL SERVICE — Energy Related Costs

**CHART 8.25**

ing and transmission equipment, and other costs associated with the generation, transmission, and distribution cost of the system.

Energy costs are those which vary with the amount of energy generated, such as the cost of fuel.

After expenses are broken down into the three parts, the analyst can further break them down into classes of service, such as residential, commercial, and industrial. Chart 8.25 shows how one analyst might decide the costs for the three main classes of service.

In making the cost allocation the analyst examines the income statement and balance sheet and forms an opinion as to which are the controlling factors.

Most rate schedules are based upon kilowatts and kilowatthours. Few make a separate charge for customer-related costs. Notice its magnitude in residential service. In many respects, the customer-related costs are fixed costs; i.e., they do not vary with the energy used. Note also the relatively small energy-related costs in residential service. Yet most residential rates are based on kilowatthours only.

The commercial service has only a relatively small customer-related cost. A two-part rate (one based on kilowatts, the other on kilowatthours) fits well. Likewise, in the industrial service, the customer-related cost is quite small. The two-part rate also fits very well here.

## Allocating the Plant

Chart 8.26 shows for the peak day the class load curves and the total system load for a typical company. The three classes do not peak at the same hour. The system peak is probably at a different hour. There is diversity among the classes. In the allocation, the aim is to assign to each class its fair share of the diversity.

Among analysts, opinions vary as to the best way to allocate the plant. Three or four methods stand out. Sometimes all are used. Examples are:

**Peak-Responsibility Method**   This method is based upon the premise that the company builds plant to serve the peak. Therefore, each class should have allocated to it that portion of the plant that is in proportion to the class demand at the time of system peak. Each class benefits by some share of the diversity. For example, for residential service, the plant investment is allocated in accordance with the ratio of the residential demand (at the time of system peak) divided by the system peak.

**Maximum Noncoincident Method**   This method is based on the premise that each class of service if served independently would require plant to meet its class demand. Under this method, all the class maximum demands are added together regardless of the time of occurrence. The allocation factor is the ratio of the class maximum demand to the sum of all the classes' maximum demands.

**The Average-Load Method**   This is a very simple method. The average load for each class is first determined by dividing the kilowatthours used by the class by the number of hours in the period under study (as 8760 h in 1 year). The allocation factor is the proportion each class is to the total

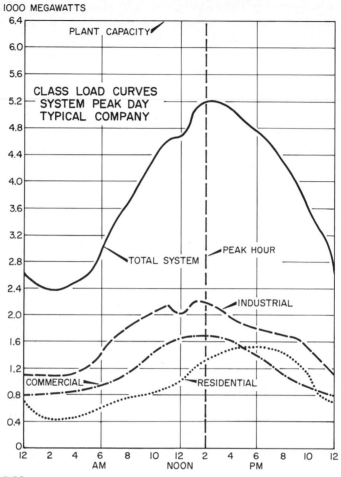

1000 MEGAWATTS

PLANT CAPACITY

CLASS LOAD CURVES
SYSTEM PEAK DAY
TYPICAL COMPANY

TOTAL SYSTEM

PEAK HOUR

INDUSTRIAL

COMMERCIAL

RESIDENTIAL

**CHART 8.26**

system average load. This method is based only on kilowatthours. It gives no recognition to kilowatt demands on the system.

All these allocations are somewhat complicated. Frequently, outside experts who specialize in this field are called in to make the studies and to testify. But it should be kept in mind that the cost studies are estimates, representing the judgment of the expert. Load conditions change from time to time. While such estimates are valuable and useful, final determination as to price should not be based solely upon them. Other factors also should be considered. Judgment is of primary importance.

## COST CURVES

The cost allocations described above can give the company's percent return by classes or for some large customer. As these studies are also used as one of the guides in rate making, it is advisable to determine the cost of furnishing service over the whole range of use. This range varies with volume and with load factor.

For the purpose of this kind of study, it is necessary to select some particular period, such as a particular year. Average costs vary from year to year. Also, according to well-established regulatory practice, the property as a whole is used in finding fair value and fair return. In keeping with this, property as a whole is used in allocating costs and in rate making.

The efficiencies and fuel costs of plants vary. Each customer is assumed to bear its share of the higher-cost plants and to receive its share of the benefits of low-cost plants. Some plants are new; some are old. The same principle applies for old and new plants.

Prior to 1968, when almost every new plant could produce energy at a somewhat lower unit cost, regulation required that all customers should share in the benefits. Prices were lowered from time to time to accommodate the lower costs. At times, a larger industry would want to locate adjacent to a new low-cost generating plant and be billed under a price schedule based on this plant alone. Regulation frowned on such a practice. All rate schedules reflected some of the lower unit cost.

Now the reverse is true. The new plants cost more per kilowatt of capacity than the average. Some claim that only increased use of service should be billed at the higher cost. But practically all customers are increasing their use of service. To maintain the principle of pricing and to avoid unjust discrimination among customers, the higher-cost plants should be averaged into the whole company. But all the cost studies show that although unit costs are rising overall, for any particular period, such as a year, the unit cost of serving a large user is lower than that of serving a small one. That is to say, the unit cost continues to go down with increasing volume.

This is reflected in Chart 8.27 which is a typical set of cost curves covering a whole range of kilowatthours and load factor. Notice that for any load factor the unit cost goes down as the use of energy goes up. Also for any use of kilowatthours, the average unit cost goes down with increased load factor.

As all unit costs are now rising, another set of cost curves would be required to reflect the costs of next year, and still another for the year after. This will probably continue until the average embedded unit cost of plant ceases to rise. It now appears that the incremental unit investment cost of plant will be above the average embedded cost through 1985, thus causing the average to rise at least until then.

A rate schedule with a kilowatt charge and a kilowatthour charge, with sliding scale for volume, can be designed to be close to the cost curve over the whole range of use.

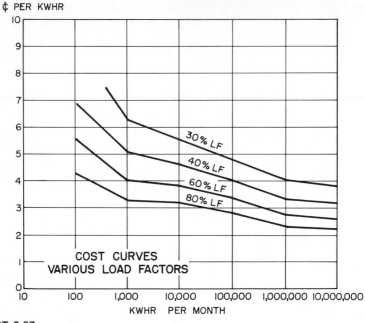

**CHART 8.27**

It is especially wise to follow the cost curve in serving large users, as they can generate their own electricity. A loss of big users would adversely affect the cost of serving the small users, thus causing those prices to rise.

**Residential Cost Curves**   At the moment practically all residential rates are of the *block* form, with a sliding-scale price based upon kilowatthours only. Such a schedule must include the fixed costs, which are higher than variable cost, in the kilowatthour charge. This is the principal reason why it is necessary for the price per kilowatthour to slide down with increased volume. The fixed cost *per kilowatthour* is much higher in the small-use range (see Charts 8.20 to 8.23).

As there is no kilowatt charge in block rates, it is necessary to base the rate schedule on the average load factor of the class. This may be around 40 percent for residential service.

Chart 8.28 shows the residential curve for three load factors, including the 40 percent class load factor. Although 40 percent is the average of the class, individual load factors, even allowing for diversities, vary over a fairly wide range. Obviously, as will soon be demonstrated in discussing rate design, a schedule based upon kilowatthours alone can only roughly approximate the cost curve.

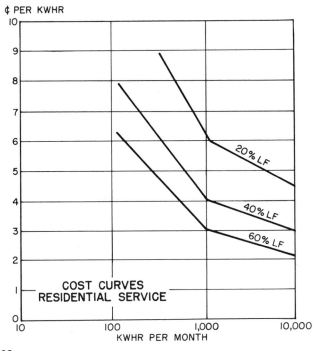

¢ PER KWHR

COST CURVES
RESIDENTIAL SERVICE

KWHR PER MONTH

**CHART 8.28**

The point here is that for each class of service, the unit cost of furnishing service decreases with increased volume and with increased load
factor. To avoid unjust discrimination, the rate schedule should, in a
major respect, allow for this characteristic.

This principle is embodied in practically all rate schedules and has
been almost from the beginning. It is in keeping with the nondiscrimination clause in most of the legislation creating regulatory bodies, and has
been upheld by a large body of court decisions.

## RATE DESIGN

### PRINCIPLES OF RATE DESIGN

The principles of rate design will now be discussed. Also, there will be
designed some of the specific rate schedules required to enable Edison
Power Company to earn the cost of money, as called for in the case study
outlined in Chapter 7.

Because of the many variations both in the use of energy and in kilowatts of demand, the industry has almost an infinite variety of cus-

tomers. Determining the right price for electricity for the various classes of customers is an intricate process. Pricing policy—rate making—must weigh all the economic and other factors entering into furnishing service to various kinds of customers so that all may be treated fairly and so that in the end the company will earn enough net income to enable it to continue as a healthy, going concern.

Here are some of the principles which are observed in rate making:

1. *The rate should be simple and understandable.*

2. *The rate should be acceptable.* The company is in business to furnish electricity. The rate must be fair, and also the customer should understand that the price is fair and reasonable.

3. *The rate should be competitive.* The electric utility may be the sole supplier of electricity in the area, but that does not mean that there is no competition. Large commercial and industrial customers could generate their own electricity.

4. *The rate should encourage higher load factor.* To the extent practical, this feature should be built into a schedule. Where not practical, customer information programs can be used.

5. *The rate should be reasonably nondiscriminatory.* All customers using service under similar conditions should be billed at substantially the same price. It is necessary and advisable to group customers by classes, but all in the class should be treated substantially the same when the characteristics of service are about the same.

6. *The rate should cover the cost of furnishing the service.* When a company's sale of kilowatthours increases, its costs also increase. Under inflationary conditions costs would increase even if sales did not. The rates charged should cover all costs, those caused by increased sales and those caused by inflation.

## KINDS OF RATE SCHEDULES

Most residential and commercial rates are *block rates*, based on the kilowatthours used. These provide for a lower unit price for larger users. In this respect they reflect the cost curve. Provisions are made in the price per kilowatthour to cover all costs including customer-related costs and demand-related costs. In some instances, there are minimum bills or "per customer service charges" in connection with the block rate. Under the block rate, the rate does not vary with demand.

Other rate schedules include a separate charge to cover the demand-related costs. These demand-energy rates can be better fitted to the cost of furnishing service over the whole range of use. Few rates make a separate charge to cover customer-related costs.

Prior to 1968 (the year of change in the economic climate) there were three reasons for the sliding-scale nature of the block rate:

1.  During any test period, the unit cost of furnishing service is less for the larger volume.

2.  The fixed charges, which are included in the kilowatthour price, decline steeply with volume.

3.  The overall cost of making and delivering a unit of energy was declining.

Since 1968, the third factor has changed to a rising unit cost. The first two factors remain. The only way to maintain the first two factors under current conditions is to increase the whole rate schedule as unit costs rise. As will be demonstrated later, the rate schedule should now slide down less steeply with increased volume.

The rates referred to herein are used for illustration purposes only. Neither the styles nor the levels pertain to any particular company.

**Residential Service**   Chart 8.29 shows a residential rate schedule for Edison Power Company. This rate was not designed by any one person

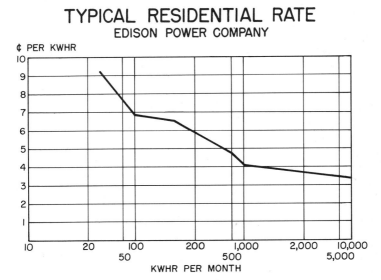

## TYPICAL RESIDENTIAL RATE
### EDISON POWER COMPANY

**CHART 8.29**

as a means of covering the cost of furnishing service for any one moment or year. Rather, the rate was developed over a period of years, with the judgment of many people entering into its design. It has the required sliding-scale feature. In this form it does not have a demand charge.

This type of rate is appropriate where the range of use falls within certain narrow limits. Until recently, this was true of residential service. Now, however, with new applications of electricity in the home—the heat pump, air conditioning, electric house heating, and other uses—a well-equipped residential customer can use a considerable amount of electricity. For all use above 500 kWh, the rate is a flat rate of 3.3 cents per kilowatthour. There is no allowance for a customer's load factor. Some companies and commissions have found it desirable to remedy this for the larger user by incorporating a demand principle or a load-factor principle in the rate schedule. By so doing, the company can give the customer credit for good load factor, thus encouraging efficient use of electricity. (This practice is discussed more fully later.)

**Water-Heater and House-Heating Rates** In some cases the company may add a lower step in the rate to apply where electricity is used for water heating. In other cases there is a lower rate for the number of kilowatthours used by the average water heater. Sometimes a separate meter is supplied for off-peak water heaters.

Electric house heating is becoming popular as more and more people come to recognize the many benefits it offers. A company may put a lower step in the rate to cover the use of either direct-panel heating or the heat pump for electric house heating. Some companies install a separate meter for house heating which is charged at a special rate. Companies with a heavy air-conditioning demand in summer are anxious to sell electricity for house heating to balance out the load, and thus make the price lower than otherwise.

Chart 8.30 shows for the period from 1930 through 1976 the price of 100 kWh, of 250 kWh, and the total average price of residential service. This illustrates the effect of the sliding-scale principle. Note that although the sliding-scale principle is maintained, the total average price has risen, as it should.

**Commercial Schedules** Chart 8.31 shows a typical commercial rate schedule. Like the residential rate, the commercial schedule is based on kilowatthours only. The commercial schedule shown is the type generally used for smaller stores and other small commercial establishments. It has longer steps than the residential rate because the use of kilowatthours in this class is somewhat higher and demands are higher. It is especially necessary to have longer blocks in this rate when no allowance is made in the rate schedule for the amount of demand.

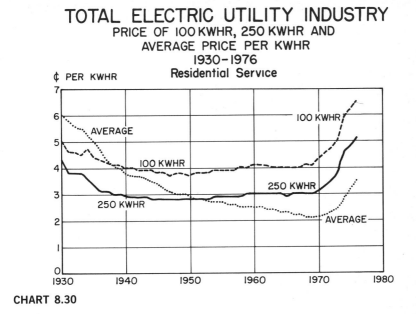

# TOTAL ELECTRIC UTILITY INDUSTRY
## PRICE OF 100 KWHR, 250 KWHR AND
## AVERAGE PRICE PER KWHR
## 1930-1976
### Residential Service

**CHART 8.30**

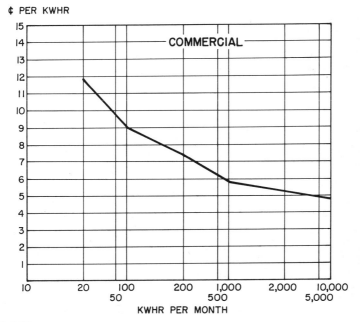

**CHART 8.31**

In some companies the length of the first few steps may vary with the kilowatts of demand. As the demand increases, the steps are longer. The reason for this is that the higher demands require more capacity and involve more fixed charges. Under this rate, a small customer at a high load factor can earn a lower rate than a large customer with a poor load factor. This is proper both from the standpoint of the customer and from the standpoint of the company. For the larger commercial installations, such as hotels, office buildings, and department stores, a demand-energy rate is often used.

**Large Light and Power Service**   This class (Chart 8.32) includes the large users of electricity, such as manufacturers, refineries, and a host of other industries. Because they use large amounts of power, it is important to have their rate fit the nature of their use and the cost of serving them. This rate takes into consideration both the number of kilowatt-hours used (the energy charge) and the kilowatts of demand they impose on the system (the demand charge).

The demand-energy rate is designed to cover the fixed costs and the variable costs in all combinations. It has a sliding-scale demand charge to which is added a sliding-scale energy charge. This type of rate gives a customer due benefit both for increased use and for higher load factor; the larger the customer, the lower the average rate earned at the same load factor.

It is possible to design a rate of this kind to cover a very wide range of use. However, some companies prefer to have separate class rates for each type of industry. The tendency is to reduce the number of rates available to simplify the whole rate structure. The trend is toward fewer rate schedules.

## OTHER FACTORS IN RATE DESIGN

**Value of Service**   Electricity has a certain value to the customer, who has a general feeling as to what it may be worth. Companies try to set the price so that the customer will be able to use it, but at the same time the price should cover the costs.

**Competition**   Competition plays a greater part in the design of rate schedules than most people realize. Rates to large industries must be competitive with other forms of power, or the customer will use the other forms, such as diesel power or steam engines to operate machinery. The customer may decide to install a generator to make electricity. Electric companies and commissions constantly study these competitive

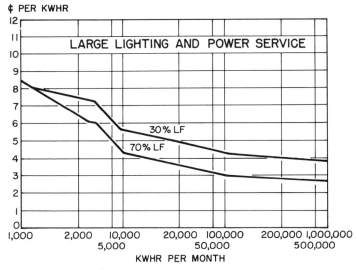

**¢ PER KWHR**

LARGE LIGHTING AND POWER SERVICE

30% LF

70% LF

KWHR PER MONTH

**CHART 8.32**

forms of energy so that the price of electricity to industry will be competitive.

Competition also comes into play in other classes of service. The commercial establishments can also use other forms of energy. They can install their own generating equipment, but few of them do because, as a rule, electric rates are competitive. The larger generating units and the diversity in large interconnected systems enable the power company to sell electricity for less than the cost of power produced by such individual plants. A commercial customer may also use gas (or other forms of fuel) for cooking, heating, or air conditioning. It is the electric industry's aim to have the commercial rates meet this competition.

Because of the importance of price, electric utility companies are striving to find better and cheaper ways to make electricity. There are no secrets in the power business. Through various committees in the trade associations, the companies exchange ideas and experiences with each other. When one company learns a better way to do a certain job, all have access to that knowledge. When one company makes a mistake, all the companies can profit by its experience.

**Cost of Furnishing Service**  While rates should be designed so that each class pays its fair share of the costs, it is not possible nor practical to design rates so that all steps in all the rates result in the same percent return on the investment. The aim is to make the prices bear some reasonable relationship to cost. The rate should not bring an unreasonable benefit or hardship to any particular customer.

With all these factors taken into account, rate making is largely a matter of judgment based on experience.

**The Demand Charge** The *demand charge*, or the separate charge for kilowatts, is frequently not understood by the customer, largely because of the lack of knowledge of the difference between a kilowatt and a kilowatthour. For a full discussion of the demand charge see *The Electric Power Business*.[1]

**Changing Rates** Rates are seldom changed radically at any one time, and there are rarely any great changes in rate form. The company generally tries to avoid any disrupting change in the billing to individual customers. It is likely to work out a long-range program of standardization, simplification, and improvement in rate form and make the changes as conditions justify.

Naturally the commission's viewpoints have a bearing on the forms of the rates. The commission's interest in this instance is the same as the company's interest—to have them simple, understandable, and acceptable; to have them cover the value of the service and reasonably cover costs.

When any rate changes are made, complicated and exhaustive studies are undertaken to determine the effect they will have on the company's revenue. Studies are also made to see how the changes will affect individual customers and classes of customers. It would be impractical and costly to calculate the bill of each individual customer on the new proposed rates. Certain shortcuts have been devised which give accurate results. One such shortcut is described below.

**Frequency Distribution of Kilowatthours—Ogive Method** In residential service, for example, very few customers may use only 1 or 2 kWh a month or 10 kWh a month. A few will use 3000 or 4000 kWh or more each month. The great bulk of customers will use something in between. To test the effect of a proposed rate it is necessary to know how many customers are using any given number of kilowatthours a month. This will give a picture of how the customers are using the service. Chart 8.33 shows the distribution of customers by their kilowatthour use for the residential class of service for Edison Power Company in 1976. This is called a *frequency distribution curve*. From this curve can be figured the number of kilowatthours being sold at each step in the rate. For example, the number of kilowatthours per month being used by customers who would fall under the first rate block of, say, 25 kWh per month can

---

[1] Edwin Vennard, McGraw-Hill, New York, 1970.

# FREQUENCY DISTRIBUTION OF RESIDENTIAL CUSTOMERS
## EDISON POWER COMPANY, 1976

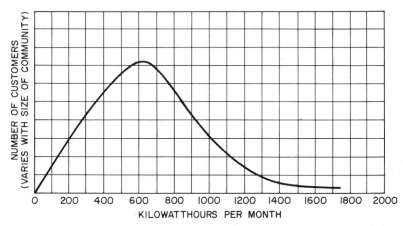

**CHART 8.33**

be read from the chart. To see the effect of a new rate, these kilowatt-hours are multiplied by the price per kilowatthour for the first block of the new rate to determine the revenue that would be produced in that block. As a check on this frequency distribution, it is tested on the existing rate schedule to see if it gives an accurate result.

In the same way, the effect on revenue can be calculated for any rate changes in the commercial or industrial class. Normally the changes applied to large industrial customers must be figured individually.

**Distribution Computations Facilitated**   There was a time when the construction of these frequency distributions of customer use meant counting individual customers from meter records. Before data processing machines were available, this was a very laborious process. Some years ago, however, it was discovered that by the use of ogives and probability paper, families of curves could be constructed empirically from which the same information could be developed as is obtained from the physical count of customers, and with much less effort and time. This method, developed by William Parkerson, is called the *ogive method* and has received broad acceptance among rate analysts in companies and on the staffs of regulatory commissions.

Not only does the ogive method facilitate the construction of customer distribution curves from current data, but, by estimating future average uses, it facilitates the construction of curves which permit the forecasting of future revenues from rates, after increased use is taken into consideration.

SAMPLE LIGHT AND POWER COMPANY
1976 OGIVE CURVE FOR RESIDENTIAL ELECTRIC SERVICE

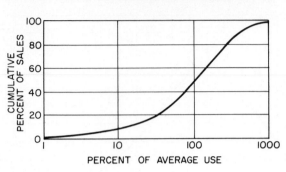

**CHART 8.34**

The shape of the ogive changes with the change in average use per customer. From a curve as shown in Chart 8.34, the ogive can be determined for any use per customer. Analysts now use computer models, programmed with these ogive data.

## CALCULATING FUTURE RATES OF EDISON POWER COMPANY

In Chapter 7 a determination was made of the revenue requirements of Edison Power Company over the next 10 years, after taking into account all practical improvements in system planning, expense control, and marketing. Now the aim is to calculate the rates needed to produce that revenue by years. To repeat for emphasis, these calculations are for purposes of illustration of methods, techniques, and tools. The rates pertain to no particular company. Local conditions determine local rates. Here, the residential rate will be designed. The same principles apply to the other rate schedules.

Chart 8.35 shows the cost curves for Edison Power Company for the years 1971, 1976, and 1982 at 40 percent class load factor. The 1971 curve is shown to indicate the increase in costs experienced since then.

It was explained earlier in this chapter why these cost curves go down for larger users. The whole cost curve rose between 1971 and 1976. The increase has been across the board, not just for increased use. In light of these cost curves, any rate schedule that rises with increased use or for larger users would violate the principles of rate making and cause unjust discrimination.

**Design of Rates in Future Years**  Knowing the revenue requirements by years for each of the next 10 years, the analyst prepares a number of

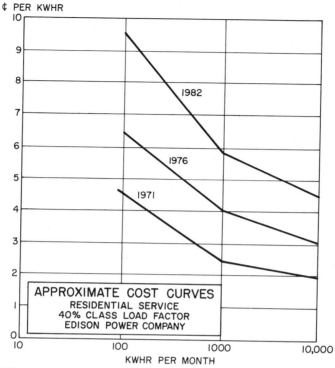

¢ PER KWHR

APPROXIMATE COST CURVES
RESIDENTIAL SERVICE
40% CLASS LOAD FACTOR
EDISON POWER COMPANY

1982

1976

1971

KWHR PER MONTH

**CHART 8.35**

alternative schedules of schedules that will produce the revenue. The term schedule of schedules is used, as the task requires a different schedule for each period, such as a year. All customers are affected across the whole range of use of each class. As each schedule should result in a lower average price with higher volume and higher load factor, the whole schedule must rise each year or periodically. This will probably continue until unit embedded costs either level off or decline.

Having designed a number of alternative schedules of schedules, each of which will produce the required revenue, management and the commission decide which one is best.

Some regulatory bodies are reluctant to look at the future, but the experience over the past 5 or 6 years indicates such a look is inevitable if the companies are to be allowed to finance the facilities required to serve those future needs. Some years ago, the Federal Power Commission decided it must look at the future.

Chart 8.36 shows the same cost curves as in Chart 8.35. Superimposed are the rate schedules of 1971 and 1976. Notice that each of these only approximates the cost curve. In each case, the rate for the

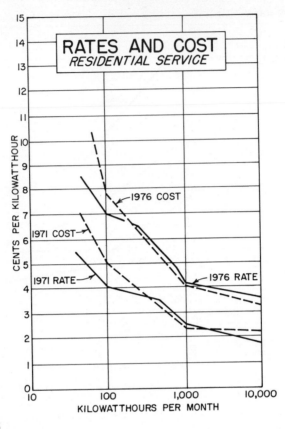

**CHART 8.36**

small user is considerably under the cost curve. To the extent the rate is under the cost curve for the small user, it must be above that curve for the large user if the schedule is to earn a fair return. In 1971 the bottom step was too low—it went below the cost curve. This needed correcting. It was corrected by 1976.

This favoring of the small user has been practiced, almost from the beginning. This gives weight to "ability to pay." But it is done in a way that does not violate the principle of lower price for larger use. The "below cost" feature to the smaller user, as shown here, has not been considered unjust discrimination for these reasons:

- Incremental costs as well as embedded costs are taken into account in rate making. The fact that the price is below the average embedded cost does not mean the company is better off without these small users. Distribution lines and the meter reader will probably pass the individual house in any event. The added cost, or incremental distribution and accounting costs, of picking up the small user may be slight.

- Most people have a habit of moving out of the "low income" bracket to a higher one.

- It is popular.

- This is where judgment enters.

But there is a practical limit to this practice. That also is where judgment enters.

Chart 8.37 shows the schedule of schedules that will produce the required revenue for Edison Power Company by years through 1985. Notice:

- All schedules follow the principle of declining price with increased volume. In the absence of a demand charge, load-factor building is done through education.

- All customers pay a proportionate share in the increased cost of service.

- The price for the average use (marked by a dot) increases each year.

- The tail-end prices are increased more than the prices for smaller users, as the downward slopes are lessened.

This price discussion completes the pricing portion of the plan called for in Chapter 7.

## INNOVATIONS IN RATE DESIGN

The literature on the electric energy business indicates that from the early years analysts have realized that it is desirable to have a pricing policy that takes into account both kilowatthours of use and kilowatts of demand. Partly because of the general lack of understanding of the difference between the kilowatt and the kilowatthour, most rate schedules for smaller users have been of the simple sliding-scale block form which is the type of rate based upon kilowatthours only. However, from the early days, the kilowatt-demand feature has been incorporated in rates for large users.

From time to time in the past, attempts have been made to incorporate the demand feature or the load factor principle into the rates for the smaller users. The author has experienced a number of these attempts and has taken an active part in initiating a few of them.

# APPROXIMATE SCHEDULE
# OF SCHEDULES
## EDISON POWER COMPANY
### 1976–1985

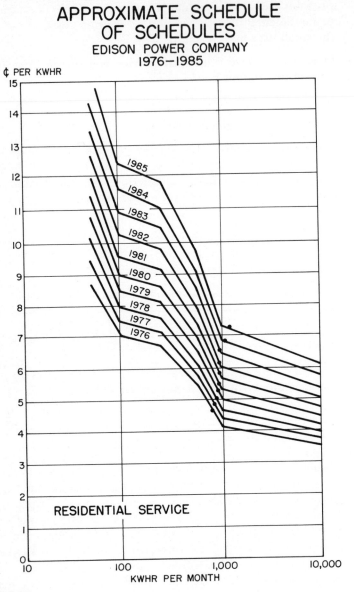

**CHART 8.37**

During the late 1920s, the so-called room-count rate schedule for residential service was quite widely used. Under this type of rate the length of the blocks was based upon the number of rooms in the dwelling. Usually there was a minimum, such as 3 rooms, and a maximum, such as 10 rooms. This rate was never popular with the electric customers.

Also, there was a kilowatt feature in the schedule for the small commercial customers. Here the blocks in the rate were lengthened depending upon the number of kilowatts of connected load. To save the expense of a demand meter, connected load was determined by a physical count of all kilowatts connected on the premises. Usually there was a minimum such as 1 kW. As can be surmised, the count of the connected load was subject to considerable error. There was endless controversy. The records indicated that an unusually large number of customers had only 1 kW connected. It was well known that they had more.

From this experience management concluded it was unwise to base any billing on an estimate, such as the number of rooms or connected load. Billing should be based upon the actual measurement by a meter acceptable to the public, the regulatory body, and the company.

In 1932 the author was transferred to a department-head position in the holding company of the parent organization. This company owned roughly 40 utilities serving about 3000 towns in 17 states. All these companies had the residential room-count rate and a commercial rate with a demand feature based upon the estimated connected load. The conclusion was that both residential rates and commercial rates should be changed back to the simple block form. Any change in the basic style of a rate is not easy. The new block rate for residential service had to raise some customers billed on the basis of a three- or four-room house and had to lower the rates to some customers having a large number of rooms. Rates for many of the commercial customers had to be raised because they were being billed on the basis of 1 kW of connected load, whereas most of them had much more.

Some years later a large percentage of the companies in the country concluded that they should again attempt to put a demand feature or load-factor principle in the residential rate schedule and the schedule for commercial customers. The problem with the small commercial customer was easily solved by adopting an optional demand-energy rate which would become available to all customers having roughly 5 kW of demand or more. Usually this demand type rate would become beneficial to commercial customers having roughly 30 percent load factor or above.

However, optional rates also have their difficulty. As the customer determines load characteristics, the option must be with the customer.

But the customer is not well informed on these matters. Consequently, the companies felt some obligation to keep the customers informed concerning their option.

In the case of residential service, most of the demand features became available to customers having a demand of roughly 5 kW or above. Demand meters were installed. Some of the rates simply added the demand feature when the load reached 5 kW. Thus, the customer's rate was increased when 5 kW of demand was reached. This was hard for the customer to understand. Actually, this practice went against some rate principles which call for a lower price with increased volume. Largely because of this practice, this type of demand feature became unpopular and was finally withdrawn in most cases.

Some companies took a different approach in the application of the demand feature. In this case all bottom steps of the residential service were raised to a certain level, such as 2.5 cents or 3 cents per kilowatt-hour. A demand meter was installed on those customers having a demand of roughly 5 kW or above. At that point the kilowatts of demand governed the length of the blocks. There were steps in the rate that went below 2.5 cents or 3 cents for those customers having good load factor. This demand feature had the advantage of making it possible for the customer to get a lower price with improved load factor. No customers were raised when the demand meter was installed.

However, even with this feature few customers understood the rate. Even among employees and some company management the demand feature was unpopular, and after a few years it was withdrawn.

**Current Considerations** In recent years there has been a renewed interest in some sort of load factor principle in the rates to the smaller users. There is discussion of *peak-load pricing* or *time-of-day metering*. The time-of-day metering concept goes beyond the simple demand charge and requires a recording device on the meter. This creates a new complication, as well as added meter expense.

For some time electric utilities have utilized recording demand meters on some of the large industrial users. Few, if any, attempts have been made to record the demand on the smaller users (except in load-research studies). Perhaps the manufacturers can develop an efficient, simple device that can allow this measurement.

There have been a few advocates of a so-called *upturn rate schedule*, i.e., schedules that slide up with increased use. This idea stems from the fact that the cost of making and delivering energy from the new power plants is higher than it is in the older power plants. Also, this idea stems from the belief by some people holding the "no growth" philosophy. However, the no-growth idea has been largely discredited. People now realize that growth in the use of energy is essential if living standards and jobs and income per capita are to be sustained and improved. Also, there is the question as to whether the regulatory body has the authority by statute to allow a particular sociopolitical bias to enter into rate-system determination.

To some, the upturn rate schedule would discourage use and thus conserve energy. Naturally it is wise to avoid all waste of energy and to conserve oil and gas. It is questionable whether it is wise to adopt a pricing policy with the idea of conserving coal and uranium (except for waste) when these materials are not in short supply. The regulatory body is empowered to ensure that the customers get the service they want and need and that the price is fair to the customer and to the investor. The upturn rates violate these principles.

The rate engineers realize that in the existing economic climate, the average price should decline less steeply than it did in the climate before 1968. However, as previously pointed out, all cost studies indicate that the unit cost goes down for the larger user. To violate this principle would lead to a charge of unjust discrimination.

Recently there has been talk of a *lifeline rate*. This concept embraces the idea that those people who are less able to pay should be billed at a price that is less than that required to cover full cost. There may be something to this. As pointed out previously in this chapter, this fact has been given consideration almost from the beginning. However, there is a practical limit as to how far a regulatory body can assist people who are less able to pay for the goods and services they want and need. Government has many avenues available to it for assisting people who need help. It is questionable whether any single commodity or service should be pinpointed for subsidy, especially one that takes such a small percentage of the family budget.

The concept of *long-run incremental cost* (LRIC) pricing has been receiving some attention. The idea here is that incremental cost, now being higher than embedded costs, should be controlling or have a strong influence on the rate schedule. This is really not a new concept. Almost from the beginning, analysts and rate people have taken incremental costs into account in making cost studies and in rate design. Consideration is given to both the embedded and the incremental costs. Naturally, the long-run incremental costs, which today are higher than average, must be taken into account in rate design. The question is whether or not the higher costs should be spread among all users of energy.

At the moment rather exhaustive studies are being made of all these various concepts. These studies are healthy. Possibly, out of them will come some practical ideas and suggestions as to how a load factor principle can be incorporated in the rate schedule for the smaller user.

Also, in many quarters today *load management* is receiving attention (see Chapter 7). This may be simply a new term for marketing to build load factor. Both practices have the same objective. Wise marketing has always had as a primary objective the increasing of load factor. Witness the increase in industry load factor from about 50 percent in 1930 to about 65 percent in 1965. This was largely brought about by a marketing practice that could very well be called load management. But since about 1968, regulatory bodies and others have discouraged marketing with the result that load factor has dropped from 65 percent to about 61 percent.

With all these new ideas, care should be exercised to ensure that they do not work contrary to the purpose for which they are designed and do not require unwarranted expense in capital and operation. Also, care should be taken to develop customer understanding and acceptance of those approaches which may prove to be beneficial.

These concerns are of particular importance in view of the Public Utility Regulatory Policies Act (PURPA), which is part of the National Energy Act passed in October 1978. PURPA sets out requirements for state regulatory commissions and nonregulated utilities to consider what are termed federal "standards" for retail rate design and policies and procedures affecting electric utilities. The standards include cost of service pricing, prohibitions on declining block rates, time of day rates, interruptible rates, seasonal rates, and load-management techniques.

Well before PURPA, Edison Electric Institute and the Electric Power Research Institute were engaged in a comprehensive electric utility rate design study on behalf of the National Association of Regulatory Utility Commissioners, with the full participation of the American Public Power Association and the National Rural Electric Cooperative Association. The first-year conclusions emphasized the need for further research and for the evaluation by regulators and utility managers of the cost effectiveness of time-differentiated rates and load controls.

During 1978, research continued, with emphasis on five areas: (1) the conduct and assessment of rate design experiments; (2) appraisal of load-management hardware; (3) evaluation of user equipment under peak load pricing and load controls; (4) assessment of customer acceptance of novel rate designs; and (5) improvement of methods for analyzing the costs and benefits of time-differential pricing and load management.

In a variety of ways, the state regulators and the utilities have been moving effectively in the rate design area. The existence of PURPA and its implementation should not be allowed to result in adoption of unproved approaches in the rate area. The test of innovations must be, as always, that the benefits to customers and utilities clearly outweigh the costs.

# REGULATION

## WHY REGULATION IS NECESSARY

In the free society, competition is generally relied upon to keep prices fair and reasonable. Anyone wanting to purchase something can buy it from the seller offering it at the lowest price. People can shop around for their groceries and clothes and automobiles. If the seller sets a price too high and others are offering the same thing for less money, the higher-priced article will not sell. The seller must cut the price in order to make a sale. Manufacturers and dealers have to maintain high efficiency in their business operations and must continue to make a quality item, reasonably priced, or they will not stay in business for long.

It is a good system and it works well in most businesses. However, in the case of the electric energy industry it was found that this type of competition did not work to the consumer's advantage.

The electric utility business was begun under this competitive system. Companies were formed and operated when investors in the industry thought they could sell electricity, leading to vigorous competition for customers. An attempt was made to rely upon this competition as a means of keeping the price of electricity fair and reasonable.

For example, in the early days of the electric industry, there were some 25 or 30 separate utility enterprises in the city of Chicago. Some were merely isolated generating plants, while others were quite large. They operated at various voltages with various kinds of equipment. There was much duplication of distribution lines. There were no large central stations, and the plants were inefficient as measured by present-day standards.

This condition at times resulted in inferior service, confusion, higher rates to electric customers, and inadequate or no return to investors for the use of their capital. It was also recognized by utility executives and government officials that the electric utility business had important differences from the ordinary business enterprises. They realized that, while direct competition was good for business in general, it was costly, impractical, and undesirable both for electricity customers and for electric utility companies. As a result, the idea of government regulation for electric companies, as a fair and workable substitute for competition, began to take form.

This concept recognized, first, that the electric utility has an obligation to serve all who apply for service at reasonable rates, second, that direct competition is not in the public interest in the case of an electric utility company, and third, that government regulation can provide the safeguard of the public interest ordinarily obtained through direct competition.

The states already had experience in regulation. Many states had for years regulated railway companies and gas and water suppliers. Many cities also had regulatory experience.

The early lighting companies operated only in cities and towns. Later it became more economical to build large central stations and to transmit the power over transmission lines. Eventually, a single company would serve large areas and hundreds of cities and towns. This made it desirable to provide statewide regulation. Later, when transmission systems began crossing state lines, federal regulation of interstate operations came into being.

## THE REGULATORY BODIES

The first states to have formal regulation were New York and Wisconsin. In 1907, public service commissions in these states were given regulatory jurisdiction over electric utilities. Other states followed suit, and today almost all have means for regulatory control of electric utilities. In 49 states, public-utility commissions watch over the operations of the electric utility companies. (Most commissions also regulate other types of public-utility intrastate operations.) In Nebraska, almost all electric utilities are owned by the state or local governments and are not regulated by the state.

The federal government has the authority to regulate electric utilities under the interstate commerce clause of the Constitution and also under its provisions for controlling public lands. The power to license water-power sites dates back to 1896.

The Federal Power Act was intended to fill a gap in the regulatory process. State commissions had jurisdiction over intrastate matters, but had no jurisdiction over electric energy delivered or received in interstate commerce. The Federal Energy Regulatory Commission (FERC), successor to the Federal Power Commission (FPC) which was created under the Federal Power Act in 1920, has the power to regulate projects in interstate commerce, on public lands, and on navigable streams and other water resources under federal control.

The Securities and Exchange Commission (SEC) has certain authority under Title I of the Public Utility Act of 1935 over holding companies. (A *holding company* is one which owns the securities of a number of electric companies. It usually operates all the companies as an integrated system, thereby taking advantage of benefits from the most modern techniques and equipment.) The SEC also regulates the sale of securities by operating utility companies under the Securities Act. The SEC's control is mainly in the financial field and deals with such things as the limit of total capitalization, capital structure, the terms of security sales, and the protection of utility assets.

The FERC has the power to regulate rates for interstate wholesale power if it finds rates unjust, unreasonable, unduly discriminatory, or preferential. It may investigate the actual legitimate cost of the property, "the depreciation therein, and, when found necessary for rate-making purposes, other facts which bear on the determination of such cost or depreciation, and the fair value of such property " [16 U.S.C. 824g (b)]. In such cases, the quality of service also comes under the commission's jurisdiction. It is empowered to regulate security issues, mergers, and interlocking directorates and managements. Other regulatory functions affecting electric energy are under the Economic Regulatory Administration, which is part of the Department of Energy, established in 1977.

State commissions regulating electric utilities operate under state laws that set forth their duties and powers. They have the authority to examine company records, question company officials, and obtain pertinent data from other sources. The commissions hold public hearings, and the operations of these electric utilities are clearly exposed to public view. State commissions generally have wide powers to decide whether a rate is reasonable.

PURPA is adding new dimensions to the regulatory task and increasing the federal role in utility regulation. Many of PURPA's provisions are ambiguous and subject to varying interpretations. Litigation will undoubtedly be required to clarify issues which will inevitably arise. However, PURPA is vastly superior to the original proposals, in that it maintains state discretion over whether to implement the federal standards which the commissions must consider under the Act. Initially, measures were proposed which would have mandated the application of certain types of rates, without due regard to the particular characteristics of the utilities in the areas under state commission jurisdiction.

## WHY DUPLICATION IS AGAINST PUBLIC INTEREST

In the normal unregulated business operation there may be as many as five or six intermediate commercial firms between the maker of a product and the person who ultimately uses it (Chart 9.1). The company that makes the product in most cases does not sell it at retail. First, the product is shipped by (1) a separate transportation company, such as a railroad or truck line, to (2) a distributor or wholesaler, who may store the

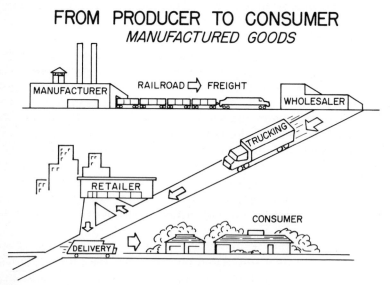

# FROM PRODUCER TO CONSUMER
## *MANUFACTURED GOODS*

MANUFACTURER    RAILROAD ⇨ FREIGHT    WHOLESALER

TRUCKING

RETAILER

CONSUMER

DELIVERY

**CHART 9.1**

product and later (3) ship by a local trucking company to (4) the retailer, who may hire (5) a firm to deliver the goods to the consumer.

In the electric power business, one company makes the product and delivers it to the consumer. Electricity is sent over high-voltage expressways to local centers from which it is shipped at a lower voltage to consumers. There is a permanent physical delivery system from the producer of electricity direct to the customer.

An ordinary manufacturer can make as much or as little of a given product as desired, and is not required to supply all the customers of any particular area. The electric utility company, however, is required by law to meet all, or practically all, the power needs of all the people of a given area.

The manufacturer can store products, can run the plant full-time, and can stockpile the product for sale later. Toy manufacturers, for example, run their plants all year round but sell most of their products at Christmas. Such a manufacturer can use machinery at a relatively steady rate all year, getting good utilization of the equipment which had to be purchased.

But electricity cannot be stored in large quantities. It is made, delivered, and used at the same instant the customer throws the switch. The company must have generators standing by at all times, waiting for the customer to press the button. Power plants, transmission systems, distribution systems, substations, and transformers must be big enough to meet the diversified demands of all customers who may want to buy electricity at any time.

## SOME ECONOMIC DIFFERENCES

As the electric utility business is physically unique, there are important economic differences between the electric utilities and other businesses, as noted in Chart 4.1. The investment of the electric utility companies must be about six times that of manufacturing enterprises, to produce the same amount of annual sales.

If two electric utility companies served the same area, the investment costs would be much greater than for a single company. But having two companies in the area would not increase the number of customers and would not bring about greater sales of electricity. So if there were two companies, the rates to the customer would have to be higher. For this reason, public utilities are natural monopolies.

## MEANING OF THE TERM "MONOPOLY"

The word "monopoly" sometimes has a bad connotation in people's minds. They usually think of it as something against the public interest. Federal and state antitrust laws forbid monopolization, but these laws do not apply to public utilities to the extent that they are regulated by government.

There are more than 200 independent electric light and power companies in the country, all under some kind of government regulation. In some states, government-sponsored suppliers are not subject to regulation. In other states, they are regulated, but generally not to the same extent as power companies. In areas served by cooperatives, municipal plants, and others, there is the same kind of monopoly as in the areas served by companies. There is seldom duplication.

## COMPETITION CONTINUES

Although there is usually only one power company from which a customer can buy electricity, this does not mean that there is no competition for business. There is competition today in most uses except for lighting. For example, in cooking, a customer can use gas or some other fuel. There is competition for water heating, clothes drying, and other domestic uses. There is competition in space heating.

With the current realization that oil and gas are in uncertain supply and should be conserved, competition in the residential and small commercial fields has not the same significance now as formerly. The remaining fuels, coal and nuclear (until others are developed), are used through conversion to electricity.

There is further competition in the electric power field in that customers are free to generate their own electricity. Some 30 or 40 years ago many industries did so. The plants were called *isolated generating plants* to distinguish them from interconnected power systems. Such isolated plants are sometimes called *on-site generating plants*. In 1920 about 30 percent of the nation's electric energy was produced by these facilities. This factor has gradually come down to about 4 percent (Chart 4.17), as electric power users have found they could obtain better service at a lower price by purchasing energy from interconnected systems.

The people who manage a company are constantly trying to demonstrate better management by keeping their rates competitive with others operating under similar conditions. The companies study new operating methods and techniques in order to improve efficiencies. When a company is able to cut costs and thereby offer lower rates, neighboring companies are eager to learn its methods. Thus there is competition among managements for excellence in performance.

Because of the higher reliability and inherent economies of the large interconnected systems, competition from on-site generation is now of less significance. Nevertheless, the possibility is still there. Loss to the utilities of industrial business results in higher unit costs to smaller customers. This fact has a bearing on judgment when a proposal is made to charge the big user at a price above cost so as to enable the small user to get a price below cost.

## THREE MAIN OBJECTIVES OF REGULATION

**Service**    Electric utilities usually operate under a franchise or an indeterminate permit which gives them the rights to serve the community. In exchange for these rights the utility undertakes the duty of rendering reliable and adequate service. The franchises are usually for a fixed number of years and from time to time they must be renewed. This generally requires the approval of the electorate. Thus the power supplier has a strong incentive to give adequate reliable service, or else the right to do business in the community could be denied.

In carrying out its duties in rendering reliable service, the company is constantly studying all causes of equipment and human failure and is designing the system so that service will not be interrupted when such failures occur. Interruptions are frequently caused by elements beyond the control of management, such as lightning, storms, and floods. Here efforts are made to restore service as promptly as possible. The state regulatory body is interested in seeing to it that all reasonable efforts are made to keep the service continuous and to restore it as promptly as possible when an interruption occurs.

In some states the state regulatory body also has authority to see to it that service is rendered within the limits of good voltage regulation. The regulatory body may also provide that a customer cannot use devices that will require a large demand for power for a very short interval of time, as this would cause a flickering of the lights in the neighborhood.

Usually the state regulatory body has the authority to see to it that adequate capacity is planned for the future so as to meet the energy demands of the area. (Later in this chapter, the problem of plant cancellations and the difficulty arising from inadequate returns allowed by regulatory bodies are discussed.) Usually a company planning to build a plant or transmission line must apply to the state regulatory body for a certificate of convenience and necessity for the building of such a plant or line. The commission will grant a permit if it decides that the line or plant is indeed necessary in the public interest. Thus the commission can prevent a duplication of facilities that will inevitably be against the public interest. A commission may find that the territory served already has adequate capacity in service or being installed and that therefore a proposed plant is not necessary. In this case the commission will deny the certificate.

**Preventing Unjust Discrimination**    It is not possible to make the rate for each customer exactly cover the cost of furnishing service to that customer at all times. There are many factors entering into rate design (see Chapter 8). As a practical matter customers are grouped into classes, such as residential, commercial, industrial, street lighting, other public authorities, and railroads. Rates are designed to meet the general characteristics of the particular class.

In determining the fair value of the property for rate-making purposes and an adequate return on the capital invested, the commission rightfully takes into account the property as a whole. It is possible to prepare an estimate of the earnings by classes of service, but this must be a separate analysis, as it would be very expensive and not practical or meaningful to keep books by classes of service. The cost analysis showing the earnings by classes is a tool that can guide management in designing the forms of the rates, but these analyses are really only estimates prepared by the experts. The return by classes will vary from year to year as the load characteristics change from year to year.

As noted, for practically all the years prior to 1968, the unit incremental costs were less than the average unit embedded costs. The price could decline, which made regulation relatively simple. Since about 1968, incremental unit costs have been above average unit embedded costs. With the need for rising prices, regulation became difficult. However, it is a general principle that all customers should pay a fair share of all the costs of doing business, both embedded and incremental, if unfair discrimination is to be avoided. In other words, all customers should share proportionately in the benefits of all the newer and more efficient stations when unit costs are declining, and all customers should share proportionately in the higher costs of all new plants in times of inflation when new plants are more expensive than existing ones.

At times in the past, when new plants produced energy at lower than average cost, a large industrial customer would feel that it should obtain all its electricity from the power plant nearest its plant site. Another would feel it would like to get all its energy from the power company's newest and most efficient plant. However, to avoid undue discrimination the commissions required that the property as a whole be taken into account in fixing rates to the customers. The same principle holds today when new plants produce energy at higher than average cost.

**Fair Return**   So that the companies can continue to provide good and adequate service, the regulatory commissions have generally approved rates which would permit them to earn a fair and reasonable return on the value of the property as determined by the commission. During the years before 1968 when there were no large-scale increases required, this allowance of the *fair return* worked well. This does not mean that the regulatory body guarantees a return to investors in all circumstances. Indeed, all during the years of the Great Depression of the 1930s most companies earned a below-standard rate of return. The commission did not order rates raised to produce what might be considered a fair return. In those years, an increase in rates might have caused a decrease in revenue. As a matter of fact, during those years of low return the companies under good business management actually reduced rates further with commission approval in their attempt to build revenue through sales. There was great excess capacity, and little money was being raised to add a plant. The existing investors had to be content with what management could earn under those unusual conditions.

But since 1968, the regulatory bodies have faced different conditions. Prices have been rising and plant expanding. Under these circumstances, regulation must assure the investors that they will receive a fair return or the needed capital will not be available to enable the commission and the utility to carry out their joint obligation of assuring the energy that people want and need.

While the economy is going through a healthy increase, as it is today, and since people want and need more and more electricity, it is in the interest of the public as well as the investor that the commission allow the utility *at least* sufficient earnings to attract in the free market the capital required to build the plants to generate the additional kilowatthours. In addition some commissions take into account some factor for good management and allow the company sufficient return to do the research and provide innovations that will enable continued improvement so as to advance efficiency in the operation.

## KINDS OF REGULATION

**The Free Market**   In the free society, people or organizations are free to invest or not to invest in securities of companies such as electric utility companies. When they do invest, they become part owners of the company.

Electric utility companies follow the securities market carefully so that they can take advantage of conditions enabling them to borrow money at favorable interest rates and sell stock on a favorable basis. The regulatory commissions also follow market prices and yields on utility securities. The reason for their interest is that the cost of money is one of the major costs of doing business in the power industry. This cost is usually considered in deciding what is a fair return to the electric utility.

As these electric companies are owned by investors, the electric companies are sometimes referred to as *investor-owned* utilities. As a whole, these companies are often called the *investor-owned electric utility industry*. However, the word "company" itself signifies ownership and operation under the free-enterprise system. The words "electric light and power companies" or "power companies" or "electric utility companies" are often used in speaking of that portion of the industry financed in this fashion. These companies serve about 80 percent of the nation's electricity customers. The rest of the electricity customers in the country are supplied by government power projects or cooperatives. There are several kinds.

**Federal Power Projects**   Federal power projects are projects which have been financed and built by the federal government. In some cases, the financing is through direct appropriations by the Congress, and in other cases, the capital is lent by the federal government. The Tennessee Valley Authority (TVA) is an example of a project which in the past had been financed by appropriations from the Congress and by investment of earnings. As of 1975, the TVA was authorized to sell $15 billion in revenue bonds to the public and to the U.S. Treasury, the proceeds to be used for future expansion.

**Public Utility Districts**   The laws of some states permit the organization of *public utility districts*. These are political subdivisions of the state. This type of organization is prevalent in the states of Washington and Nebraska. In Washington the district usually encompasses a county. In Nebraska the size may vary from that of a single county to a large regional area.

**Municipal Ownership**   Since 1882, 4362 municipalities have at one time or another established their own electric utility systems. By the end of 1976, only 1907 of these were still in operation. The number has gone down mainly because most municipal plants are limited in size; usually each plant serves only the town or city in which it operates. Because of their small size, most of these municipally owned plants cannot make electricity as cheaply as the large interconnected systems of the power companies. Also it has been found that the small municipal power plants often do not have the power capacity necessary to serve large industries.

The municipal power plants are financed through the sale either of bonds which are the general obligation of the community or of bonds supported by the revenue of the municipal power system.

**REA Cooperatives**   In 1935 the federal government began a rural electrification program designed to hasten the extension of electricity to America's farms. In 1936 the Rural Electrification Administration (REA) was created. The purpose of this program was to make federal money available at low interest rates in order to encourage the extension of service into thinly populated rural areas. The federal government lends money at 2 or 5 percent interest with a condition that the loan be repaid within 35 years. At the end of 1976, 927 rural electric distribution cooperatives were active borrowers from the REA.

As of September 30, 1976, 98.6 percent of all the farms in the United States had central-station electric service available. Of these farms, about 43 percent are served by the electric light and power companies, 51 percent by the rural electric cooperatives, and 6 percent by other suppliers, such as public utility districts and municipal plants. The cooperatives bought about 36 percent of their total power requirements from electric utility companies during fiscal year 1976. The remainder was generated by cooperatives or bought from some government power project or system.

To summarize, in 1976 America's electricity was generated by the various types of suppliers in the percentages shown in Table 9.1.

## VALUATIONS

The regulatory bodies of the nation have an intricate and complicated task to perform in looking after the interest of the public—as the consuming public and as the investing public. Generally speaking they are well staffed, and they carry out their functions with skill. The most exacting task is to find the combination of fair value of property and fair return on that property. There are people who have devoted their lives to these matters. Many valuable books have been written on this subject. There are university scholars who testify on these subjects and express their views in textbooks. The state regulatory bodies have at their command this wealth of knowledge and experience as well as the experience and judgment of other regulatory bodies meeting similar problems. Added to this, the commissions have as their guide a great wealth of court decisions. The rate case when decided can be appealed to the

**TABLE 9.1**  PROPORTION OF ELECTRICITY GENERATED
BY VARIOUS SUPPLIERS

| *Supplier* | *Percent Electricity Generated* |
| --- | --- |
| Electric power companies | 77.7 |
| Federal projects | 11.6 |
| Municipal systems | 3.8 |
| Public utility districts | 4.9 |
| Cooperatives | 2.0 |

lower courts by any party. These lower-court decisions frequently are appealed to a higher court. Some cases reach the Supreme Court.

The lay student and public may find it difficult to judge the efficiencies of these regulatory bodies without intimate knowledge of the intricate procedures in each case. Possibly the best measure of the efficiency of the regulatory procedure is that the electric power industry has met all demands of the American people for electric service in times of war and peace and that the average price of electricity has continued generally downward until the unusual inflation beginning around the years 1968–1970 (see Chapter 4). The regulatory lag since about 1970 is discussed later in this chapter and in Chapter 7.

The best overall measure of operating results of the Edison Power Company is the percent return on the total value of the property used in serving the public. The *amount available for return* is the amount left after all operating expenses and all taxes and depreciation are paid. This money is used to pay bond interest and preferred stock and common stock dividends; whatever may be left is retained in the business and reinvested.

When the Edison Power Company is called before the commission to answer questions as to why the rates should not be reduced, or when the company finds it necessary to go to the commission for a rate increase, the commission sets a fair and proper value on the company's property for the purpose of determining a fair rate for electric service. In determining this value, the commission takes many factors into account. It may consider the value as recorded on the books, the value of securities outstanding, the amount set aside for depreciation, and in some cases the present value of the property or the cost of building the same plant today. Following are some of the aspects of "value" that the commissioners may consider.

**Original Cost**   Edison Power Company, like most electric utility companies, keeps its books in accordance with the uniform system of accounts established by the FPC (now the FERC). These records give the *original cost* of the company's facilities.

Most power companies, including Edison Power Company, also keep what is called a continuing property record of all the items, large and small, that go into the making of the electric utility system and the year in which these various items were installed. Edison Power Company's electric system has been built in phases over the years, with additions and betterments made to meet the growing demands of its customers. The continuing property record shows, for example, that in 1950 the company installed so many poles of a certain classification, so many transformers of a certain size, so many pieces of hardware properly classified, so many pounds of wire of a certain size, and many other items. It also includes a record of power plants installed, with each plant broken down into its appropriate parts. For 1951 there is a similar record, and so on for each year. The cost as shown in this record is the original cost of the property. Of course, as a result of price rises since the equipment was installed, the cost of replacing the item in a current year may be considerably higher. If prices had gone down, it could well be that the reproduction cost of the property might be less than the total value shown on this record. Present-day costs of building electric utility systems are generally far in excess of original cost.

**Net Original Cost**   The depreciation reserve reflects the charges to current income estimated to be required in order to amortize the cost of the plant and equipment by annual amounts over the expected useful life of the equipment. When a unit of property is retired from service, the amortized amount less net salvage is then charged to the depreciation reserve. The annual charges to income permit the company to recoup the cost of the equipment over its useful life.

The amount of depreciation is an accounting entry based on the best judgment of experts of the commissions and the power companies. Edison Power's engineers study the life histories of various kinds of equipment. Certain items may be expected to last 20 years before wearing out; others, 30 years. With these studies as a guide, management records certain annual amounts in the depreciation reserve.

A number of factors contribute to a decision to replace a piece of equipment, including physical wear and tear, inadequacy, economic obsolescence, needs of public authorities, and accidents. Obsolescence is an important factor in deciding to replace equipment. For example, new boiler-plant equipment and generating equipment which are more efficient than that which had previously been available have been, and are being, designed. There comes a time in the economic life of a power-plant unit when the company would do best to take it out of use and replace it with a unit that can make electricity more efficiently.

As replacements are not made every year for most items, the depreciation reserve normally builds up over a period of years. For Edison Power Company, the reserve for depreciation on electric utility plant as of December 31, 1976, was $340,306,785, which was almost 20 percent of the gross electric plant account.

Many regulatory bodies deduct the depreciation reserve from the total original cost of the utility to arrive at what is called the *net original cost* of the property.

In some cases, it is argued that the net original cost more nearly represents the value of the property, for the reason that the company used the cash in the depreciation reserve to build new plant and therefore it does not have to earn a return on this property. When the depreciation reserve is deducted from the total original cost of the property, the result is that the company earns no return on that portion of the property built with cash from the depreciation reserve.

Others argue that the depreciation is a prepayment for the replacement of property when it wears out or becomes obsolete, both of which are inevitable. Management could put this cash in a savings bank and earn interest on it, but that interest could be something less than a return on utility property, as the return takes into account some risk. If the reserve is placed in a savings bank, the interest of course would be credited to net income. The new money needed to build the property would be all from the market, and the company would be allowed a full return on this, as on any other property. Those who hold this view argue that the company obligates itself to furnish continuous and adequate service to customers receiving service from property built with the depreciation reserve, as it does on all other parts of the business. The company must maintain the property and keep it in good operating condition.

**Value of the Securities**   Edison Power Company, like all power companies, has a record of the book value of all its securities outstanding. The commission may give some consideration to this value in finding the appropriate value for rate-making purposes.

**Reproduction Cost New**   In the earlier years, many commissions relied upon a physical inventory and valuation of the property in determining its value for rate-making purposes. To make this valuation an appraiser would be hired to make a detailed inventory of the whole property, counting and listing all items of equipment by size and classification. Then the cost of building everything at current prices was figured. This was referred to as the *reproduction cost new* of the property. In finding fair value under the reproduction cost theory, it is generally considered proper to deduct the observed depreciation in order to find the present depreciated value of the property.

**Trended Present Cost**   As most companies now have fairly complete property records, a close approximation to this reproduction cost can be found by what is called the trending process. Indexes such as the Handy-Whitman index are used in finding the present value, working from the original cost figures. (The Handy-Whitman index is a refined estimate by utility engineers and appraisers of the cost of basic items used by the utilities, by years. It is roughly equivalent to a "cost of living" index for energy companies.)

By way of example, the property records may show that the company purchased a certain number of 5-kVA transformers in 1955 at a certain price. This figure is recorded for that year. Using the Handy-Whitman index for transformers, one can convert that cost into the current cost of those particular transformers. In similar fashion, all items of property can be trended forward to determine their present value. Since there is a record of what property is in service, it is not necessary to have engineers go over the property to count the items. The current value of the property, arrived at in this way, is referred to as the *trended value* of the property. It approximates the reproduction cost of the property. This could be called "original cost expressed in the current dollar value."

Besides this trended value, the observed depreciation is sometimes determined by outside engineers, as is done in the case of reproduction cost. The trended value less the observed depreciation becomes then a close approximation to reproduction cost less observed depreciation.

A regulatory body may take into account all these elements of value and then arrive at its own opinion as to what is the proper and fair value of the property to use as a base for the establishing of rates. Before finally determining which method or combination of methods it will use in determining fair value, the commission will hear expert witnesses on the cost of money. The commission may be inclined to give most of the weight to net original cost as the rate base. But this value may result in insufficient net revenue to attract needed capital from the market. If this is so, the commission's views on proper rate base may be influenced in some respects by its views as to what is a proper return.

**Values for Rate Making**   For many years the courts have held that the value of the property and equipment used was the basis for determining the allowable return. As the commissions must follow the rules laid down by the courts, there are many commission decisions which state that value is the proper basis of return.

Frequently courts and commissions, instead of using the term *value,* have held that utility companies are entitled to earn a return on the *fair value* of the property devoted to public service. In the leading case on this subject (*Smyth* v. *Ames*), decided in 1898, the Supreme Court of the United States said:

> The basis of all calculations as to the reasonableness of rates to be charged by a corporation maintaining a highway under legislative sanction must be the fair value of the property being used by it for the convenience of the public. And in order to ascertain that value, the original cost of construction, the amount expended in permanent improvements, the amount and the market value of its bonds and stock, the present as compared with the

original cost of construction, the probable earning capacity of the property under particular rates prescribed by statute, and the sum required to meet operating expenses, are all matters for consideration, and are to be given such weight as may be just and right in each case. We do not say that there may not be other matters to be regarded in estimating the value of the property. What the company is entitled to ask is a fair return upon the value of that which it employs for the public convenience.

In the Natural Gas Pipeline case in 1942 the Supreme Court seemed to abandon its former position in the field of public utility regulation. Upholding a rate-fixing order of the FPC, the Court said:

> The Constitution does not bind rate-making bodies to the service of any single formula or combination of formulas. Agencies to whom this legislative power has been delegated are free, within the ambit of their statutory authority, to make the pragmatic adjustments which may be called for by particular circumstances. Once a fair hearing has been given, proper findings made, and other statutory requirements satisfied, the Courts cannot intervene in the absence of a clear showing that the limits of due process have been overstepped. If the Commission's order, as applied to the facts before it and viewed in its entirety, produces no arbitrary result, our inquiry is at an end.

In 1944 in the Hope Natural Gas Company case, the Supreme Court swung even further from its former position when it said: "Under the statutory standard of 'just and reasonable' it is the result reached, not the method employed, which is controlling. It is not theory but the impact of the rate order which counts."

In recent years, most states have given the greatest weight to net original cost. As this is about the lowest rate base, and as there has been a regulatory lag, utility earnings have been depressed, with the result that financing has become difficult. This does not appear to square with the Supreme Court statement, "It is not theory but the impact of the rate order which counts."

Many scholars have different views as to the proper values for rate-making purposes, and these views vary among the states. There is no specific formula for finding the answer to these complicated matters. The process requires wisdom and judgment on the part of the regulatory bodies which analyze each case on its merits after hearing all the appropriate facts and opinions on each case.

**Return**  Having found the value of the property for rate-making purposes, the commission then usually sets out to find the proper or fair return to be allowed on that value. Commissions try to find a return which is fair to existing investors, will enable the company to attract the required new capital in the free market, and enable the company to maintain a good credit position.

The commission will examine values of securities listed on the stock exchanges. Very probably an expert in the market will be asked for an opinion as to what is a fair return, and the expert may also be asked what the minimum return should be in order to enable the company to raise money in the free market for expansion to meet the customers' demands.

After weighing all these factors, the commission reaches a decision, states the value of the property for rate-making purposes, and sets a certain percentage as a return to be allowed on that value. This percentage multiplied by the valuation results in the earnings or return which the commission has allowed.

Having thus set a figure for these earnings, the commission then orders an adjustment in rates which will produce these earnings. If the earnings set by the commission are higher than the company is currently making, a rate increase is in order and the company is asked to prepare new rate schedules which will produce the higher revenue. If the earnings allowed are less than current earnings, the company is asked to file new rate schedules lower than those currently in effect.

During the 1930s and early 1940s the Edison Power Company, like most of the electric utility industry, was reducing its rates on a voluntary basis. During World War II the company was unable to buy new equipment, and in the period following the war it experienced higher rates of inflation. Consequently it could not realize the normal benefits of economy from installing new machines, and a rate increase was inevitable. As the company began to feel the effect of the new larger and more efficient generators, of increased sales, and of the greater degree of interconnection and coordination, it was able to resume the pattern of rate reductions.

As noted in previous chapters, the enormous change in the economic climate, which began during the period 1968–1970, has made it necessary to almost continually raise rates, with a rate case almost every year. This trend is expected to continue at least through 1985 and probably beyond.

## REGULATORY PROCEDURE

The Edison Power Company continually files with its regulatory body all its appropriate financial statements and operating reports. Thus the regulatory body has a continuing knowledge of all the company's operations and procedures. The commission is aware of proposed plans for construction, for it must issue appropriate certificates showing their convenience and necessity. At any time the commission can, and does, ask for pertinent information and receives it.

Any time the company feels a need for a change in any pricing schedule, it must have commission approval and show justification for making the change. The change requested may be for a reduction in rates, in which case a formal proceeding may not be necessary. If the commission feels a rate reduction of substantial amount is in order and the company questions the advisability of the amount of the reduction, a formal rate proceeding might ensue. If the company feels that a rate increase is in order, a formal proceeding would be required.

In such a formal proceeding the company presents its whole case in testimony before the commission. Management people and appropriate experts testify. The company may employ outside experts to give independent judgment as to value of property, justification for the proposed change, and rate of return. The witnesses may be cross-examined by the commission counsel with the advice of the commission staff. Interested parties may intervene, submit testimony, introduce expert witnesses, and cross-examine the company's witnesses. Often the proceedings go on for weeks and months while the commission weighs all the evidence and issues an order giving its findings and conclusions. These may be accepted by the parties concerned or reviewed by the courts.

All the company's records of revenue and expenses are open to inspection by the commission. The expenses are reviewed by the commission as a matter of regular routine and specifically inspected in rate cases.

State laws vary and the practices of state regulatory bodies vary in the treatment of some of these expenses. Among the items frequently debated are the treatment of taxes in connection with (1) accelerated amortization, (2) liberalized depreciation, (3) investment tax credit, and (4) guideline depreciation.

**Construction Work in Progress (CWIP)**    During a period of expansion, as at present, construction work in progress becomes a sizeable item. A power plant may cost $1 billion and be under construction for 10 years. There are two views on accounting procedure.

(1) CWIP is not included in the rate base, the company earns no return on the capital representing construction, and the company is allowed as an expense the interest on the money borrowed to carry on the construction. The plant investment is included in the rate base when the plant goes into service.

(2) The CWIP is included in the rate base as construction progresses. The company is allowed to earn a return on the capital representing plant under construction. There is no credit for interest during construction.

Plan 1 results in rather large accumulations of interest during construction, which may reflect in the books a distorted position of the company. Also, there is a sudden large increase in book value and rate base the year a large power plant goes into service.

For these and other reasons, in recent years, more and more commissions have been allowing CWIP to be included in the rate base.

## DOES THE SYSTEM WORK?

Described in Chapter 6 were the two basic systems that people employ to get their work done when living in societies. One is the free society and the other the government planned economy. The system, as applied to electric utilities in America, provides (for about 80 percent of the business) that the invested capital be obtained from the free market. There are millions of individual investors who are direct owners. Almost everyone who has an insurance policy or a savings account or who is in a pension plan is an indirect investor. The people are free to invest or not to invest. Furthermore, they can invest their savings in electric utilities or other businesses or investment trusts. Government acts as the regulator of the price of service to the public. By statute, the regulatory body is empowered with the authority (1) to prevent duplication of property, (2) to ensure a fair price to the public, (3) to ensure reliable and adequate service, (4) to ensure that all the needs and wants of the people are met, (5) to ensure an adequate return to investors so that they will continue to put their savings in electric utilities to enable the companies to build the necessary plants, and (6) to avoid unjust discrimination.

Has the system worked? Generally speaking, yes: (1) there is almost no duplication; (2) the price record compared with other commodities and services has been excellent (see Charts 4.8 and 4.9); (3) the reliability record has been outstanding; (4) all demands for service have been met; (5) by and large, except for recent years, the investors have been treated fairly; and (6) unjust discrimination has been avoided.

**More Recent Experience**   In recent years difficulties have arisen with respect to the regulatory process. Here are the principal reasons:

1.   For practically all the years prior to 1968, the public and the regulatory bodies had been accustomed to a declining average price of electricity in contrast with a rising price of practically all other commodities and services.

2.   It was generally known that the electric utility business was one of constantly improving efficiency in that practically every new power plant built could produce electric energy at a somewhat lower price than the previous ones. These improved efficiencies of converting energy coupled with the improved efficiencies in overall operation enabled the companies to lower the price when inflation was roughly 2 to 3 percent

per year. Since 1968, improvements in the efficiencies have continued (although to a lesser degree), but these improvements could not possibly offset the other increases in the cost of doing business, including inflation, cost of money, unusual environmental pressures, difficulties in getting approvals to build plants and to locate them, and the unusual increase in the cost of fuel. All these factors made it necessary to increase prices. These increases will not be for any one year, but must be on a continuing basis—as long as the incremental unit cost of building plant is higher than the average unit embedded cost.

3. During the period when prices were going down, new rate schedules could be based upon the rate base of the previous year with reasonable assurance that the coming years would result in better earnings or still lower prices or both. Investors' confidence was sustained. They were assured satisfactory earnings for future years. Since 1968, the situation has completely changed. No longer can last year's rate base be counted upon to be a satisfactory base for the establishment of rates for next year or the year after. Any rate schedules (except a schedule of schedules, as mentioned in Chapter 7) designed to provide a fair return on last year's rate base cannot possibly provide a fair return on next year's rate base when it is known that practically all unit costs will be higher next year. This effect is part of what is termed *regulatory lag*.

4. The result of this dilemma is illustrated in Chart 7.18. Regulatory bodies have known that the cost of money was rising. Actually, on the average, they have been allowing a percent return somewhere near the bare cost of money. However, as they were basing their calculations on a previous year's experience, and as they did not sufficiently account for the appropriate rate base, the effect has been that the utilities experienced an inadequate rate of return.

**The Results**    The results have been damaging to the utilities, the investors, and the public.

1. The bond rating agencies have downgraded the ratings of the bonds of many companies. This lowering of the ratings has caused the cost of new bond money to utilities to rise. As the return allowed by the commissions is largely based upon the cost of money, this means that electricity rates to the public, in the long run, will be higher than they would be otherwise.

2. As this regulatory practice discouraged investors during the period 1967–1974, the utilities faced difficulty in financing. In the public interest, regulation should encourage investors. Furthermore, the cost of capital has been higher than it would have been if the proper percent return had been allowed.

3. In many cases the market value of the utilities dropped con-

siderably below book value, although there has been some recovery since 1974.

4. A number of companies were required to cancel the building of power plants that had been in the construction schedule. To be sure, many of these actions were postponements, and not true cancellations. Also, some of the postponements were caused by a slowdown in the rate of growth of the economy in 1974 and 1975. Nevertheless, the financial difficulties had their influence. Such slowdowns are against the public interest for a number of reasons. Jobs, wages, and income per family are dependent upon continued increased production and productivity, which in turn are dependent upon an adequate supply of energy. Unfortunately, as it takes 8 to 12 years to build a power plant, the results of this problem may not be felt by the public and industry for some years to come. In recent months, some correction has been made in the regulatory lag.

**Remedies** A number of remedies have been suggested. Some have been put into practice, with the result that in more recent months improvements have been experienced.

1. The use of a future rate base is almost essential. During a period of rising prices, when it is generally realized that the price of electricity service will probably continue to rise for some years to come, it is axiomatic that the best way to provide the proper return is to use as a rate base the year or years for which the rates are designed. Obviously, this involves some forecasting. It means estimating the future years. However, these estimates, especially as they apply to the electric utility business, are more accurate than most people realize. Witness the results of the forecast described in Chapter 7. With the large capital investment per dollar of revenue, fixed charges on plant investment are equal to something like 40 percent of the utility's gross revenue. The plant investments are committed 5 to 12 years in advance. They are fairly well known. Fixed charges can be calculated with a fair degree of accuracy. Labor may be about 15 to 18 percent of gross revenue. Most of the balance goes to pay for fuel. This variable is usually covered by a fuel clause. It is possible to calculate the price of the energy by years for about 5 years in advance with a fairly high degree of accuracy.

The regulatory body need not take undue risk in basing the price upon a future rate base and an estimate of the future. If necessary, a condition could be contained in the order respecting the return on common stock. If the future earnings go above some predetermined allowance on common stock, the utility could be required to reserve these earnings. If the future rates fall under some predetermined common-stock allowance, the company could apply for appropriate adjustments.

The FERC and about half the state commissions have established a policy of basing the rates upon some form of future rate base.

2. The New Mexico state commission has adopted the concept of "cost of service indexing." An allowable range of 1 percent in the rate of return on the utility's common equity was fixed. Then a comprehensive adjustment clause was instituted, under which, if a utility's return is below the allowed range, it can adjust its rates upward to the lower end of the range, while if the utility's return is above the allowed range, it must adjust its rates downward to the upper end of the range. Future rate cases for the utility are necessary only if the return on equity needs to be changed, rate structure needs to be revised, or some similar problem arises. As the utility can raise its return above the minimum allowed only by achieving greater efficiency and economies, the system has a built-in incentive for the utility to resist cost increases and effect economies. At the same time, the plan effectively eliminates most of the problems caused by regulatory lag.

3. Regulatory lag can be shortened. The average regulatory lag before state commissions in 1971 was 9.5 months. This average increased to 10 months in 1972 and to 10.5 months in both 1973 and 1974. In 1975, it dropped to 9.7 months; increased again, to 12.1 months, in 1976; and declined once more to 9.7 months in 1977. Unfortunately, the average state commission regulatory lag for the first half of 1978 has again risen, to 11.7 months. Many rate cases are repetitious. The time required for such cases can be shortened appreciably by a concerted effort, which has been made in the past in some jurisdictions, especially during 1975. Sadly, as the average regulatory lag for 1978 indicates, such improvements in the regulatory process tend not to be permanent.

4. One method of minimizing the deleterious effects of long regulatory lag is the use of interim increases, put into effect while the rate case is still in process. Such increases have been allowed in recent years in 45 states. Often, a utility is allowed by law to put an increase into effect automatically if a commission has not acted on its rate request by the end of a specified period. In other states, a commission order is required. Often, such rates are subject to refund if their amount exceeds the increase finally held allowable by the commission. Interim increases were allowed in 27 percent of the cases decided by state commissions in 1973. This percentage increased to 38 percent in 1974 and 50 percent in 1975, then decreased to 35 percent in 1976. In 1977, the percent of cases in which interim increases were allowed fell further, to 20 percent, but during the first half of 1978 the percentage increased to 30.4 percent.

5. Many companies realize that education may be an important factor in bringing about an appropriate remedy. Very complicated economic and financial considerations need to be simplified so that they can

be readily understood by the regulatory bodies and the general public. With better public understanding, it is probable there will be greater public acceptance of the necessity for the increases. The customers can judge to what extent they want to avail themselves of the service at the price necessary to provide the facilities. With this better understanding, there would probably be less public objection.

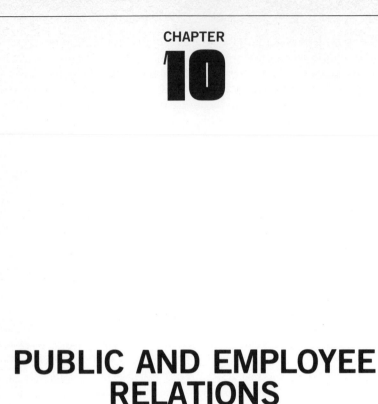

CHAPTER

# 10

# PUBLIC AND EMPLOYEE RELATIONS

## IMPORTANCE OF PUBLIC AND EMPLOYEE RELATIONS

In a modern corporation operating in an industrialized society, public and employee relations are managerial functions ranking in importance with engineering, accounting, and finance. In most large companies in the United States and Europe these functions have been in operation for many years. They are still relatively new compared with some other functions, but they have grown in stature so that the heads of these functions are usually officers and frequently members of the board of directors.

In some companies both of these functions are dealt with under one vice president. In other companies there is one department head who handles public relations and another who handles employee relations. As these two functions have something in common, they will be treated jointly in the early part of this chapter. Then they will be treated separately.

In both functions, management is dealing with people—their hopes, desires, beliefs, attitudes, knowledge, and ability to understand. These things have to do with the behavior of people, and they vary over a wide range. The behavior of people is affected by: (1) how the company functions; this will be referred to as *functional public relations* and *functional employee relations;* and (2) knowledge of company affairs. There is a direct correlation between knowledge of company affairs and functions, and attitude toward the company. To a large extent management has control over its functions relating to both the public and the employees. Also, to a large extent, it has control over how well the customers and employees are informed respecting the company's affairs and its problems. Experience has shown that there is much that management can do to build and to hold good relations with the public and the employees.

As touched upon briefly in Chapter 6, the sampling of opinions of people has been developed to the point where it is almost a science. By dealing with a very small, well-balanced sample of people, it is possible to measure their attitude and their knowledge. These measurements serve as guides to the establishment and the carrying out of well-designed public- and employee-relations programs. The programs call for doing specific things regarding the functions of the company as they relate to the attitude of customers and employees. Periodic measurements are made of the progress being made.

Similarly, specific programs are designed for public and employee information, dealing with those facts where ignorance has a bearing on attitude. Management's aim in dealing with the public and employees is (1) to analyze the problem, (2) to draft a specific plan, (3) to test the plan, and (4) to go ahead with the plan decided upon. Once a program is agreed upon, management then allocates sufficient money and personnel to carry out this function.

As public and employee relations are functions that are somewhat more nebulous than engineering, construction, or accounting, management sometimes does not allocate sufficient resources to carry out the specific program needed. At times, management may issue a news release when it wants to inform the public of certain facts. Such a release may be read by a very small percentage of the people. If it is desirable to have all employees or all customers know, understand, believe, and retain a fact, it is necessary to go through a certain procedure, just as the construction engineer must go through a certain procedure in building a power plant.

## GENERAL DESCRIPTION

Every function of the company that touches on a customer or an employee leaves an impression for good or ill. Every contact by person, letter, or advertisement with a customer or employee makes a positive or negative impression. Management's aim is to examine all functions with the idea of doing more of those things that are pleasing to people and fewer of those things that are displeasing.

Good management carries on continuing measurements of attitude and knowledge of customers and employees to determine those areas of ignorance having a bearing on attitude. "A person's judgment on any subject is no better than that person's knowledge." Management carries on a continuing program of public and employee information, always attempting to dispel ignorance which has a bearing on attitude. Continuing measurements are conducted to determine the number of people who have heard, believed, and understood the facts as presented. Knowledge on the part of the public and employees breeds understanding. Understanding breeds trust. The essence of good public relations is the building of trust.

The growing recognition of public relations as an essential part of a company's process of doing business is due in large measure to the fact that many of the techniques involved in public- and employee-relations practices depend heavily on modern methods of communication. With the limited techniques and avenues available before the turn of the century, little was done in this field. Furthermore, businessmen during the period of industrial infancy were generally too busy with other aspects of the business to concern themselves with the attitude of the public and employees toward them. With few exceptions, they had no accurate means of knowing what that attitude was.

Electric utilities have a special stake in good public relations. The utility is sometimes described as being "vested with public interest." The service rendered is so essential to modern living that the utility is required by law to furnish adequate service to all who request it within the service area, and to furnish service without discrimination. The utility has practically the entire public as customers. Furthermore, the customers can go nowhere else to buy the service. If they want it, they must take it from the company serving the area, whether or not they are displeased with management. Thus, to a unique degree, the electric utility is dependent upon favorable public opinion.

**Definitions**   There are a number of definitions of public relations. *Public Relations News* describes it as the management function which evaluates public attitude, identifies the policies and procedures of the individual or an organization with the public interest, and executes a program of action to earn public understanding and acceptance.

Another definition is this: the function which, through positive action based on measurements, strives to so operate the company and so inform the customers as to cause the public to like the company.

Likewise, employee relations may be defined as: the function which, through positive action based on measurement, strives to adopt and carry out policies and to so operate the company and to so inform the employees as to cause good employees to want to work for the company and new employees to want to join the company.

It cannot be overemphasized that the key to good public and employee relations lies in the hands of the chief executive and the principal people making up the managerial team. Obviously, management employs experts to carry out these functions. However, management lets it be known by all that it is taking a direct interest in these affairs and is following the progress of the various programs.

The chief executive's policies and approach inevitably will be reflected by the members of the organization in their contacts not only with customers but also with each other, thereby affecting the entire character of the utility. Administrative confidence, engineering excellence, sound planning, and other attainments may go unappreciated both by the public and by employees themselves if there is an appearance of indifference at the top level.

Some companies have experimented with having the various departments handle their own public and employee relations. With few exceptions, this has proven to be unworkable and ineffective. The lack of uniformity and the poor coordination that result from this setup are undesirable under any circumstances. Obviously, the functions of public relations and employee relations are inevitably carried on through all the various departments. It is necessary, however, to assign one or two departments with specific responsibilities to see to it that the functions are carried out, just as it is important to have a department in charge of engineering. The benefits that result from having someone in charge of these activities outweigh any advantages to be found in a diversified operation.

Any and all contacts or relations a company has with people—customers, employees, or stockholders—are part of the fabric of public relations. Extending like the roots and branches of a tree, they are varied and sometimes not immediately apparent, but they all deserve the attention of management and the public relations executive. One of the chief barriers to an effective and all-encompassing public relations program is the tendency on the part of company personnel to concern themselves only with the immediate and superficial aspects of their primary jobs.

The meter reader who thinks his or her sole job is to read meters, to the exclusion of the many other means of serving the customers and gaining goodwill, shows a misunderstanding of the utility's aim and urgently needs correction. Here is a situation where "if the pupil hasn't learned, the teacher hasn't taught." Tact, ingenuity, skill, and perseverance are required to make all company personnel broaden their outlook so that the forest, not just a particular tree, can be seen. When that is accomplished, a sense of corporate unity, purpose, and cooperative action will be achieved, together with awareness of the public relations importance of individual action.

Many of the specific problems facing management stem to a large extent from necessary price increases and the effect of these on public attitude. Today, the problems involving increased prices and environmental concerns weigh heavily upon management. Management now asks itself: Has the company prepared a program to take the public into its confidence concerning everything the company is doing in its endeavor to supply service to the public? To make it reliable? To restore the service after an interruption? To economize so as to keep the price as low as practical consistent with the value of service to the public and with the reliability required? Does the public really understand that management must commit itself 5 to 12 years in advance to the building of the power plants so essential to meet the needs and desires of the public? Has the company reminded the customers of the relatively small percentage of the family budget that is required for electric service? Has the customer been reminded of the many services that are rendered in the home for the convenience of the customer? Has the customer been continually reminded of the value of the service to well-being, comfort, and convenience? When people are uninformed and are dissatisfied for any reason, they tend to think of the price as being exorbitant.

Management is now asking itself whether the customers are fully informed as to how electric service has already helped materially in cleaning up the environment. Are the customers fully informed as to all the things that management is doing to help further improve the environment? Do the customers understand the cost of performing each of the environmental functions and that this cost must of necessity be borne by the customer? Have the customers been so informed that they can reach a judgment as to whether the benefits of certain further improvements justify the costs?

## STEPS TAKEN IN ESTABLISHING A PUBLIC RELATIONS PROGRAM

1. Management first establishes the desired objectives that it wants to accomplish. These include short-term, immediate, and long-term objectives. They are set down concisely and clearly so that there can be no misunderstanding of the intended goals. In setting up these goals or objectives, the aim is to have each one benefit the public and win public approval. These are goals toward which employees will want to strive. Further, it should be possible to state each objective in terms of the self-interest of the various segments of the public, including customers, government, investors, employees, and management.

2.   Next, management takes stock of its present position. After the objectives have been formulated, the present policies and practices of the company are critically reviewed to determine whether each is an aid or a detriment toward the accomplishment of goals. To the extent possible and practical, those policies not in the spirit of the new objectives should be discontinued. New policies and practices, in keeping with the aims, should be established.

3.   In the third step, management determines and evaluates, by measurement, current public attitudes toward the company. Simply stated, this is a matter of finding out what the public likes about the company, and also what it dislikes. This can best be done by means of a survey, preferably made by an outside organization. The survey should be well designed and conducted with great care.

Attention is given to employee-attitude surveys as well. Employees are the source of a large part of the information which the public receives. It has been estimated that each employee has some kind of contact with roughly 50 to 60 families. To many of these families or to most of them, the employee, whose knowledge and behavior have a bearing on what the public thinks, *is* the company.

A detailed study of the survey results is made in order to derive the most value from them. The results will probably show that, as is the case with a large number of utility customers, the public is not well informed about the business. Probably there are wide areas of misinformation. Some company policies may not bear the stamp of approval of employees or customers. The interpretation of the survey results indicates the direction in which action must be taken to emphasize and expand good practices, and to eliminate or minimize those which are bad. It may be necessary to continue some practices even though the survey reveals they are disliked by the public; this then becomes a problem of explaining why they are necessary.

4.   The development of a program or plan of action is the fourth step. Probably no organization does all it could do for better public relations, as there are always limiting factors such as time, money, and personnel available. A definite plan of action helps to ensure that efforts are expended first on goal-directed activities certain to strengthen goodwill, rather than on non-goal-directed activities, the value of which may not be clear. The plan includes broad details of the various programs, the different segments of the public toward which these programs are directed, the means of communication, and the points of contact to be used in executing the programs. A manual of policies and practices is prepared so that everyone concerned is completely informed and there is a ready reference to the planned activities.

5.   Step 5 is the securing and training of personnel responsible for

carrying out the plan of action. In a well-organized company there is a vice president in charge of the public relations function, which includes advertising, equal in rank to any other officer of the company below the president or managing director. Even the best program will have only limited success unless it is administered with skill and enthusiasm. Frequently found in this same department is the function of employee relations. But in many companies, there is in addition to a vice president in charge of public relations, another vice president in charge of employee relations. If the function of dealing with people is centered in one individual, that person must have highly skilled and broad-gauged department heads responsible for public relations and employee relations. Each of these areas is broken down into two or more parts, such as functional public relations and informational public relations, functional employee relations and informational employee relations.

In this well-organized company there is a public relations committee, composed of top-management representatives from each of the major departments. The vice president in charge of public relations serves as head of this committee, which is responsible for formulating public relations policies. The company chief executive sits in on the meetings.

6.   The sixth step is a periodic review of public attitudes and of a program of action. The program is revised whenever the analysis shows this to be advisable. Situations change. What may once have been a good program may produce different results a year later.

## OUTLINE OF PUBLIC RELATIONS

Because public relations and employee relations are closely related, the following brief and extended outlines cover some aspects of both, as they would apply to a typical company.

### BRIEF OUTLINE
A vice president–board member is in charge of the whole activity. Under this executive fall the following functions:

1.   Public relations
    a.   Functional
    b.   Informational
2.   Employee relations
    a.   Functional
    b.   Informational
    c.   Selection and training
    d.   Salary administration
    e.   Union negotiations

3. Investor relations
   a. Functional
   b. Informational
4. Government affairs
   a. National
   b. State
   c. Local
5. Public affairs
   a. Environment
   b. Civic affairs
   c. Conservation

## EXTENDED OUTLINE

I. Public relations
   A. Functional: one person in charge full time, with the following duties:
      1. Serve as head of a committee of all department heads, with the chief executive officer taking part.
      2. Do periodic public opinion surveys to determine customers' attitude—identify customer likes and dislikes. A programmed effort is made to do fewer of those things not liked and more of those which are viewed favorably.
      3. Bear in mind that every contact between company and customer makes an impression for good or ill. *All* company functions having a bearing on attitude are listed and examined through the committee of department heads. The list is practically endless, and includes:
         a. Telephone demeanor
         b. Courtesy of *all* employees: training in courtesy, courtesy of letters
         c. Handling of complaints
         d. Cutoff notice and policy
         e. Deposit policy
         f. Penalty policy
         g. Applications for service
         h. Language of service agreements and contracts
         i. Clarity and simplicity of rules
         j. Tone of letters and advertisements
         k. Noise
         l. Conditions of company equipment
         m. Tree trimming
         n. Neatness of meter readers
         o. Training of drivers to rigidly practice courtesy and safety
      4. Set goals to be checked by measurement of the number of customers who say, "nothing," when asked, "Please name those things you don't like about your electric company and its service." Good managements have attained responses as high as 92 percent of the customers saying "nothing."

B. Informational: One person is in charge full time, with the following
   duties:
   1. Periodically measure public opinion to determine customers'
      a. Attitude
      b. Knowledge
   2. By trial and error, develop test questions which disclose whether
      customers have good or bad attitudes toward the company and
      the reasons for such attitudes. To test the reliability of the test
      questions, first ask them of one or two hundred people known to
      have a bad attitude and then put the questions to the same num-
      ber of people known to have a good attitude. Once the pretested
      questions are predesigned to disclose these two attitudes, these
      same questions are asked of a representative cross section of all
      customers. After each question relating to attitude, ask "Why?"
      Other questions have to do with fact, not opinion, about the com-
      pany.
   3. In analyzing the survey, separate all responses indicating bad atti-
      tudes from those exhibiting good attitudes. Determine the num-
      ber and percentage of people with bad attitudes who could iden-
      tify *facts* correctly, as compared with the number in the favorable
      category. This kind of research usually discloses that those peo-
      ple with unfavorable opinions show a high percentage of igno-
      rance as to the facts. The people with good attitudes usually score
      high in answers to questions of fact.
   4. This kind of research will disclose *what* facts to communicate to
      people in order to create good attitudes. There is no need to
      waste money presenting facts already known. Communicate facts
      which customers are unaware of and which have a bearing on im-
      proving attitudes.
   5. Prepare informational programs designed to reach each segment
      of the overall public:
      a. Employees
      b. Opinion leaders
      c. General public
      Different media and techniques are used in reaching the various
      segments of the public.
   6. Always pretest all communications before spending money on
      wide use. Select 100 people at random, or a balanced cross sec-
      tion. Give them a proposed advertisement or piece of literature
      and ask them to read it. More often than not, these pretests show
      that the readers do not understand the message, for which they
      should not be blamed. The purpose of communication is to com-
      municate, which means it is necessary to use language which the
      reader demonstrably *can* understand—otherwise the money is
      wasted. The communication should be (a) heard (or seen), (b) un-
      derstood, (c) believed, and (d) retained. Always measure the per-
      centage of the public who hear, see, or read the communication.

If an ad is read by 10 percent of the people, it must be run 10 times to reach all the customers once.

7. After 3 or 4 months, again make a survey to measure the retention of the information communicated. Research shows that in most cases a fact must be seen or heard or read about four times for it to be retained for any length of time. If the purpose is to communicate a fact, to have it reach all the people and be retained, put in the budget the amount of money required to carry out the purpose. Management will decide whether or not it wants to inform the public on this or that fact, but be sure the money is appropriated to do what management wants.

8. Periodically call on all editors and top executives of magazines, newspapers, and radio and television stations. Particular effort should be made to keep these highly important opinion leaders thoroughly informed on all company facts, policies, and plans.

9. Carry on research to determine relative cost "per recall" and efficiency of various means of communication, such as television, radio, direct mail, bill inserts, newspaper ads, magazine ads, releases, and speeches.

10. Set measurable goals as to the percent of the customers desired to have knowledge of certain facts in a certain length of time. Prepare required budget.

11. Judge performance in public relations by *measurement* at periodical intervals.

II. Employee relations
   A. Functional: One person is in charge full time, with the following duties:
   1. Screening new employees
   2. Recruitment
   3. Interviewing graduates
   4. Job training
   5. Job specifications
   6. Job classification
   7. Work rules
   8. Salary administration
   9. Labor negotiations
   10. Grievances
   11. Employees' information courses
   12. Vacation policy
   13. Sick leaves
   14. Working conditions
   15. As in public relations, make measurements of attitudes and knowledge of employees to determine those things liked and disliked about the company.
   16. As in public relations, hold periodic meetings of department heads to discuss all employee dislikes, and make an effort to reduce the causes of dissatisfaction. As to those things employees

do like, make an attempt to increase or improve them. Make a list of all company functions and policies that might have a bearing on the attitudes and behavior of employees. These include:

   a. Working conditions
   b. Fairness of pay scale
   c. Fringe benefits
   d. Vacation policy
   e. Attitude of supervisors
   f. Lighting conditions
   g. Indoor climate conditions
   h. Noise level
   i. Relative intelligence of supervisor as compared with supervised
   j. Handling of complaints
   k. Handling of grievances
   l. Promotion policy
   m. Washroom conditions
   n. Transportation conditions
   o. Medical examination
   p. Health treatment
   q. Lines of communication, etc.

B. Informational: One person is in charge full time, with the following duties:

1. As with public relations, make measurements of employees' attitudes and knowledge of facts. Design and pretest test questions to disclose the number of employees who have unfavorable attitudes and behavior, as distinguished from those having good behavior and attitudes. The survey also asks questions of fact about the company. To aid in getting correct readings of attitudes, ask employees *not* to sign their names. No employees are identified.

2. As in the public surveys, separate the employees into groups of "bad behavior" and "good behavior" and analyze the knowledge of each group. Again, it will probably be found that those employees having bad behavior and attitudes are largely uninformed as to the facts. Those of good behavior are better informed. These measurements disclose the nature of the informational program that should be carried on among the employees.

3. Design a comprehensive employee information program. Communicate those facts, the ignorance of which has a bearing on attitude. Include the following course information:

   a. How Americans improve their living standards
   b. The use of machines, energy, people, capital, and material
   c. The electric energy business
   d. Company affairs
   e. Special current subjects such as: the environment, the energy crisis, conservation, regulation, and nuclear power

4. Divide the employees into groups of 25. Research shows that "group discussion" is by far the most effective means of communicating. Put each group of 25 under a trained conference leader, of a level no higher than an immediate supervisor. Give the leader a 2-week training course on the material to be covered and how to conduct a conference.

5. Conduct these conferences on company time, an hour a month— forever. They are a way of life in a well-managed company. By the time one group is informed, there will be a turnover in employees, meaning a new group must be schooled.

6. Set goals for the percent of employees showing good behavior and good attitudes and the percent having the required knowledge on each important fact.

7. Budget the required money. Consider the program satisfactory when the goals have been reached—as shown by *measurement.*

Selection and training, salary administration, and union negotiations, vital though they are, will not be discussed in this brief treatment. The aim here is to emphasize communication needs, which too often are given short shrift in employee relations.

III. Investor relations

In this area, many of the basic principles of public relations apply. Investors are a special audience of great importance to the ability of the electric utility to meet its responsibilities. Although space limitations do not permit a discussion of the particular techniques which are used to keep shareholders and potential investors well informed, they are worthy of careful consideration by management, and should extend beyond quarterly letters and annual reports.

IV. Government affairs

Government affairs is an activity of considerable complexity, involving as it does the interaction of administrative, legislative, and regulatory considerations at all levels of government—national, state, and local. The subject is mentioned here only to highlight its increased importance to the success or failure of utility enterprises. Management should develop the governmental-affairs area with great care and attention, to ensure the highest possible effectiveness of its efforts in each of these arenas.

V. Public affairs

A. Environment

1. Functional

a. Carefully follow all functions of the company having any effects for good or ill on the environment. List all of these.

b. Within practical economic limits and where the benefits justify the cost, strive to do all things necessary and advisable to maintain a clean, healthful, pleasant environment. Weigh the demands and needs of the people for good and improved living standards (requiring more machinery run by energy and, therefore, the burning of fuel) on the one hand, and the demands for pure environment on the other hand.

c. On the major issues, survey the public. Tell the customer the cost and the benefit. Ask for a choice.

d. Check company vehicles for pollution and noise.

e. Reduce noise where possible.

f. Carry on continuing research for ways to improve the environment.

g. Take an active part in community affairs.

2. Informational

a. Measure public attitudes and knowledge.

b. Inform the public of:

(1) All that has been done, is being done, and is planned.

(2) The company's contribution through fly-ash removal, better burners, and the conversion of inefficient, air-polluting engines of commerce and industry to clean electric energy.

(3) Playgrounds and recreation areas.

(4) Beautification of substations.

(5) Facts on water temperatures.

(6) Research.

(7) Benefits of nuclear power.

(8) Research on the electric vehicle.

(9) Air-pollution programs, including tall stacks, scrubbers, controls, etc.

(10) Cost impact vs. benefits, and alternative methods.

B. Civic affairs

1. Take an active leadership role in community affairs:

a. Urban renewal

b. School problems

c. Unemployment

d. Transportation

e. Fund raising

2. Inform customers of what is being done.

C. Conservation

1. Expand activity.

2. Tie into national, regional, and local programs.

3. Help customers to avoid waste.

4. Inform customers about company.

a. What has been done.

b. What is being done.

c. What is planned.

# RESEARCH

## THE VALUE OF RESEARCH

From the very beginning management considered it wise and essential to utilize research as an important element in the development, growth and improved efficiency of the electric energy business. The development and improvement of machinery play a large part in research activities. As mentioned previously, the electric utility business requires a capital investment of about $4 for each $1 of annual revenue. (The ordinary manufacturing business requires about 50 cents in capital for each $1 of annual gross revenue, or about one-eighth as much.) This investment ratio of the electric energy business is one of the highest of all industries. This means that there is an unusually large aggregate of machinery required to produce annual utility revenue. Machinery requires continuing research for its development and improvement.

Owing to the importance and magnitude of research activities, modern management usually assigns this function to a particular department. It is the purpose of this department to continually keep itself fully informed on all research matters—projects currently being conducted and those contemplated. All management should have a general knowledge of research. The purpose of this chapter is not to give a full treatise on research, but rather to offer a general summary.

A few of the values of research in electric energy supply are to:

- Ensure an adequate supply of economical fuel.

- Ensure the highest practical efficiency in converting primary fuel to electric energy and to operate the business economically. This improvement in efficiency is reflected in the price to the public and the earnings to investors.

- Help increase overall productivity in the national economy.

- Help raise living standards.

- Help create jobs.

- Develop new, better, and more efficient methods of getting a job done.

- Improve the availability and efficiency of the various appliances and applications used by consumers.

- Help improve the environment.

- Promote safety.

- Develop computer models to improve efficiency of operation and to assist in wise management decisions.

- Learn more and more about the complicated economic characteristics of the business so as to enable the development of sound policies in load management and pricing.

- Develop improved skills, methods, and tools of management.

- Develop improved and efficient methods of dealing with human desires, knowledge, attitudes, and behavior.

## EVOLUTION OF RESEARCH

In the early stages, research was carried on largely by individuals searching for new inventions and better tools and methods. The classic example is that of Thomas Edison in building the first electric utility company.

This feat required many phases of research carried on in his laboratory.

With the evolution of modern management, the research process has passed through evolutionary changes. Beginning in the early part of this century, research methods evolved into concentrated efforts by particular manufacturers and groups in developing new products and improving existing ones. Competition among the manufacturers in the United States stimulated this research. Each manufacturer endeavors to produce a superior, safer, and more efficient product so as to obtain and to hold the particular business. Manufacturers carried on most of the research during the earlier development years of the electric utilities, and the utilities purchasing the equipment closely followed and stimulated development. Each step upward in the size of the generators required additional exhaustive research. The utilities decided upon the size of the units that would best fit their method of operation. The same has been true in the development of higher voltages in transmission and distribution.

Prior to the years when the EEI became active in carrying on a research program, the Institute committees closely followed the research and development activities of manufacturers. Then as now, whenever a turbine or a relay or other equipment failed, a research project was formed to determine the causes and the improvements required. This has been a continuing process. Utility management has been motivated by the desire to provide electric service of the highest practical reliability at the lowest practical cost consistent with the cost of capital. Through its many committees, the electric utility industry has promoted the necessary research to aid in realizing this objective.

Has the process worked? How does one measure the efficacy and value of research? The record may provide a few hints. In Chapter 4 mention was made of a few of the primary improvements in efficiency which enabled improvements in reliability and lower price. A review of some of those steps may be appropriate from the standpoint of the research.

**Larger and Larger Generating Units**   Each step upward in size required managerial decisions as well as new research and development. With increased size came increases in temperature and pressure of steam. Efficiency goes up with a rise in temperature and pressure. Each step upward required research and development in metallurgy, valves, piping, controls, vibration, costs, and all the factors required to enable reliable commercial use under the new conditions.

**Increased Voltage in Transmission**   Each increase in voltage required managerial decisions accompanied by the appropriation of funds and years of research and development to enable the efficient and safe use of higher voltages.

**Interconnections and Pooling**  Continuing research has been necessary to enable the transmission systems to operate on an interconnected basis and in pooling arrangements. Further research was required to form the large coordinated areas now in operation throughout the country.

**Improved Load Factor**  For many years research has been conducted on the load characteristics of individual appliances, classes of service, and systems. This research has enabled more accurate and efficient load management so as to increase load factor. The research also enabled a wiser and more beneficial pricing policy designed to encourage the increase of load factor. Improved load factor—from about 50 percent in 1930 to about 65 percent in 1965—shows the result of this activity. Research has disclosed the diversity available among various appliances, among classes of service and among interconnected systems. In this way, research has enabled the companies to realize the benefit of practically all the diversities available. Further research can be expected to yield additional improvements.

**Fuel Conservation**  Possibly the best example of the results of research is found in the reduction in the amount of coal (or coal equivalent) to produce 1 kWh. This amount was about 8 lb in 1900. Today it is roughly $^2/_3$ lb of coal per kilowatthour. Present research may reduce this still further.

**Nuclear Energy**  The field of nuclear energy has required major, concentrated, coordinated research for many years, resulting in significant benefits to the public. Research on the breeder reactor is now going through the same process, although delays are being experienced in the United States. Fusion is the next step.

**Productivity**  Research contributes to the advance of productivity in all manufacturing. Chart 4.18 illustrates the improved productivity in the electric utility industry as compared with all manufacturing. Research was required for each step in this record of progress.

**Reliability**  Research has made possible the 99.98 percent "on" service record for the average customer during a 10-year period.

**Price**  Chart 4.9 shows the consumer price index for all business compared with the electricity component for the period 1930–1976. Managerial decisions to carry on the required research contributed in large measure to this record for the electric utility industry.

**Public Attitude and Behavior**   Chapter 6 (in part) and Chapter 10 discuss public, employee, and government relations. In dealing with human desires, beliefs, knowledge, attitudes, and behavior, management has many skills, methods, and tools at its disposal. The efficient and effective use of these depends upon research—different from that involving machinery—but important research nevertheless. The value of such research lies in good employee and public attitudes, so essential for successful management and the survival and progress of the enterprise.

## EXAMPLES OF THE NEW METHOD

With the idea of concentrating research efforts on accomplishing a specific objective, there was an increase in the entire research program. Here are but a few of the specific objectives:

- Control the nuclear reaction.
- Build a commercial, competitive nuclear power plant.
- Develop a 1 million kW turbine generator.
- Develop the breeder and make it commercially feasible.
- Develop a lightweight battery.
- Clean up the stack gases.
- Develop a transmission system of 1 million and 1.5 million V.
- Develop high-voltage underground cables.
- Control fusion.
- Convert coal to oil and gas.
- Develop the electric vehicle.
- Develop automatic meter reading.
- Develop computer models to assist in management decision making.
- Develop magnetohydrodynamics.
- Develop the large-scale fuel cell.

- Make solar energy reliable and economically feasible.

- Determine load characteristics of appliances.

These and other projects are put into a specific schedule and time frame.

## RECENT EVOLUTION

Prior to about 1960, most research in the electric utility business was conducted by manufacturers in cooperation with the utilities and by individual electric utility companies. In 1952 the EEI launched a coordinated research effort with the formation of the Research Projects Committee. In 1961 the EEI formed a research division for the purpose of stimulating and coordinating research. A survey was made among all the sectors of the business, including all the manufacturers, to determine the extent of their research in total volume—without disclosing any particular company's projects. An independent accounting firm was employed to conduct the survey. Simultaneously, a survey was made among all the electric utility companies to determine the specific research being carried on by them and the volume. This enabled closer coordination of research activities.

At that time, consideration was given to the wisdom of conducting all research of manufacturers and utilities through a central organization such as the EEI. The conclusion was that utility companies and the public benefited from the competitive research activities of individual manufacturers. Obviously, the electric utilities paid for the research in the price of the equipment. The EEI research committee acted largely as a catalyst and a coordinator to avoid unnecessary duplication.

About the same time, there was formed an international research organization to enable the exchange of knowledge on research among the various members of the international association. Shortly thereafter, in 1965, the EEI took the initiative in forming the Electric Research Council (ERC), which coordinated the research of all parties. This permitted contributions to research activities from government-owned utilities and some of the manufacturers.

In June 1971 the ERC issued a report entitled *The Electric Utilities Industry: Research and Development Goals through the Year 2000.* Table 11.1 shows a summary of the research and development costs to utilities, manufacturers, and government by major classifications of research through the year 2000. Notice that total research and development costs were estimated to be $32 billion, expressed in 1971 dollars. This was the estimated budget of the ERC, to be financed by utilities, manufacturers, and government.

**TABLE 11.1**  SUMMARY OF TOTAL R&D COSTS TO UTILITIES, MANUFACTURERS, AND GOVERNMENT
*(Millions of 1971 Dollars)*

| | 1972 | 1973 | 1974 | 1975 | 1976 | 1977 | 1978 | 1979 | 1980 | 1981 1985 | 1986 1990 | 1991 2000 | Total |
|---|---|---|---|---|---|---|---|---|---|---|---|---|---|
| Energy Conversion | 500 | 619 | 701 | 799 | 841 | 922 | 873 | 857 | 806 | 3875 | 4098 | 8430 | 23,319 |
| Transmission and Distribution | 110 | 123 | 147 | 157 | 156 | 160 | 160 | 167 | 168 | 913 | 1036 | 2474 | 5772 |
| Environment | 47 | 91 | 108 | 103 | 104 | 107 | 109 | 107 | 105 | 503 | 503 | 1006 | 2893 |
| Utilization | 8 | 8 | 8 | 8 | 15 | 15 | 15 | 15 | 15 | 67 | 64 | 128 | 369 |
| Growth and Systems | 2 | 3 | 4 | 5 | 5 | 5 | 5 | 5 | 5 | 22 | 21 | 51 | 133 |
| TOTALS | 667 | 844 | 968 | 1072 | 1121 | 1209 | 1162 | 1151 | 1099 | 5380 | 5722 | 12,089 | 32,486 |

*Source:* R&D goals task force report to the Electric Research Council, June 1971.

# THE ELECTRIC POWER RESEARCH INSTITUTE
## (EPRI)

During the latter part of the 1960s, many factors contributed to the desirability and necessity of further coordination and increases in research activities. Some of the contributing factors were:

1. Managements and others associated with the energy business had realized for a long time that gas and oil reserves were rapidly diminishing and that there was need for a national energy policy calling for stepped-up development of coal and nuclear fuels (see Chapter 13). However, only in recent years has the problem reached the consciousness of the Congress and the public. These shifts in use of primary fuels require large research activities that will be described later.

2. The whole economic climate changed around 1968, causing a reversal in the long-term downward trend in the cost of producing and delivering a unit of electric energy. Rate increases became essential. This called for stepped-up efforts in all areas in which any improvements in efficiency were possible.

3. As will be pointed out in Chapter 12, electric utilities began their efforts in meeting environmental problems during the early part of the century. However, beginning around 1965 to 1970 there developed a greater awareness on the part of government and the public of the importance of more active attention to all environmental matters. Electric utilities were called upon to greatly step up their research and performance.

When the ERC developed the outline of the required research effort to the year 2000, it became apparent that there was need for a new and expanded organization in research, supported by all segments of the industry including the government. The result was the Electric Power Research Institute (EPRI) which succeeded the Electric Research Council and assumed responsibilities for the research and development programs of the Edison Electric Institute and the Electric Research Council.

EPRI appointed Dr. Chauncey Starr as its first president on January 1, 1973. He was formerly Dean of the School of Engineering and Applied Science of the University of California, Los Angeles. The headquarters of EPRI are in Palo Alto, California. There is an office in Washington, D.C., to provide coordination with such parallel federal efforts as those in the Department of Energy, the National Science Foundation, the Environmental Protection Agency, the Nuclear Regulatory Commission (NRC), and others.

As of 1978 EPRI had about 1300 research projects under management or in contract negotiations. The total value of these projects is about $1,500 million. EPR's share of this amount is about $800 million.

# ELECTRIC POWER RESEARCH INSTITUTE OBJECTIVES

The major objectives of the Electric Power Research Institute, as set forth in the Articles of Incorporation, include the following:

- To promote, engage in, conduct, and sponsor research and development for electricity production, transmission, distribution, and utilization and all activities directly or indirectly related thereto.

- To provide a medium through which investor-owned, government-owned, and cooperative-owned power producers and all other persons interested in the production, transmission, distribution, or utilization of electricty can sponsor electricity research and development for the public benefit.

- To discover, through study and research, ways to improve the production, transmission, distribution, and utilization of electric power in order to insure the adequate power supply vital to the progress of the nation and the world community.

- To seek, through scientific research and development, solutions to the environmental problems related to the production, transmission, distribution, and utilization of electric power.

- To provide a medium for coordination and cooperation and for the exchange of information among all organizations, public or private, concerned with electric power research and development.

- To develop, prepare, and disseminate information and data on scientific research and development activities in the field of electric power.

EPRI does not make operating decisions for the utility industry or for government bodies. Its mission is to focus national utility R&D resources on continuously providing the technological options needed for insuring that future electricity demands can be met in a manner that best serves the overall public good.

The technical staff of EPRI is composed of four divisions:

- The nuclear power division

- The fossil fuel and advanced systems division

- The energy systems, environment, and conservation division

- The transmission and distribution division

## Funding

EPRI is supported by all segments of the electric utility industry, including investor-owned companies, government-owned agencies, rural electric cooperatives, the TVA, and the U.S. Department of the Interior. In most cases, the investor-owned companies participate through the EEI, which provides about 90 percent of EPRI's total funding. The government-owned organizations contribute through the American Public Power Association. The rural electric cooperatives contribute through the National Rural Electric Cooperative Association. There are more than 500 member organizations supporting the EPRI program.

## Industry Advisory Groups

More than 250 utility-industry executives and engineers serve on EPRI's industry advisory committees. The 21 members of the Research Advisory Committee work with EPRI's president and board of directors to identify R&D needs of the utility industry and to coordinate the activities. There are four divisional managers, corresponding to the four EPRI technical divisions.

## SUMMARY OF EPRI ACTIVITIES

The following is a brief outline of some of the principal research projects being accepted by EPRI.

1. Basic systems
   a. Nuclear plants
   b. Fossil fuel and advanced system objectives
   c. Fossil fuel
      (1) Low- and intermediate-Btu coal gasification
      (2) Coal liquefaction
      (3) Direct utilization of coals
   d. Environmental control technology
   e. Basic research and resource extraction and utilization
2. Advanced systems
   a. Electrochemical energy conversion and storage
   b. Electromechanical energy conversion and storage
   c. Fusion
   d. Solar energy and geothermal energy
3. Transmission and distribution objectives
   a. AC overhead
   b. AC underground
   c. DC systems
   d. System planning, security, and control objectives
4. Energy systems, environment, and conservation objectives
   a. Energy demand and conservation
   b. Energy supply
   c. Environmental impact of energy production, transportation, and use
   d. Energy system modeling

## SUMMARY OF EPRI PROGRAM EMPHASIS

**Cooperation with Federal Agencies**   A number of federal agencies are actively engaged in matters relating to electric energy. These include:

- Department of Energy (DOE)
- National Aeronautics and Space Administration (NASA)
- National Bureau of Standards (NBS)
- Research & Development Division of the Environmental Protection Agency (EPA)

It is the purpose of EPRI to coordinate its research efforts with all these agencies.

In 1977 there was close coordination with federal agencies in the following areas:

- Near-commercial LMFBR
- Battery Energy Storage Test (BEST) facility
- Solar energy
- Coal gasification and liquefaction
- Transmission and distribution
- Fuel cells
- Environmental issues
- Fusion
- Safety and reliability of nuclear power plants

Research covers the fields of:

- Primary energy resource processing
- Energy conversion methods
- Electrical systems
- Environment and conservation

**Distribution of 1976 R&D Funding**   Chart 11.1 shows the percentage of the funding required in each of the primary areas of research. This chart also shows the percentage of funding that is motivated by concerns

# DISTRIBUTION OF 1976 R&D FUNDING
## *ELECTRIC POWER RESEARCH INSTITUTE*

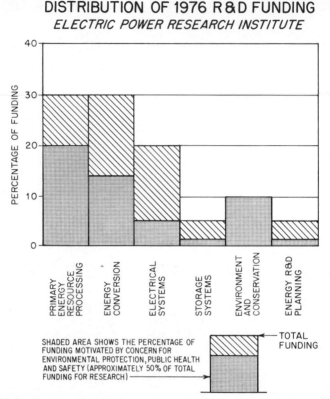

CHART 11.1 (Source: Electric Power Research Institute. A Summary of Program Emphasis for 1976.)

for environmental protection, public health, and safety. About 50 percent of the 1976 funding has been devoted to projects in these areas.

**Key Technological Options for the Utility Industry**  Chart 11.2 shows some of the new technological options that are expected to be available during the next few decades.

**Primary Program Plan**  Chart 11.3 shows the principal research activities classified according to short-range, mid-range, and long-range significance. This also indicates the approximate time of scientific feasibility, engineering, and commercial production.

## NUCLEAR RESEARCH

A major research effort on splitting the atom was conducted by the U.S. government during World War II, based on Einstein's formula $e = mc^2$,

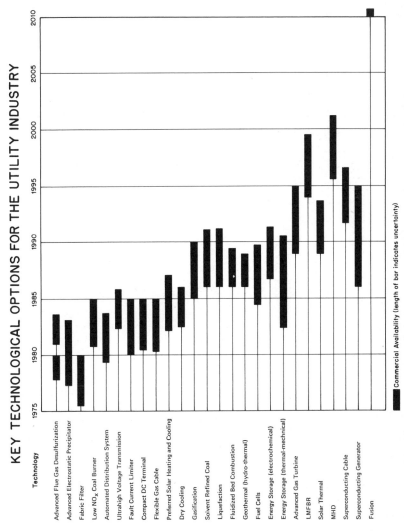

KEY TECHNOLOGICAL OPTIONS FOR THE UTILITY INDUSTRY

Technology

Advanced Flue Gas Desulfurization
Advanced Electrostatic Precipitator
Fabric Filter
Low $NO_x$ Coal Burner
Automated Distribution System
Ultrahigh Voltage Transmission
Fault Current Limiter
Compact DC Terminal
Flexible Gas Cable
Preferred Solar Heating and Cooling
Dry Cooling
Gasification
Solvent Refined Coal
Liquefaction
Fluidized Bed Combustion
Geothermal (hydro-thermal)
Fuel Cells
Energy Storage (electrochemical)
Energy Storage (thermal-mechnical)
Advanced Gas Turbine
LMFBR
Solar Thermal
MHD
Superconducting Cable
Superconducting Generator
Fusion

1975  1980  1985  1990  1995  2000  2005  2010

 Commercial Availability (length of bar indicates uncertainty)

**CHART 11.2** (Source: Electric Power Research Institute. A Summary of Program Emphasis for 1976.)

335

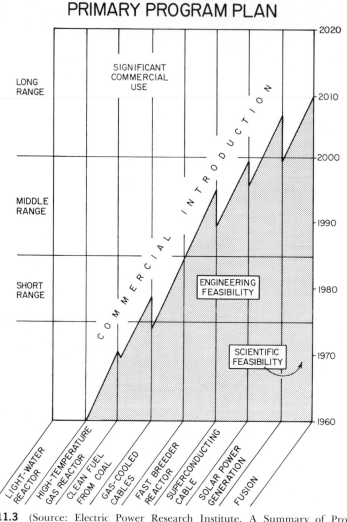

# PRIMARY PROGRAM PLAN

**CHART 11.3**   (Source: Electric Power Research Institute. A Summary of Program Emphasis for 1976.)

where $m$ represents the mass, $e$ is the energy, and $c$ is the speed of light. Light travels at a speed of 186,000 mi/s. $186,000^2$ is roughly 35 billion. In other words, the energy released by the splitting of the atom is roughly 35 billion times the weight of the atom.

President Roosevelt ordered the Manhattan Project to build the atomic bomb. A management expert of high rank in the army was appointed manager. He was given clearance for the money, manpower, material, and supplies required to do the job. The allies' best available talent was employed. All skills, including resources of leading manufacturers and their research staffs, were concentrated on one objective. In 2 years the bomb was built, and it ended World War II. Furthermore, there was developed an enormous amount of knowledge that could be used for peaceful purposes.

Shortly after World War II, the United States government made this knowledge available to those who wanted to use it for peaceful purposes. Electric utility management and others recognized nuclear energy as a new and abundant source of fuel for producing electric power. Through the EEI, electric utility management formed a Task Force on Atomic Power, which included leading nuclear scientists and utility executives who were most familiar with the new process. The Task Force was instructed to develop and report upon the most practical and feasible methods for controlling nuclear fission so that the resulting heat could be expended over a long period of time and not in a fraction of a second as in the case of the bomb. After about a year of study the Task Force issued its report listing the feasible methods and pointing out certain priorities. The LWRs, both BWRs and PWRs, were high on the list (see Chapter 3).

The industry committees on atomic power (now referred to as nuclear energy) together with the AEC (now NRC) of the federal government and the leading manufacturers and consultants considered it wise to proceed on the development of an LWR.

With guidance from the AEC, certain research projects, in the form of prototypes, were proposed. The first major project was at Shippingport, Pennsylvania, on the property of the Duquesne Light Company. The AEC helped in the financing. The company contributed approximately the amount of money required to build the equivalent capacity in a conventional steam plant. This project was successful in doing that for which it was designed—although, as expected, the cost per kilowatthour was high.

As part of the overall national research effort, the AEC proposed from time to time slightly different concepts and possible larger prototypes. Those dealing with electric energy supply were invited to carry on a particular research project. This required the establishment of groups that would enable the financing.

Many such groups were formed, with each project promising a higher efficiency and lower cost than the previous one. President Johnson set a goal of 10 years for achieving the production of electric energy from nuclear fuels that would equal the economy of conventional fuels. Thanks to many research projects throughout the country, financed largely by the AEC, by manufacturers, and by electric utility companies, economic feasibility was reached about 2 years ahead of schedule. The

LWRs were used to produce heat and steam. In this way the conventional boilers that produced steam from fossil fuels were replaced.

Simultaneously other countries in the world were engaged in similar research efforts to develop economical nuclear fuel for the generation of electric energy. The managements of electric utility companies in the United States coordinated their efforts through the committees of the EEI. In this way all knowledge was shared, and duplication was reduced to a minimum. Furthermore, knowledge was shared with other countries. Again, management had concentrated knowledge, talent, and effort on one objective. A new, clean, and economical fuel was brought into being.

**The Breeder**   Uranium, which is the basic fuel for the LWR, is composed of two atoms of different weight, uranium 235 (0.7 percent) and uranium 238 (99.3 percent). In the United States, nuclear fuel is utilized in which the uranium 235 content is raised to a 2 to 4 percent concentration. Thus, the LWR, while contributing greatly to the energy supply,

## EFFECT OF BREEDER INTRODUCTION ON URANIUM REQUIREMENTS

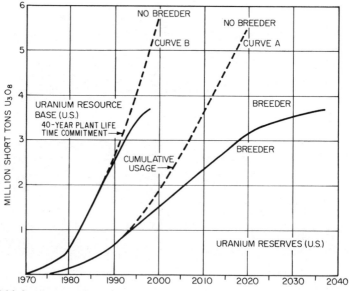

**CHART 11.4**   (Source: Fig. 9, p. 2442. Pt. 4, Authorizing Legislation Fiscal Year 1976 for Energy Research and Development Administration. U.S. Senate Joint Committee on Atomic Energy.)

utilizes only a small fraction of the usable fuel in uranium. The remainder contains more usable fuel but it must be reprocessed before being employed to make energy.

The breeder reactor is based upon a different and more complicated process, which enables the reactor to "breed" fuel for use either in the LWR or in the breeder reactor itself. Through this process, roughly 70 to 80 percent of the real fuel in uranium can be realized. Two charts illustrate the value of the breeder from the fuel standpoint. Chart 11.4 illustrates how the uranium resources in the United States will be exhausted at a fairly early date if the breeder reactor is not available to enable the replenishment of the fuel. Under curve A the cumulative usage of uranium will exhaust the uranium resources by the year 2020. Under curve B, with a 40-year plant lifetime commitment, the uranium resources would be exhausted by the year 2000. Note how uranium will be extended when the breeders become available for the furnishing of fuel.

Chart 11.5 shows the reserves available from coal as compared with those in oil and gas. Also of particular significance is the indication of the uranium reserves available with LWRs alone as compared with the fuel

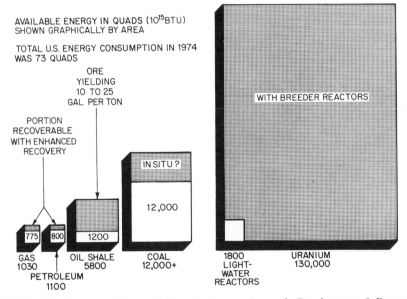

## AVAILABLE ENERGY FROM RECOVERABLE DOMESTIC ENERGY RESOURCES

AVAILABLE ENERGY IN QUADS ($10^{15}$BTU)
SHOWN GRAPHICALLY BY AREA

TOTAL U.S. ENERGY CONSUMPTION IN 1974
WAS 73 QUADS

ORE YIELDING 10 TO 25 GAL PER TON

PORTION RECOVERABLE WITH ENHANCED RECOVERY

IN SITU ?

12,000

WITH BREEDER REACTORS

775   800   1200

GAS 1030   OIL SHALE 5800   COAL 12,000+   1800 LIGHT-WATER REACTORS   URANIUM 130,000

PETROLEUM 1100

**CHART 11.5**  (Source: A National Plan For Energy Research, Development & Demonstration, ERDA–48, 1975.)

availability with breeder reactors. It has been estimated that the breeder reactor will extend the availability of reasonably priced uranium reserves from less than 100 years to over 1000 years. In addition, the breeder will have desirable environmental qualities.

It is difficult to calculate the approximate production cost of the breeder reactor. This will depend to some extent upon the credit to the breeder production for the fuel that it will supply to other reactors. The present thinking is that the production cost will be approximately 1 mill per kilowatthour. Thus, there is a potential for saving billions of dollars in fuel. More important, however, the nuclear fuel available from the breeder reactor is a necessity if the world is to continue its endeavors to improve living standards and to create jobs, both of which are closely related to the use of energy. Oil and gas will be available only for a relatively short time. Furthermore, the breeder reactor will be able to operate at higher temperatures and pressures and will therefore have a thermal efficiency considerably higher than that of the LWR.

The principles of the fast breeder have been known almost from the beginning of nuclear research. However, owing to technological complexities, scientists considered it advisable to first develop the LWR—although it was known that the LWR would be less efficient in the use of uranium.

Research on the breeder reactor has been moving forward, although not as extensively as that on the LWR. The Enrico Fermi plant (see Chapter 3) was financed by a group under the leadership of Walker Cisler. It was an LMFBR of approximately 60,000 electric kilowatts in Detroit. This project experienced the usual difficulties expected in all major research projects, but it produced energy and operated according to the purpose for which it was designed. It demonstrated the feasibility of using liquid sodium as a coolant in an LMFBR. This information contributed to the knowledge of the breeder reactor not only in the United States but throughout the world, for simultaneously with the research in the United States, other nations notably England, France, Canada, and the Soviet Union have progressed in research to enable the building of a commercially feasible breeder reactor.

In the United States the next step in the development of a commercial breeder calls for the building of a reactor of roughly 300,000 electric kilowatts. President Carter has proposed that the breeder research effort be postponed on the grounds that it results in the production of plutonium, the principal ingredient of the bomb.

Those in favor of continued research argue:

1.  A nuclear power plant is only one of eight possible ways to produce weapons-grade fissionable material. Most of the other

ways are far less expensive and time-consuming. The American people should not be denied the benefits of the breeder under such circumstances.

2.   Other nations, such as Britain, France, Belgium, West Germany, Japan, and the Soviet Union, are going ahead with breeder research.

3.   The best, and possibly the only, realistic way to prevent proliferation is through the political process.

4.   While there are other possible avenues for taking the next step after the LWR, most scientists consider the breeder the best.

5.   Postponing research while nuclear fuel is in short supply may place the United States in an inferior position in this important technology.

6.   Most Western nations look to the United States for leadership in nuclear energy.

Research on the breeder is already a major national effort. Expenditures to date have exceeded $2 billion. Breeder research in the United States is carried on in a number of locations and is financed by many sources. The Department of Energy and EPRI are playing a leading part. A large fast-flux test facility (FFTF) is scheduled to be in operation in 1980 at Hanford, Washington.

A 350-MW electric demonstration reactor, known as the Clinch River Breeder Reactor (CRBR), has been under development in Tennessee, and about $500 million has already been spent. The total expected construction expenditure will be more than $1 billion. This is an LMFBR, cooled with liquid sodium. Until it was postponed, initial operation of this reactor was scheduled for 1983. The total LMFBR program, extending beyond the year 2000, was anticipated to cost some $10 billion. The design, construction, test, and operation phases of this project are illustrated in Chart 11.6.

Prior to the completion of the Clinch River project, work was scheduled to begin on the *prototype large breeder reactor* (PLBR). This would be the first large LMFBR plant in the United States. Chart 11.6 shows the period for design, construction, and testing. Notice that the testing would have been done in the latter part of the 1980s.

This PLBR project was expected to operate as part of a large, interconnected electric utility system. Operation was expected to begin about 1988. As costs have risen sharply in recent years, the PLBR project was expected to have a cost that would make its energy production somewhat

# SCHEDULE OF DESIGN, CONSTRUCTION, TESTING, AND OPERATION OF BREEDER REACTORS

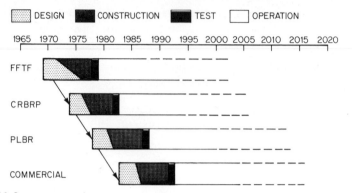

# GENERATION COST OF BREEDER REACTORS COMPARED TO LIGHT-WATER REACTORS

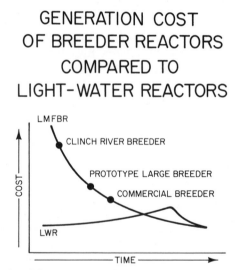

more expensive than that of the LWR. Therefore, this would be followed by further research in the development of the commercial breeder. Chart 11.6 shows the schedule from design to expected operation in 1995.

Chart 11.7 gives an estimate of the generation cost of the breeder as compared with the LWR. It is likely that the commercial breeders that could compete economically with the LWR could become available around the year 2000. Naturally, price is an important consideration in the development of a breeder reactor. But more important, it is necessary as a fuel supply to replace rapidly diminishing gas and oil reserves.

**Fusion**    The splitting of the uranium atom led to the development of the atomic bomb—the release of an enormous quantity of energy in a fraction of a second. The control of that process enabled the controlled release of the energy over a long period of time. Thus, managements in the field of energy had available to them a new fuel resource to supplement fossil fuels.

The hydrogen bomb, enormously more potent than the atomic bomb, is based upon a different atomic principle—fusion, the combining, or fusion, of certain elements. Here the nucleus of heavy hydrogen (deuterium) combines with one of tritium. This union releases very large amounts of energy, as in the hydrogen bomb.

Now the problem is to control the process so that the energy can be dissipated gradually over a long period of time. Scientists throughout the world agree that the control can be accomplished although it is more complicated and difficult than control of the atomic bomb. Temperatures of about 100 million degrees Celsius are required to initiate the fusion reaction. This means that a "vessel" must be designed that can contain this high temperature. Research indicates two possibilities:

1.  Build an electromagnetic field. This is known as a *controlled thermonuclear reaction* (CTR). The largest test facility in the United States is the Tokamak Fusion Test Reactor (TFTR) being built at Princeton, New Jersey, with federal funds. The estimated cost is $215 million. This facility is scheduled to be in operation in 1981. The rating would be about 6 kW. However, there would be no net kilowatts gained, as the research project itself is operated by another machine requiring 35,000 kW. If this study proves the concept to be feasible, it will be followed by a second power reactor which will be designed to produce a net power output. The estimated cost is $500 million.

2.  The second concept for the release of fusion energy is called *laser-pellet*. A laser beam is focused on a fuel, causing it to compress and attain high temperature.

Both research projects are being accelerated at a contemplated research cost of about $2 billion through 1980. Some scientists estimate that it will require about $20 billion to bring this source of energy to a commercial scale, if research indicates that this development is economically feasible.

Fusion has another advantage as compared with fission. There would not be the waste-disposal problem associated with fission. Ultimately, the fusion reactors may be fueled with deuterium, which is derived from seawater—a resource which would be practically unlimited.

The scientists and research directors of the fusion research projects indicate that, if fusion continues to appear feasible, it can be brought into commercial operation sometime between 2000 and 2040. The process is so promising that energy managers throughout the world are pressing forward with the research. The question is what fuels should be used during the transition from oil and gas to fusion and/or solar power. Most experts believe the transition will be accomplished with the help of coal and nuclear fuels.

# CHAPTER

# 12

# ENERGY AND
# THE ENVIRONMENT

## INTRODUCTION

One of the most pressing problems facing electric utility management relates to the environment. Indeed, a large amount of time of top management is devoted to this problem, which is quite involved and has many ramifications.

Naturally, good management, and especially management in the electric energy field, is interested in making its proper contribution toward the desired end of environmental improvements. In a free society, government sets the proper standards in all matters relating to environmental protection and safety. Citizens do all within their power to abide by these standards. Unfortunately, as is probably the case in any great movement, the pendulum tends to swing too far from one side to the other. There are those who feel that the laws setting the standards have swung too far in too short a time. Adding to the problem is the fact that most of the laws were written before there was time to really weigh the costs against the benefits. Fortunately, laws and administrative practices can be modified when proper authorities are convinced that changes are advisable. Here, electric utility management can and does contribute.

Most people agree that in matters of this kind, the benefits to be derived should at least equal the cost of providing those benefits. Often, especially in the early stages, the costs are not known, and it is difficult to even estimate them. Furthermore, there is a tendency on the part of advocates to overestimate the benefits and underestimate the costs.

Now that companies have had some experience, the managers of the electric energy business and their staffs are in a good position to provide intelligent estimates of costs. Management's viewpoint is valuable, particularly as it involves the presentation of facts and opinions to the regulatory bodies and to the public.

Customers should be and are being reminded that it is they who derive the benefit from environmental measures and they who pay the cost. More often than not, the cost is higher than most people think.

The cost to a company of further improvement in the environment is an expense contributing to the total cost of manufacturing a product or furnishing a service, just as material and labor are costs. Money spent on environment frequently contributes little to increased production and productivity. As was noted in Chapter 6, a limited amount of money is available for increased salaries and wages, and this in turn is dependent upon increased production and productivity. In other words, money spent on environment (although needed) usually means less money for increased salaries and wages. When labor demands (and gets) the "normal" increases without taking this factor into account, increased prices are inevitable. This is inflation. As money spent for improved environment increases the "good life," people must be satisfied with a corresponding slowing of the rate of increase of money available to buy more goods and services. Inflation is the factor which compensates for the violation of the laws of economics and the laws of mathematics. This principle holds, no matter what laws people pass or actions they take.

When all costs are known and when consumers realize that they are paying the cost, frequently there is less public demand to go to the extreme in environmental protection, particularly if something short of the extreme, as a practical matter, is satisfactory. Therefore, a large part of the problem regarding the environment relates to public understanding. Wise management realizes this and conducts appropriate programs to properly inform the customers.

## OBJECTIVES OF ELECTRIC UTILITY MANAGEMENT

As stated in Chapter 6, there is a direct relationship between the use of energy and the provision of good living standards. It is recognized that the use of energy has some impact upon the environment. Thus the aim is to reduce these detrimental effects to a practical minimum and to use energy to help clean up the environment.

While environmental protection and improvement are important, management has other and equally important objectives and responsibilities, ranging from furnishing all the energy the customer wants and needs at the lowest practical price to paying good wages. All these objectives are interrelated. Management addresses itself to the environment problem in this context.

The cost of providing satisfactory environmental protection is quite high. This cost becomes part of the price of electricity, which management is trying to keep as low as practical. This is another reason why management tries, by all appropriate means, to avoid any undue or extreme environmental measures that would of necessity result in still-higher costs to the public without providing proven health benefits.

## ELECTRICITY'S CONTRIBUTION TO A CLEAN ENVIRONMENT

Further, concerning the growth in electric energy as it relates to the environment, it should be more widely understood that the use of electric energy has already contributed immeasurably to the improvement of the environment, especially in relation to air pollution. During the early 1920s, most of the energy used to operate factories came from either primary fuels (coal, oil, gas or wood) or local on-site generating plants. Only about 20 percent of the energy for manufacture was obtained from the large interconnected electric systems, where energy is produced from a much smaller number of large, highly efficient generating stations. One can imagine the air pollutants from the thousands of manufacturing plants with little or no control of air pollution as compared with the highly controlled electric generating plants, from which most industrial plants now obtain their energy.

## TO AID IN FUTURE IMPROVEMENT OF THE ENVIRONMENT

Practically all the improvements in environment in the years to come will be accomplished through the use of machinery. This machinery will be run by clean electric energy. Thus, it could well be that the continued growth in electric energy is an important factor that will lead to the desired ends in the improvement of environment.

Here are a few examples of beneficial ways in which electric energy may be applied:

- Electric transportation, including cross-country rail, high-speed interurban rail, intracity mass transit, and the electric automobile

- Water purification

- Water recycling

- Desalting seawater

- Waste removal

- Using waste to generate electricity

- Recycling waste material

- Cleaning of stack gases

- Converting coal to oil and gas

- Further electrification of manufacturing process

- Electric heating

## MORE WORK IN PROBLEM AREAS

**Air Pollution**   This involves removing particles such as fly ash from stack gases, removing smoke caused largely by the inefficiency of the burners, and removing or reducing the content of those gases considered harmful.

**Water**   Large amounts of cooling water are required in the steam generation of electric energy. The water is circulated through the condenser and then reenters the stream or lake from which it was first obtained. This does not cause water pollution, as the public generally understands the term. However, the temperature of the water is raised about 15 to 20°F in the area immediately adjacent to the plant. Part of management's problem is having the public understand that water use by the utility does not result in the pollution that the public normally associates with that term. Another part of the problem is how to reduce the temperature increase to that amount required by regulatory bodies.

**Plant Location**   Most people want the required number of power plants built so that they can have the electricity they want and need. Frequently, however, they do not want electric power facilities built nearby—which may be understandable, but creates a number of problems.

**Nuclear Energy**   Because nuclear energy was an outgrowth of atomic bomb research, there have been concerns expressed by some people as to potential hazards.

**Aesthetics**   The appearance of distribution lines, transmission lines, substations, and power plants is another subject of criticism by some members of the public. A good example of what can happen when people, even those in high positions, are not well informed occurred during the early years of the stepped-up environmental program. President Johnson had formed the White House Conference on Natural Beauty. His committee was composed of cabinet members and state governors. Following the first conference, the President outlined on television a five-point program. Point number two was: "All transmission lines must go underground." The managing director of the EEI was aghast. He knew roughly the cost involved and knew that the technology was as yet unavailable to transmit large amounts of power long distances underground. At that time the investment in transmission lines was about $9 billion. To go underground, even for short distances in the cities, cost approximately 20 times as much. If technology were available for the building of underground long-distance transmission lines, it would cost about $180 billion, or $2000 per family. A public opinion survey made a few years previously indicated that only a small percentage of the American people were aware of ever having seen or heard of a transmission line. One can imagine the reaction of a family living on the fifth floor of a walk-up apartment in New York City when being told that they must pay $2000 to put transmission lines underground. Obviously the benefits could not equal the cost.

The EEI Managing Director wrote a one-page memorandum giving the principal facts, called a White House staff member, and asked that the memo be placed on the President's desk. The next day the staff member called and said: "The President said: 'that's Lady Bird's (Mrs. Johnson's) department. I showed the memo to Lady Bird and will read what she wrote on the memo: 'Try harder.'"

Fortunately, the chairman of the Federal Power Commission issued a statement pointing out the reasons why it should not and could not be done. While all cases are not quite like this, it is a fact that a large body of opinion, even among some opinion leaders, is formed without adequate knowledge of the facts. Management's obligation is to present the facts so that proper judgments can be formed.

## WHAT HAD BEEN DONE PRIOR TO THE RECENT PROGRAMS

Probably a large part of the public and Congressional consciousness of the environmental problem began with the so-called smog in Los Angeles. At first the cause was unknown. Research indicated that the primary cause was the large number of automobiles and that certain air formations above Los Angeles tended to keep the polluted air from moving out. Other big cities such as London and New York began to experience similar difficulties.

Here is another example of the results of lack of public knowledge on such an important subject. During the late 1960s there was formed the Citizens Advisory Committee on Recreation and Natural Beauty. Laurance Rockefeller, one of the nation's leaders in the field of environment, recreation, and natural beauty, was the chairman. The committee employed an engineering firm to begin the investigation, and the electric utility industry was chosen as one of the first to be investigated. The engineers called upon the managing director of the EEI to begin the investigation. Before their arrival, a multitude of reports on research projects and other activities pertaining to environmental matters that had been going on for many years were placed on the table. The engineers were presented with a summary of what had been done prior to and following the White House Conference on Natural Beauty and with some of the long-term goals. The reaction of the engineers was one of surprise at the comprehensiveness of the accomplishment. Next was the usual question following such a summary: "Why haven't the utilities told this story to the public?"

As a result of that investigation, the committee called upon the experts on environmental matters in the utilities for factual advice in the preparation of their report. A fair, practical, and workable report was the result.

## FIRST AIR POLLUTION COMMITTEE

The predecessor association of the EEI was called the National Electric Light Association (NELA). It functioned in much the same manner as the EEI, with a large number of committees, one for each of the principal functions in the electric utility business. The first air pollution committee was formed in 1905. It was then that the industry began a study of air pollution on a national basis with the aim of bringing about appropriate improvements in the generation of electric energy. This committee also sought improvement in the efficiency of the burners used in the boilers, where inefficiency results in waste of fuel and is manifest through smoke coming from the stack. The aims were the conservation of fuel as a means of lowering the price of energy and also the removal of smoke as one of the causes of air pollution.

During the 1920s, the industry had attained a high degree of efficiency in new burners and had practically eliminated smoke from newer plants. In 1928 the author was a department head of an electric utility with headquarters in Shreveport, Louisiana. The chief engineer had just completed a new generating station in Shreveport with three "huge" 10,000-kW units. He was proud of the plant and its efficiency, and employed a photographer to take a color picture.

When the picture was delivered, the engineer discovered that the photographer had painted in some smoke which he thought should be there. Naturally, the engineer was upset and explained that there was no visible smoke for the reason that the burners had a high efficiency—on the order of 99 percent.

About the same time, the company had built a substation in a residential section. A picture was taken of this. It was enclosed in a small bungalow similar to the small houses in the neighborhood. The point is that many years ago electric utilities were conscious of the need for the protection of the environment.

During the early years, the industry began its systematic approach for removing particles—principally fly ash—from the stacks. Various kinds of equipment, including electrostatic precipitators, had been used. So much fly ash was removed that the handling and disposal of it became a problem. There followed research to find useful purposes for fly ash.

Research to find some way to remove sulfur from coal began in 1958, initially in partnership with the Bituminous Coal Institute. Research is still being conducted, with some promise.

Some years prior to the stepped-up environmental endeavors, electric utilities in some major industrialized nations began their research in the building of tall stacks as a means of dispersing gases over a wide area so that there would be little concentration at the level at which people live. In some countries, as in England, a major part of the problem has been handled in this fashion. Some claim, however, that tall stacks do not solve the basic problem.

In other cases, companies established the practice of monitoring stack emissions. An operator was stationed in an adjacent building with instructions to watch the stacks for any excess smoke. This operator had a direct telephone line to the plant operator, who could either adjust the burner or transfer the load to another machine or plant.

In 1968 the EEI formed the Electric Vehicle Council for the purpose of promoting the required research and development of the electric vehicle. It was known then that the primary cause of air pollution in the city was the automobile. The electric vehicle does not pollute the air.

Only 15 percent of the air pollution is attributed to the generation of electric energy. This pollution can be controlled much more easily than that from millions of gasoline-driven automobiles.

## AESTHETICS

During the 1960s, when it became known that the industry's transmission-line towers left something to be desired from the standpoint of aesthetics, but when there was still not sufficient knowledge to put transmission lines underground, to any great extent, the industry employed one of the nation's leaders in industrial design. Understandably, perhaps, a number of utility engineers did not share the views of others that the towers were not things of beauty. To them the towers were like those supporting a beautiful bridge. Nevertheless, the industrial designer was instructed to design about 100 different types of towers that would meet the required electrical specifications and have an appearance that would be considered pleasing, if not beautiful. These towers were designed, and models were built of them. They have been used as models to aid engineers in improving the aesthetics of transmission towers.

For some time the industry was conscious of the fact that in many parts of the cities and in some suburban residential sections, customers preferred that the distribution lines be placed underground. Extensive research was begun on finding a way to do this more economically. At first the cost of underground lines was 8 to 10 times the cost of overhead lines. For ordinary soil conditions, this figure has since been reduced to approximately $1^{1}/_{2}$ times. Many companies have gone to underground lines in the big-city areas, especially the centers of town. Also many have gone to undergound lines for new construction in residential areas. Most companies either are making improvements in the appearance of distribution systems or placing them underground.

For a long time electric utilities endeavored to beautify their power plants. Frequently the lakes and ponds adjacent to them were landscaped and used for recreation. The same is true of hydroelectric plants and pumped-storage plants.

## WHAT IS BEING DONE

In this discussion no attempt will be made to give a full description of all that is being done to meet environmental problems. There is ample literature available for this purpose. Rather, a few of the typical problems facing management will be treated.

Problems arise when, in the judgment of management, there are alternative ways to accomplish substantially the desired purpose at far less cost to the public than the government-mandated programs would require. Management believes that it has the responsibility to present these viewpoints to the appropriate regulatory bodies and the customers.

### AIR QUALITY

The Clean Air Act of 1970, as amended in 1977, sets the standards for air quality control, and the EPA establishes emission standards.

The sulfur content of some fossil fuels is the principal cause of the problem. As has been mentioned, roughly half the fuel used by electric utilities is coal. There is an abundant supply in the United States, but Eastern coal has a higher sulfur content than Western coal.

When it is necessary for Eastern utilities to use Western coal because of lower sulfur, more fuel (energy) is required to transport the coal. The fuel generally used in transportation is oil, which is in short supply. Eastern coal miners may lose jobs. Management and regulators should aim to consider all factors before forming a judgment.

For the long term, the fuels available are coal and nuclear fuels. Therefore, the aim is to utilize these fuels without undue ill effect on the environment.

## TALL STACKS

For the purpose of enforcing clean-air standards, questions arise as to where the pollution measurement should be made. Should it be at the top of the stack, or should it be where people breathe the air? Some experts claim that tall stacks enable the dispersing of the gases over a wide area at a high level, where they are diluted in the upper air. The result is the proper level of clean air when the emissions reach the space where people live. Tall stacks in new plants are as high as 500 ft and more.

On the other hand, neighboring states or countries may complain about having pollutants dropped upon them. The questions are: "how much is bad?" and "how much are people willing to pay?" Should the "best" be attempted now at high cost, or should something "less than best" be used until technology finds a more economical solution?

For example, two research projects show promise in removing sulfur from coal. One is being carried on by a U.S. utility with two European manufacturers. The project focuses on a coal-burning technique known as *pressurized fluidized bed combustion* (PFBC). This may be developed into an effective and economical alternative to the use of scrubbers for flue gas desulfurization. There will be combustion testing of eastern Ohio higher-sulfur coal in a PFBC pilot plant at Leatherhead, England.

Another U.S. company in the Southeast reports promise in the development of solvent refined coal. In the most recent test, solvent-refined coal burned so nearly pollution-free that it more than met the current clean-air requirements for new power plants.

Also the question arises as to whether it is wise to set emission standards on a national basis. Is the proper goal to control the air quality at the top of a stack, or is it the aim to provide a good quality of air where people breathe it, under different conditions around the country?

## SCRUBBERS

Although the chemistry associated with the removal of sulfur oxides from stack gases is well known, the engineering technology has not yet been fully developed. The various systems for removing sulfur oxides are called *scrubbers*.

In some respects scrubbers create new environmental problems. For example, scrubbers using limestone are being used in experimental installations. The use of limestone with a 1000-MW boiler will produce about 600 tons a day of a thick, wet sludge, like toothpaste, that must be carried away. Furthermore, there are environmental problems and energy costs associated with the obtaining and processing of limestone and transporting the material to and from the plant.

Scrubbers are quite costly and add materially to the customer's cost of electricity. They cost as much as $70 million each to purchase and install. For example, the TVA, a very large power system, would have to install scrubbers on 63 units at 12 plants at a cost of more than $1 billion. The annual cost would be $225 million. Alternative controls could be used for $17 million per year. The TVA has stated that if it were forced to install scrubbers, electric rates would increase 15 to 30 percent.

For these and other reasons, managements are suggesting that it would be in order to have a continuing review of provisions in the Clean Air Act and its administration so that new knowledge can be taken into account regarding both costs and possible alternative methods.

At the time this was written, EPA had proposed a rule which would require that all new coal-burning utility plants remove 85 percent of potential sulfur emissions, regardless of the amount of sulfur in the coal being burned. The principal alternative being backed by the electric utility industry and by the Department of Energy would allow a "sliding scale" under which a plant burning low sulfur coal could use less control equipment. If the EPA proposal turns out to have been the one adopted, the cost to the American electric consumers in the year 1990 alone will be $3.19 billion, some $2^{1}/_{2}$ times the figure which would result from the industry's approach.

## WATER QUALITY

The problems pertaining to water quality, while simpler than those relating to air quality, are sizeable and costly. The utility uses water for condensing steam. The water is passed through the condenser and does not come in contact with substances that can cause pollution in the ordinary sense. This water which is used for cooling purposes while condensing the steam absorbs some of the heat. Government regulatory bodies, national and local, set the limits on the increase in temperature.

Some ecologists believe that the raising of the temperature by this amount, although for only a small area in the neighborhood of a power plant, tends to upset the ecological balance. Studies show that there may be some damage to aquatic life in that area. On the other hand, research indicates that some increase in water temperature may be helpful. This appears to be the case with oysters and other aquatic species.

Engineering technology is available for lessening the increase in water temperature when government regulations require that this is advisable. However, as mentioned later, the increased cost, and therefore the rise in price, of electricity is considerable.

Here are some of the various techniques in lessening the increase in water temperature:

- Cool water may be mixed with the warm water.

- Cooling ponds are rather popular. These are shallow lakes with a large surface area. Sometimes a spray system is used to facilitate the cooling.

- Cooling towers are better known, as they are quite large and visible to the public. These are towers in which water is sprayed into the air which pass through the towers. There are two kinds of towers: those having a natural draft, which require a very large structure, some 300 ft in diameter and 350 ft high; and those with motor-driven fans that move the air, which are much smaller.

The investment cost of a cooling pond may be about 2 percent of the cost of the plant. A pond usually has other uses, such as recreation. The investment in a cooling tower is about 5 percent of the total cost of the plant and has no other uses. Studies have shown that the benefits of cooling towers can be quantified. However, the evidence on a nationwide basis shows that the *costs* of towers far exceed the benefits.

The Federal Water Pollution Control Act provides the authority for the EPA to set the standards of water pollution. One of the difficulties for electric utilities is that it appears that Congress considered heat as one of the categories of pollutants. Now that more experience has been gained and some of the problems studied, there may be less-rigid requirements and a lower cost to the public as far as utilities are concerned.

## NUCLEAR ENERGY

If people are going to continue to maintain and improve their living standards and have greater job opportunities, energy will be needed to drive machines. The major fossil fuels, gas and oil, are being depleted, probably to be exhausted shortly after the year 2000. Coal will last longer, but it is not inexhaustible. Energy from uranium is essential at least until other prospects for large-scale energy become feasible technically and economically. Solar energy may become practical on a large scale, but not until after the turn of the century. Even then, many experts doubt that the cost can compete with the breeder reactor. As shown, uranium can last hundreds of years with the breeder. Scientists and engineers say that few major advances have had as much research and development devoted to them as nuclear energy, especially from the standpoint of safety.

For the very long term, solar energy and fusion appear to be best to meet all requirements. They are clean and practically inexhaustible. But practical application on a large scale will not begin until about 2000. In the meantime, there is need for a transition phase, which will require the use of coal and uranium.

From the standpoint of air pollution, nuclear energy is clean because there is no combustion of fuel. There are no stacks. The nuclear reactor replaces the fuel supply of an ordinary fossil-fuel plant. From there on, the steam cycle is the same as that of a conventional plant. The nuclear power plant requires cooling water. As the efficiency of the LWR is not as high as that of a modern conventional steam plant, the cooling-water temperature is raised slightly more than that from a conventional plant. A fossil-fuel plant may raise the water temperature on the order of 15°F, whereas the LWR raises the temperature about 20°F. Both these temperatures can be controlled as desired. From the aesthetic standpoint a nuclear plant not only has a pleasing design but also eliminates the large stockpile of coal needed for a coal-burning plant.

## SAFETY

Most of the concern for the nuclear plant relates to safety. Possibly this is a carryover from the experience of the atomic bomb, although there is no scientific connection. The bomb was designed for one purpose, the nuclear power plant for another. It cannot explode. Other concerns have to do with radiation and the problem of waste disposal.

Table 3.10 shows the number of megawatts and the percent generation of nuclear plants by countries. Chart 12.1 shows the cumulative reactor experience since 1954. The safety of nuclear power plants has been demonstrated through more than 300 reactor-years of operating experience in the United States. No member of the public has experienced radiation injury or death from the operation of a licensed nuclear power facility. There have been no radiation fatalities among employees in nuclear power plants. In addition, more than 1400 reactor-years of U.S. naval ship propulsion have been accumulated without injury to the public. The matter of safety of nuclear plants is under the control of the experts in the federal government. Extensive hearings are required before a utility is granted a license to build a nuclear plant.

# CUMULATIVE REACTOR EXPERIENCE SINCE 1954

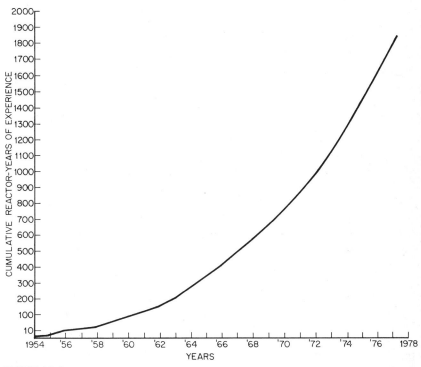

**CHART 12.1**

There must be application to the NRC for two separate licenses. One license allows the plant to be constructed. Another license permits the plant to be operated after it is built. Also, the utility must obtain many other licenses and permits, such as those from state and local governments. There must be a public hearing before the Atomic Safety and Licensing Board (ASLB). Any individual or group may object to the issuance of a license. The NRC grants a permit for construction only after the staff has given all matters complete study. Much of the process must be repeated in order to get the permit to operate the plant after it is built.

During the entire construction period, inspection is insisted upon to ensure compliance with the provisions of the permit. Furthermore, when a plant begins operation, it is carefully supervised. There is additional public assurance of safety in that the plant must be operated and maintained by skilled and experienced personnel licensed by the NRC.

There is no operation in any industry requiring more careful, painstaking insistence upon the maximum in safety. This is one of the reasons why the safety record has been excellent.

**The Work of Dr. Rasmussen**   Dr. Norman C. Rasmussen has been associated with the Massachusetts Institute of Technology (MIT) for the past 25 years. Currently, he heads MIT's Department of Nuclear Energy. Dr. Rasmussen was asked to make a study and then report on nuclear safety. "Reactor Safety Study: An Assessment of Accident Risks in U.S. Commercial Nuclear Power Plants" was the result. The study was made for the AEC, whose functions are now divided between NRC and the research and development functions of the Department of Energy. The team was composed of an independent group of nuclear scientists, including Dr. Rasmussen.

Some $4 million and 70 work-years were spent on the study in an attempt to provide the government with a "realistic estimate" of the risk posed to society by a wide spectrum of potential accidents that might have even a remote chance of occurring with commercial reactors.

A few quotations from Dr. Rasmussen may be of interest. These are from an article, "Nuclear Power: Safe by Any Measure," which appeared in the February 1976 *Electric Perspectives,* a bimonthly publication of the EEI. The report publishes a table showing the chances of an individual living near a nuclear power plant being fatally injured by a possible nuclear accident as compared with established statistics of the probability per year of other types of fatal accidents (see Table 12.1).

Then, there is this significant statement by Dr. Rasmussen:

> Of course, the most commonly expressed concern is that even though the odds of a nuclear accident may be small, the consequences may be unacceptably large. To investigate this problem, we calculated and plotted the probability and magnitude of nuclear accidents of various sizes (Chart 12.2). For comparison, similar plots for some of the more common accidents were also made.
>
> We see that fatal nuclear accidents of any size are at least several hundred times less likely to occur than equally serious fatal air crashes, fires, or explosions, and are about 10,000 times less likely to occur than the total for the risks plotted (Total Man-Caused Accidents).
>
> Furthermore, the maximum nuclear accident of 3,300 fatalities is no larger than what can be expected from these other risks. This clearly indicates that nuclear accidents will not be larger, in terms of fatalities, than other kinds of accidents society is currently exposed to. In fact, certain risks such as dam failures and airplane crashes into crowded areas have the potential for producing considerably more deaths.

**TABLE 12.1**  RISK OF EARLY FATALITY* BY VARIOUS
TYPES OF ACCIDENTS FOR THE TOTAL
U.S. POPULATION

| Accident Type | Probability per Year |
|---|---|
| Motor vehicle | 1 in 4000 |
| Falls | 1 in 10,000 |
| Fires and hot substances | 1 in 25,000 |
| Drowning | 1 in 30,000 |
| Poison | 1 in 50,000 |
| Machinery | 1 in 100,000 |
| Air travel | 1 in 100,000 |
| Falling objects | 1 in 160,000 |
| Electrocution | 1 in 160,000 |
| Railway | 1 in 250,000 |
| Lightning | 1 in 2,000,000 |
| Hurricanes | 1 in 2,500,000 |
| All others | 1 in 25,000 |
| All accidents | 1 in 1,600 |
| Nuclear accidents | 1 in 5,000,000,000† |

* Refers to fatalities occurring within a short period of time of an accident.

† Based on an industry of 100 reactors and on a population at risk of 15 million persons, which was calculated from population and weather data for regions surrounding U.S. nuclear plant sites. The nuclear probability was estimated, since there have been no nuclear-related fatalities.

**Radiation**  Dr. Rasmussen has this to say concerning radiation:

As to the normal releases of radioactivity from operating plants—all commercial-type reactors release very small amounts of radioactivity into the environment—the dose received by even the closest inhabitants to nuclear power plants is but a small fraction of what they receive from natural sources. In fact, the doses received from medical X-rays, high-altitude flights and the building materials of some homes greatly exceed this small reactor-produced dose (Table 12.2).

The Federal Radiation Council has recommended that the incremental whole-body radiation exposure for an individual member of the general public should not exceed 500 millirems per year. (A *rem* is a unit of measurement of the biological effect of radiation; a millirem is one one-thousandth of a rem.)

Dr. Rasmussen closes his article with this comment: "Nuclear power obviously brings with it certain important problems and responsibilities to those nations that choose to exploit it. I believe that the past 30 years'

# PROBABILITY AND MAGNITUDE
# OF NUCLEAR ACCIDENTS
# COMPARED WITH MORE COMMON ACCIDENTS

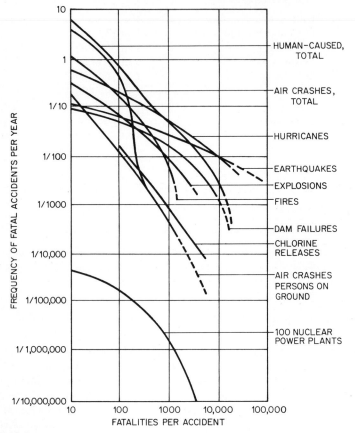

**CHART 12.2**

experience with this form of power has clearly demonstrated that responsible nations can enjoy its several advantages without exposing the public to undue risks."[1]

---

[1] On January 19, 1979, after this discussion of the Rasmussen report was set in type, the NRC announced it no longer regarded as reliable the study's numerical estimate of the overall risk of a reactor accident. An NRC spokesman said, however, according to the *New York Times* of January 20, that "this means the chances

**TABLE 12.2** ESTIMATED WHOLE-BODY-DOSE RATES OF
RADIATION FOR U.S. POPULATION IN 1970*

| Source | Millirems per Year* |
|---|---|
| Natural | 102 |
| Medical | 73 |
| Fallout | 4 |
| Occupational | 0.8 |
| Nuclear reactors | 0.003 |
| Miscellaneous | 2 |

*Source:* "Effects on Populations of Exposure to Low Levels of Ionizing Radiation," "Biological Effects of Ionizing Radiation" (BEIR), findings from the National Academy of Sciences.
"Nuclear Power: Safe by Any Measure:" by Dr. Norman C. Rasmussen published in *Electric Prospectives,* Feb. 1976, Edison Electric Institute.
* Units of radiation dose per year.

**Nuclear Waste** One of the problems associated with nuclear energy is how to handle, dispose of, and store the radioactive waste so that it will not be a hazard for the present or for future generations. For the present the government policy is to store the wastes in storage centers above ground. Naturally, these are closely guarded.

It is not generally realized that the waste product will be relatively small, just as a pellet of uranium fuel is quite small as compared with the amount of coal providing equivalent energy. For example, one of the government centers for temporary storage utilizes as little as 350 acres. This area is capable of containing all the high-level wastes generated by all the nuclear reactors in the United States up until the year 2000.

---

of accident could be greater or could be less than was stated in the report" and that "the action does not mean that the commission now considers nuclear power plants to be dangerous." EEI commented that "the Rasmussen study was a valuable and still useful effort to estimate risk under extreme, hypothetical conditions which never have happened. It was not an assessment of nuclear reactor safety as it has occurred in actuality."

In its statement on the report, NRC said: "With respect to the component parts of the (Rassmussen) study the Commission expects the staff to make use of them as appropriate, that is, where the data base is adequate and analytical techniques permit. Taking due account of the reservations expressed in the review group report and in its presentation to the Commission, the Commission supports the extended use of probabilistic risk assessment in regulatory decision-making." ["NRC Statement on Risk Assessment and the Reactor Safety Study Report (WASH-1400) in Light of the Risk Assessment Review Group Report," NRC, Office of Public Affairs, Washington, D.C., release no. 7919, Jan. 18, 1979.]

Considerable research and study is going forward in the endeavor to locate and decide upon a permanent storage. Following are the viewpoints of John W. Simpson:[2]

> A first step would be to reduce the size of radioactive waste by separating out the radioactive materials. These materials would be mixed with a suitable binder, such as boron silicate glass, and processed into a solid cylinder as impervious to corrosion and the leaching out of radioactive materials as possible. The cylinder also must be capable of being cooled by natural air circulation.
>
> This cylinder would then be put inside another container made of metals known to be most resistant to corrosion (titanium and even gold are possibilities) and then again encased in an outer container of great mechanical strength (stainless steel or reinforced concrete). The resulting cylinder would be self-shielding, thus reducing the radiation level adjacent to it to below the level considered safe for short time exposure of human beings. The cylinder would then be stored in an area known to have been geologically stable for hundreds of millions of years and predictably stable for millions of years to come. It should be stored in an area in which no water has existed for millions of years, where the possibility of water even reaching it is almost nonexistent.

Quoting Mr. Simpson further:

> It has been calculated by the Atomic Energy Commission that the probability of water flowing through such a storage in a thousand years is from one in a million to one in ten billion. If water should reach the storage site and then flow through the earth to a populated area, the radioactivity would probably be less than 1% permissible dosage. So you end up with one chance in a million of receiving an exposure of $1/10$ of 1% of the permissible dose in a thousand years.

## COSTS OF PROTECTING THE ENVIRONMENT

Properly, the benefits of environmental protection should be evaluated in dollars so that care can be exercised in keeping the costs to a level not in excess of the benefits.

On the matter of costs, the following may be of interest:

---

[2] Director of Westinghouse Electric Corporation and Chairman of the Energy Committee. Chairman of the Atomic Industrial Forum (at the time of this statement) he is also past president of the American Nuclear Society. From a reprint of a number of *Position Papers* on nuclear energy which appeared in *Fortune* magazine.

**TABLE 12.3**  ESTIMATED COSTS FOR THE ELECTRIC
UTILITY INDUSTRY FOR CLEAN-AIR AND
WATER-QUALITY LAWS

|  | Capital Costs 1976–1990 (billions of dollars)* | Consumer Costs in 1990 (billions of dollars)* | Equivalent Annual Cost per Household in 1990 (dollars) |
|---|---|---|---|
| Industry base (no controls) | $384 | $117 | $1200 |
| Clean-air controls: | | | |
| State implementation plans and new source performance standards | $21 | $11 | $110 |
| No significant deterioration (EPA supports, Senate passed) | $22 | $ 5 | $ 48 |
| Water-quality controls: | $21 | $ 3 | $ 33 |
| Total of control costs | $64 | $19 | $191 |
| Control costs/base costs | 17% | 16% | 16% |

*Source:* National Economic Research Associates, 1977.

* All dollars are in constant 1975 dollars. Current dollars would of course be much higher because of inflation.

1.  *Costs*  Capital costs in the electric utility industry for air and water pollution are estimated to be at least $64 billion for the period 1976–1990. (All costs mentioned here are in 1975 dollars.) With inflated dollars, these costs will be much higher.

- $21 billion will be for requirements under the Clean Air Act of 1970.

- $22 billion will be for no significant degradation under the 1977 amendments as supported by the EPA.

- $21 billion will be for controls under the Federal Water Pollution Control Act.

The annual cost in 1990 is equivalent to $191 per household. Consumer electric bills will be 16 percent higher (see Table 12.3).[3]

2.  *Energy Impacts*  6 to 8 percent of the energy produced by a new coal-fired generating unit will be needed to operate scrubbers for sulfur

---

[3] Lewis J. Perl, "The Cost of Clean Air Legislation." Hearings before the Senate Committee on Public Works, Feb. 22, 1977, and unpublished materials of Dr. Perl.

oxides and cooling towers for thermal discharges.[4] In effect, for every 14 new coal units, a fifteenth must be built for environmental controls.

Nearly 500 million extra barrels of oil-equivalent will be needed to comply with the Clean Air Act of 1970 and the "no significant deterioration" amendments of 1977, supported by the EPA.[5] This is for the period 1976–1990.

3. *Benefits* According to a study by Dr. Lewis J. Perl of National Economic Research Associates, the benefits to be derived from 1977 legislative amendments supported by the EPA are one-fifth of the costs. These amendments provide that the industrial process shall result in "no significant deterioration" in air quality. For example, the agency may set a certain standard of good quality for a certain state. Another state may have air quality that is better than the standard. The amendment requires that this higher quality should not be disturbed, which may be desirable. The question: is it worth $22 billion? Are people willing to pay the excessive cost for the extraordinary quality when the lesser standard is satisfactory for health?

These are samples of the studies and deliberations of the government regulatory bodies and company managements in their endeavor to provide electric service with a minimum detrimental effect upon the environment. The aim is that the benefits derived from the improvements should at least equal the cost of obtaining them.

In the face of numbers such as these, policy makers and regulators must carefully evaluate the consequences of environmental overregulation in their effect on the personal finances of the American taxpayer and consumer as well as on the overall economy of the nation.

---

[4] Testimony of Donald G. Allen before the House Committee on Public Works and Transportation on Apr. 19, 1977.
[5] NERA.

# THE ENERGY CRISIS AND THE NEED FOR A LONG-TERM ENERGY POLICY

As noted in previous chapters, the primary sources of energy in recent years have been fossil fuels (namely coal, oil, and gas), falling water, and nuclear energy. Most of the available sites for economical hydroelectric power have been developed. This leaves only fossil fuels and nuclear energy as sources until research can make other forms feasible.

Also, roughly 80 percent of the primary energy used in the United States in recent years has come from oil and gas, which are in relatively short supply. The U.S. production of domestic crude oil has not increased, because old wells were depleted as rapidly as new wells were developed. Consequently, to meet its growing energy requirements, the United States has been required to rely more and more upon imported oil, a large part of which has come from Southeast Asia and the Middle East.

These facts have been known to energy people for some time. However, owing to a lack of an overall energy policy in the United States, the public did not become fully aware of this problem until the oil embargo

THE ENERGY CRISIS AND THE NEED FOR A LONG-TERM ENERGY POLICY

cut off practically all oil imports. This brought on what is now generally referred to as the "energy crisis."

The shifting from one form of primary energy to another is not a new phenomenon. Terming the current energy problem an energy "crisis" tends to confuse the issue, causing many people to believe there is a shortage of total energy. While there may be a public awareness that oil and gas are in essentially short supply, there is no immediate shortage in the supply of coal or uranium if these sources are both wisely developed. The problem arises because it was not anticipated that imports of foreign oil could or would be cut off.

The embargo was followed by a large increase in the price of imported oil. This factor plus all the other inflationary factors caused the price of other primary fuels to rise precipitously. As pointed out in previous chapters, this development came as an unpleasant surprise to a public which had become accustomed to an abundant supply of low-cost energy.

In Chapter 11, there is an outline of the plans being carried on to develop other sources of primary energy and to make coal more generally useful. The energy crisis will be discussed here under the following headings:

- Review of energy in relation to living standards
- Fuel resources
- U.S. trends and patterns
- World trends and patterns
- Conservation and the efficient use of energy
- How to become independent of oil imports

In order to discuss this problem as a whole, brief reference will be made to previous charts which have a bearing on the subject.


## REVIEW OF ENERGY IN RELATION TO
## LIVING STANDARDS

In all national efforts of this kind where a change in basic policy is needed, the aim is always to ensure that the benefits to be derived from the change are at least equal to the cost of making the change. It is assumed that the overall goal is to improve the well-being and living standards of the greatest number of people. Previous chapters have illustrated the close correlation between living standards and the use of energy.

Refer to Chart 6.54 for the relationship between the percentage change in energy use and the percentage change in employment. Chart 6.55 shows the result of studies by Fremont Felix involving all the nations on earth. This chart shows the standing by nations in the relationship between energy use per capita and national income per capita.

Chart 6.56 shows, for the United States, the relationship between the use of electric energy per capita and the per capita personal income in constant dollars. For example, notice from this chart what would happen if suddenly there were no further growth in the kilowatthours per capita. Correspondingly, there would be no further growth in per capita national income. It is difficult to imagine how the benefits from a "no growth" policy in electric energy could equal the enormous costs in employment and personal well-being of a no growth policy on income per capita. In this connection, see also Chart 7.7 for the relationship between total energy consumption and gross national product.

This review leads to the observation that, whereas it is wise to conserve energy by reducing waste to a minimum, care should be exercised to avoid a policy calling for curtailment of *useful* energy. Such a policy would have a detrimental effect upon overall living standards.

## FUEL RESOURCES

For many years the U.S. Department of the Interior has maintained continuing studies of the various fuel resources. A leader in the field is Dr. M. King Hubbert of the Bureau of Mines. Chart 13.1 shows a review by Dr. Hubbert testifying before the Committee on Interior and Insular Affairs, U.S. Senate (1974). This chart indicates why it is now necessary to import large amounts of oil and why these imports will probably increase in the future. Chart 13.2 shows the change in the projection after taking into account the Alaskan North Slope oil and enhanced recovery methods. Chart 13.3 shows the domestic oil production in a slightly different way. Notice that there has been almost no increase in production from 1967 through 1978. The new wells were offset by the depletion of the old ones. Domestic production drops even after the proven reserves are considered and the North Slope of Alaska is included. With new fields, new pools, and extension of old fields, and with proven reserves of the North Slope of Alaska, domestic production will remain about steady through 1990. With those conditions (which may or may not be realized), all additional oil for the United States will have to be imported.

# U.S. PETROLEUM LIQUIDS PRODUCTION CYCLE
## (1972 ESTIMATE)

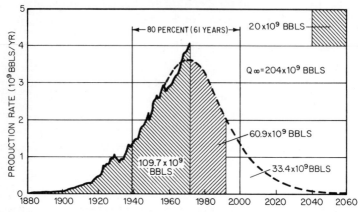

**CHART 13.1** (Source: U.S. Energy Resources: A Review as of 1972, M. King Hubbert, Committee on Interior and Insular Affairs, U.S. Senate, 1974.)

A similar "bell shaped" curve has been developed by Dr. Hubbert for natural gas. This is shown in Chart 13.4. Chart 13.5 shows a projection of the natural-gas supply in the United States, including the Alaskan North Slope and with stimulation techniques.

All studies of oil and natural gas indicate that the supplies will be exhausted sometime shortly after the year 2000. Conservation of these valuable fuels is essential. Substitute fuels, such as coal and uranium which are in much longer supply, should be replacing oil and gas as early

# U.S. PRODUCTION OF PETROLEUM LIQUIDS

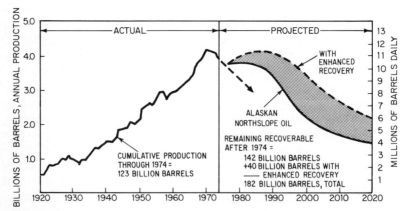

**CHART 13.2** (Source: A National Plan for Energy Research, Development and Demonstration, ERDA–48, 1975.)

# PROJECTED U.S. PETROLEUM LIQUIDS SUPPLY
## (PROJECTION SERIES C)

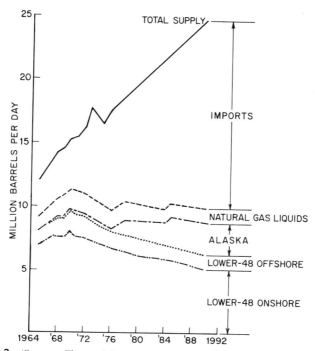

**CHART 13.3**   (Source: Figure 6.1, p. 136, Energy Information Administration, U.S. Department of Energy, Annual Report to Congress, Vol. II, 1977.)

# U.S. NATURAL GAS PRODUCTION CYCLE
## (1972 ESTIMATE)

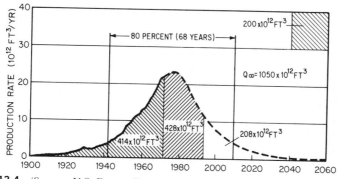

**CHART 13.4**   (Source: U.S. Energy Resources: A Review as of 1972, M. King Hubbert, Committee on Interior and Insular Affairs, U.S. Senate, 1974.)

# U.S. NATURAL GAS SUPPLY

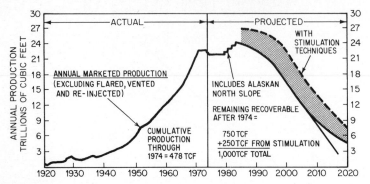

**CHART 13.5**   (Source: A National Plan for Energy Research, Development and Demonstration, ERDA–48, 1975.)

and as expeditiously as practical. Again examine Chart 11.2 for a picture of the available energy from all recoverable domestic energy resources. Note the volume of available coal compared with gas and petroleum. Then compare these with the volume of uranium available with breeder reactors.

## U.S. TRENDS AND PATTERNS

For a pattern and forecast of the U.S. electric energy requirements to the year 2000, review Charts 7.10 and 7.12. A close examination of Chart 7.12 indicates that the pattern is one of a gradual decline of the rate of increase per year as the use of energy per capita increases. The author has made no attempt to forecast energy requirements beyond 2000. However, others have done so. Practically all forecasters believe that the long-term trend is one of a gradual decline in the rate of increase, which also appears to be a world pattern. Chart 7.14 is not a forecast but simply the application of the mathematical formula derived from the trend from 1900 to 2000.

Review Chart 4.19, which shows the percent of primary energy in the United States that is used for conversion to electric energy. This is now just under 30 percent and is forecast to be about 50 percent in the year 2000. There appears to be a pattern that will continue beyond the year 2000, with the percentage of primary energy used for conversion to electricity increasing beyond 50 percent after the year 2000.

From Charts 4.19 and 7.12, it is possible to find a point at the year 2000 representing the forecast of total primary energy. An efficiency of conversion from primary to electric energy may be assumed to be about 38 percent. A reexamination of Chart 4.10 indicates that this pattern will continue to the year 2000 and probably beyond, with the rate of growth of electricity continuing to be about two to three times the rate of growth of primary energy.

Review Chart 3.3 for a picture of the manner in which the pattern of primary fuels has changed over the years. From 1850 through 1870 or 1880, most of the fuel was wood. Bituminous coal and anthracite coal reached their peak after 1900 and then began to decline. The large use of oil and gas began during the 1920s and accounts for a very large percentage of the total fuel used today. Bituminous coal is only about 20 percent. Efforts should have begun 25 years ago to make the gradual shift back to coal.

This pattern is further illustrated in Chart 3.4, which also shows the forecast to the year 2000. This forecast is based upon a judgment as to what will happen if present trends continue. Naturally, by the year 2000, coal should be a much higher percentage and oil a much smaller percentage of the total. Chart 3.5 should be reviewed because it shows how primary fuels are used for conversion to electric energy, with a forecast to the year 2000.

Chart 13.6 shows for the United States the forecast of the U.S. De-

## PROJECTED CHANGES IN USE OF ENERGY IN THE UNITED STATES

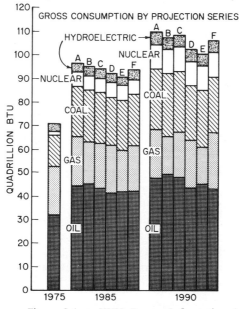

**CHART 13.6** (Source: Figure S.4, p. XXV, Energy Information Administration, U.S. Department of Energy, Annual Report to Congress, Vol. II, 1977.)

# ANNUAL GROWTH RATES OF OIL AND GAS PRODUCTION AND THE ECONOMY, 1975-1990 (PROJECTION SERIES A THROUGH F)

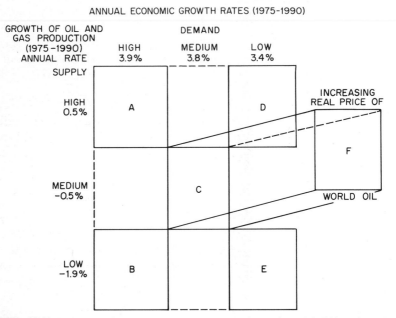

ANNUAL ECONOMIC GROWTH RATES (1975-1990)

**CHART 13.7** (Source: Figure 1, p. 4, Energy Information Administration, U.S. Department of Energy Annual Report to Congress, Vol. II, 1977, Executive Summary.)

partment of Energy, indicating energy by source through 1990. This chart shows the results based upon a variety of assumptions regarding annual economic growth rates and annual rates of growth of oil and gas production. These assumptions are shown in Chart 13.7.

Chart 13.8 shows the forecast of the United States Department of Energy indicating the U.S. energy use by major consuming sectors. This chart also assumes the projection series indicated in Chart 13.7.

Chart 13.8 does not show electric energy as a separate consuming sector. In each of the consuming sectors shown, part of the primary energy is first converted to electric energy for end use.

These forecasts anticipate a continued high use of petroleum, which means that either the high imports will continue or getting oil from coal will prove to be economically feasible so that oil can be so produced domestically in very large quantities.

# PROJECTED CHANGES IN U.S. ENERGY USE BY MAJOR CONSUMING SECTORS (BY PROJECTION SERIES)

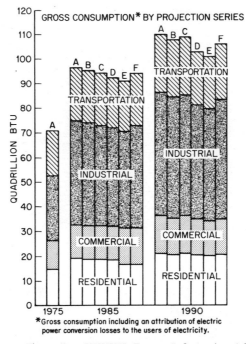

**CHART 13.8**  (Source: Figure 5, p. XXXVII, Energy Information Administration, U.S. Department of Energy, Annual Report to Congress, Vol. II, 1977.)

## WORLD TRENDS AND PATTERNS

For a number of years Fremont Felix, consultant, Gibbs and Hill, Inc., New York, New York has made exhaustive studies and published reports on the pattern and growth of world energy and related subjects. Chart 13.9 shows his forecast to the year 2020.

Note the decline in gas and oil beginning about 1985, with these resources becoming exhausted around 2020, and the continued increase in coal and shale use and the large increase in nuclear energy. The same article gives the Felix forecast to the year 2020, indicating world electricity generation and the primary fuels that will be used for that purpose (Chart 13.10).

Chart 13.11 shows the world energy forecast to 2020 by Dr. Ralph

# WORLD ENERGY USE, 1950–2020
## *FREMONT FELIX*

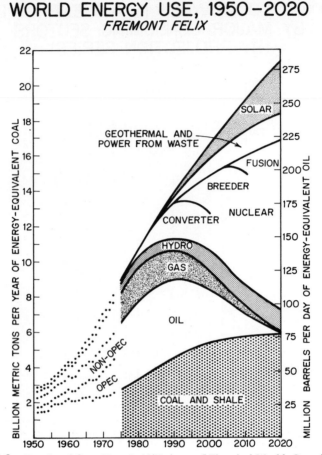

**CHART 13.9** (Reprinted from Nov. 1, 1975, issue of Electrical World. Copyright 1975, McGraw-Hill, Inc. All rights reserved.)

E. Lapp. Lapp shows somewhat the same pattern, with a decline in the fluid hydrocarbons beginning about 1995 and going to almost zero in the year 2040. The solid fossil fuels, largely coal, continue to increase. Lapp also indicates a high percentage of energy coming from nuclear fuels.

## CONSERVATION AND THE EFFICIENT USE OF ENERGY

While it is and should be the aim to conserve energy that may now be wasted, the basic goal is to utilize energy in the most efficient fashion in

# WORLD ELECTRICITY GENERATION, 1950–2020
## *FREMONT FELIX*

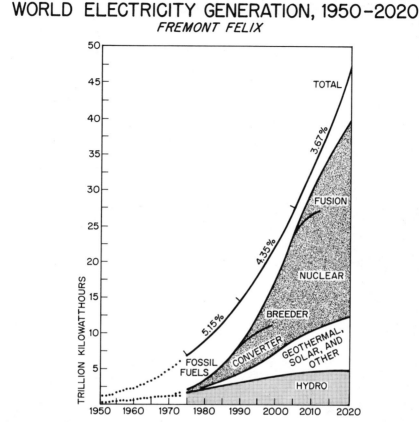

**CHART 13.10**   (Reprinted from Nov. 1, 1975, issue of Electrical World. Copyright 1975, McGraw-Hill, Inc. All rights reserved.)

improving living standards. Chart 13.12 is one measure of the actual improvement in the efficiency in the use of total energy from 1947 to 1972, with a forecast to 2000.

Fremont Felix has made some interesting studies to determine the relative efficiency of various nations in the world in their use of energy. One result is shown in Chart 13.13, where the base is expressed in 1972 per capita GNP (U.S. dollars) and the vertical scale expressed in 1972 energy use in kilograms of equivalent coal per U.S. dollar of GNP. Those nations having a higher GNP use a higher amount of energy to accomplish their purpose. The measure of efficiency is the manner in which the energy is used in relation to the GNP. Most of the principal nations are shown on this chart. The curve on the lower right-hand side is the lowest use of energy in relation to the GNP. The United States, Sweden, Switzerland, Bermuda, France, and Oman are approximately on this

# WORLD ENERGY CONSUMPTION, 1970–2040
## RALPH E. LAPP

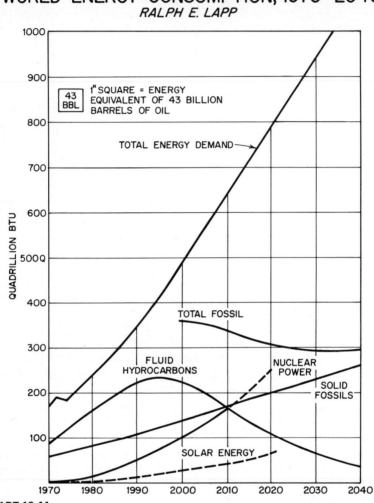

**CHART 13.11**

line with Denmark and Israel close to it. This chart refers to total primary energy use, and indicates that the United States (with all its wastes) is among the most efficient nations on earth in the use of total energy. A similar study is shown in Chart 13.14, indicating the efficiency of the various nations in their use of electric energy. The horizontal scale is 1972 per capita consumption of kilowatthours, while the vertical scale is 1972 consumption per U.S. dollar of GNP. This chart also shows the curve of the lowest use of electricity in relation to the GNP, again demonstrating that the United States is among the group of nations showing the highest efficiency in the use of electric energy.

# GROSS ENERGY CONSUMPTION PER DOLLAR OF 1958 GROSS NATIONAL PRODUCT
## *U.S. DEPARTMENT OF THE INTERIOR*

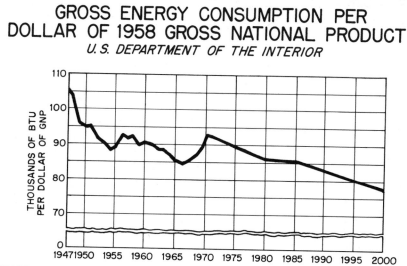

**CHART 13.12**   (Source: United States Energy through the Year 2000 by Walter G. Dupree, Jr. and James A. West, U.S. Department of the Interior, December 1972.)

These charts are shown because one frequently hears the statement that the United States uses energy in an inefficient fashion. When American usage is compared with that of other nations, however, this statement does not hold true. Obviously, this does not mean that there is no room for conservation and for avoidance of waste. Indeed, there is much that can be done to further improve efficiency.

For example, there is waste in the heating and cooling of buildings that are not properly insulated. More mass transportation in the larger cities would conserve energy. More suburban mass transportation would help. Further attention to the efficiency of appliances and equipment will help in conservation. Automobiles can be more efficient in the use of fuel.

However, when one speaks of conservation, it should not be assumed that little has been done so far in the conservation of energy. Most energy is used by business concerns which have a primary interest in making a profit. This encourages holding expenses to a minimum. Much work has been done to improve the efficiency of burners and machines. More can be done and should be done.

However, it would be unwise to assume that there is sufficient waste to enable the elimination of oil imports by means of conservation alone. The problem is so large that it will require a great deal more than could possibly be attained through conservation. The basic fact is that the United States has placed too much emphasis on the use of oil and gas, which are in scarce supply, and too little emphasis on coal, which is in

# ENERGY USED PER U.S. DOLLAR OF GNP IN KILOGRAMS EQUIVALENT COAL, 1972

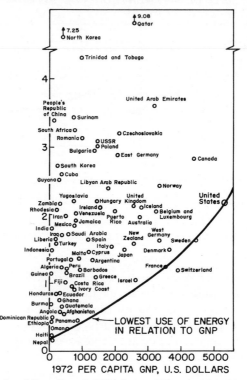

**CHART 13.13** (Reprinted from Dec. 1, 1974, issue of Electrical World. Copyright 1974, McGraw-Hill, Inc. All rights reserved.)

long supply. Furthermore, although nuclear energy has made a good start, it is not receiving nearly the amount of attention that is required if the nation is to become independent of imports in any reasonable length of time.

## HOW TO BECOME INDEPENDENT OF OIL IMPORTS

If the United States is to continue its growth in jobs and living standards, as most people desire, it must continue with a corresponding pattern of growth in the use of energy to drive the machines to produce the goods and services. As oil and gas are in short supply, this means that, until some other new fuel is developed, reliance must be upon coal and nu-

# PER CAPITA KILOWATTHOUR CONSUMPTION VS. CONSUMPTION PER DOLLAR OF GNP, 1972

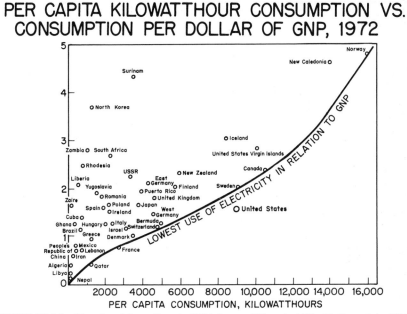

**CHART 13.14**    (Reprinted from Dec. 1, 1974, issue of Electrical World. Copyright 1974, McGraw-Hill, Inc. All rights reserved.)

clear energy to replace gas and oil. The only practical way to use the coal and nuclear energy for the end product is through the electric process; thus the trend toward electricity will not only continue as it has in the past but should accelerate. In 10 or 15 years it may be possible to get oil and gas from coal, but the economics at the moment are not certain. The processes of converting coal and nuclear energy to electric energy are well established and practical.

A good illustration of the problem can be seen from a set of charts prepared by the Joint Committee on Atomic Energy in 1973. These charts are interesting in that they graphically show the amounts of the various fuels being used and their relative efficiencies. For example, notice Chart 13.15, which represents the year 1970. The figures on the chart are expressed in quadrillion Btu per year. A quadrillion is $10^{15}$. It is a standard unit expressing a very large quantity of energy.

The chart is drawn to scale so that the distance between the lines represents the quantity in quadrillion Btu. For example, the distance between the lines showing domestic oil represents 21.6 quadrillion Btu. The distance between lines of oil imports represents 7.3 quadrillion Btu.

The chart also shows how the various fuels flow into the economy. A certain portion goes for electric energy generation and takes 14.8 qua-

# U.S. ENERGY INPUT, REJECTED ENERGY, AND USEFUL ENERGY IN QUADRILLION BTU PER YEAR, 1970

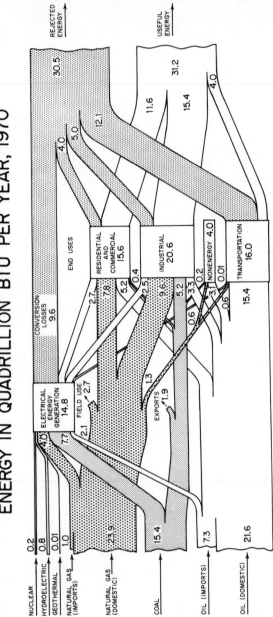

CHART 13.15

drillion Btu. The next segment to absorb energy is the residential and commercial area, which takes 15.6 quadrillion Btu. Next is the industrial area, then the nonenergy group, and finally the transportation area, taking 16 quadrillion Btu. Thus, one can see the amount of each fuel going into each division of the economy for end-use purposes.

The chart also shows for each of the areas of end use the amount of energy that is useful and the amount that has been wasted. For the year 1970, the rejected or wasted energy amounted to 30.5 quadrillion Btu. The useful energy amounted to 31.2 quadrillion Btu.

The conversion losses in connection with electrical energy generation amount to 9.6 quadrillion Btu, representing about 65 percent of the energy. This means that the average efficiency of electrical generation is 35 percent. Parenthetically, this is the highest mechanical efficiency of any of the commercially feasible large-scale engines in converting primary fuels to electric energy. It is also the highest efficiency in converting primary fuels to mechanical work. Now look at the transportation area, which shows 12.1 quadrillion Btu as rejected energy, representing about 76 percent of the 16 quadrillion Btu used in transportation and reflecting an overall efficiency of about 24 percent. The automobile engine is probably the most inefficient of the group in transportation.

Chart 13.16 reflects a similar picture for 1980. Notice the large increase in total amount of oil used. Domestic oil use remained almost the same—an increase from 21.6 to 23.9 quadrillion Btu, but oil imports increased from 7.3 to 22.4 quadrillion Btu. The oil imports in 1980 are forecast to be approximately equal to the total domestic production of oil.

Coal use increased, but not nearly enough if coal is to take its rightful place. Notice the decline in the natural gas input. Electric generation continues to be the large user of coal. Relatively little oil was used for this purpose in 1970. It is expected that a lower percentage will be used in 1980.

The 9.4 quadrillion Btu forecast for nuclear energy used for generation of electricity in 1980 is probably on the high side. There has been some slowing down in the installation of new nuclear power plants, which is unfortunate because it means prolonging the time during which the United States will be dependent upon imports of foreign oil.

Chart 13.17 for 1985 shows the still-larger increase in the use of oil, with domestic oil actually decreasing somewhat from 1980. Oil imports will be roughly 60 percent of all the oil used, and in fact the figure may be higher than that. Notice the 19.5 quadrillion Btu in this study which represents the nuclear energy to be used for electric generation. This

# U.S. ENERGY INPUT, REJECTED ENERGY, AND USEFUL ENERGY IN QUADRILLION BTU PER YEAR, 1980

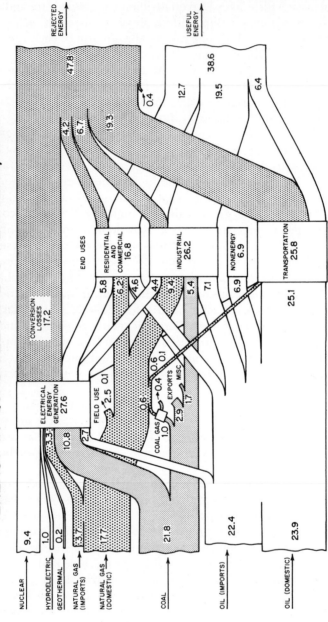

**CHART 13.16**

# U.S. ENERGY INPUT, REJECTED ENERGY, AND USEFUL ENERGY IN QUADRILLION BTU PER YEAR, 1985

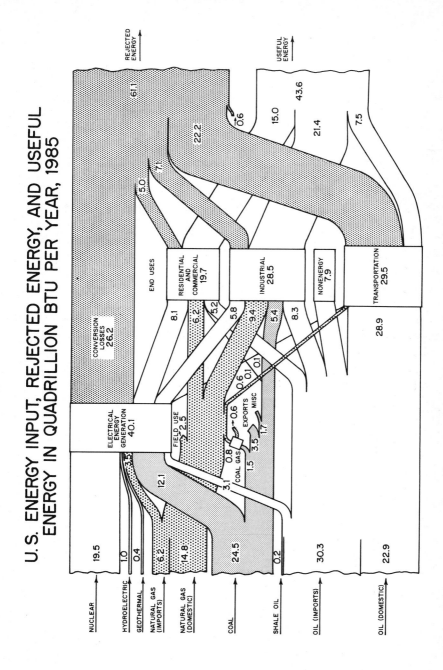

CHART 13.17

383

represents about half of all the fuel used for electric generation. It now looks as though nuclear fuel should account for about 40 percent of total electric generation by the year 2000. Coal should account for nearly half. In both instances, the achievement of these results is going to require a change in national policies.

A review of Chart 13.18 may help the reader to visualize the severity of the problem. This chart shows for 1972 in the United States oil and gas usage by categories. The largest category is transportation, with slightly over half this fuel being used by automobiles and the balance being used by trucks, trains, and planes. The next largest category is space heating for homes and commercial buildings. Next is processed steam, followed by direct heating. Consider first the two largest uses of oil and gas, namely for space heating and transportation.

In 1972, space heating required roughly 11 quadrillion Btu. Chart 13.17 indicates that oil imports will account for 30.3 quadrillion Btu in 1985. The only practical way to convert space heating from oil and gas to coal and nuclear is through the electric process. If all homes and office buildings in the United States were electrically heated, they would save the equivalent of something like one-third of the oil imports for 1985. To electrify all homes and office buildings in 10 years would be a massive undertaking, if it is at all possible. In 1976 about 13 percent of dwelling units were electrically heated. It has taken about 15 years to reach this point. Estimates are that by the year 2000 approximately half the homes and office buildings in the United States will be electrically heated.

In the transportation area, there is no practical way at the moment

# OIL AND GAS USAGE IN THE UNITED STATES, 1972

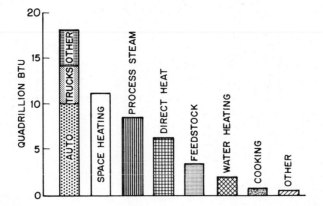

**CHART 13.18**

to electrify planes or trucks. There could and should be more electrification of trains. The big area for the use of oil is in the automobile, and there is no practical way at the moment to electrify cars for intercity operation. However, it is possible and practical to use electric energy in automobiles for intracity operations.

The lead battery is satisfactory for town use. The car could travel 50 to 75 mi or more on a single charge. depending upon the design. The battery can be recharged overnight. Surveys have revealed that the average automobile trip is 7.3 mi. The average car makes 1318 trips annually, or an average of 26.3 mi per day. Only 4.25 percent of all trips in personal cars are over 20 mi, and only 0.4 percent are over 75 mi.[1]

Assume for the moment that all automobiles are electrified and thus change from the use of petroleum to the use of coal and nuclear fuel through the electric process. Notice from Chart 13.18 that automobiles used about 10 quadrillion Btu in 1972, about one-third of the oil imports forecast for 1985. Obviously, the attainment of independence of oil imports in any reasonable length of time is going to take a combination of many and varied efforts, including all the stepped-up research for electrification and gasification of coal and the development of a lightweight battery. In addition, there should be stepped-up efforts toward legislation that would enable both the development of coal mines at reasonable cost and a method of strip mining that is both economical and environmentally safe. There must be a balance between air-quality goals and energy needs to facilitate the burning of coal and simplification of the regulatory processes that will enable more expeditious installation of nuclear power plants without the sacrifice of any safety provisions. There should be examination of the railroads in order to ensure that coal can be delivered to the power plants. Electric utilities need the allowance of a return on capital that is not less than the cost of money, or else the power plants cannot be built to enable the transition from oil to coal and nuclear energy.

A year and half after President Carter set forth his energy proposals, he signed into law the National Energy Act in November 1978. In many ways, Congress improved upon the initial recommendations, but the Act does not represent the comprehensive and effective national energy policy which the American people need. Basically, the Act emphasized conservation, but it provided little to enhance the supply side of the picture.

Conservation is, of course, a key element. However, it cannot provide the solution by itself. Even with the best in conservation, electricity

---

[1] Stanley J. Kalish, *A Study of the Market Potential for On-the-Road Electric Vehicles,* Electric Vehicle Council, 1971.

generation will have to be three to four times the 1976 level to meet U.S. needs at the turn of the century.

New sources of energy, such as solar and geothermal, must be developed. But despite accelerated research and development efforts in alternative energy forms, sources other than fossil and nuclear fuels will be supplying only about 10 percent of electricity generation by the year 2000—and about half of that will be hydro.

To avoid adverse effects on energy supply, sound regulations implementing the National Energy Act are required. In addition, proposals being put forth by the Department of Energy for the second National Energy Plan (NEP-2), focusing on supply, merit careful consideration.

Other legislation is also needed to remove the many existing restraints on the timely development of U.S. energy resources. Although government has recognized that coal and nuclear power must be used to meet future energy demands and reduce dependence on overseas oil, steps have been taken which make their utilization more difficult.

Coal has suffered through the enactment of two significant measures. The Surface Mining Act passed in 1977 has resulted in regulations restricting the opening of new mines and threatening the continued operation of a number of mines now in existence. Coal supply will be adversely affected and its cost will increase. The Clean Air Act Amendments, also passed in 1977, are adding greatly to the problems of siting and constructing coal-fired power plants. In fact, depending on the final regulations which are imposed, it may become impossible to build a power plant or any other industrial installation in many areas of the nation. Billions of dollars of added costs to utilities and their customers can be anticipated. Inflation, a priority concern of the American people, will be unnecessarily pushed if a more balanced approach to environmental controls is not taken. Further, the amendments hold a grave potential for increased unemployment, dislocation of commercial and industrial enterprises, and consequent social disruptions.

Where nuclear energy is concerned, the regulatory process has become a maze of unpredictability and delay, largely as a result of the tactics of antinuclear activists. In other nations, a nuclear power plant can be completed in about 6 years—in the United States it now takes some 10 to 12 years to license and build such a plant. Unnecessary cost increases of some $120 million per year constitute the price tag for delay of a large nuclear installation. Understandably, such conditions have made utilities wary of planning new nuclear units.

In 1978, the Administration proposed a bill to improve the regulatory process and shorten lead times. This was vigorously opposed by antinuclear forces and died with the expiration of the Ninety-fifth Congress. It can only be hoped that perseverance will eventually produce the necessary legislation.

As of October 1978, there were 72 licensed nuclear units, of which 67 were in commercial operation. If no new nuclear units were added through 1990 to that total, there would be a gap of 20 percent between the projected supply of electricity and the demand for it. Closing that gap would require the use of coal or foreign oil. The coal alternative would mean utilizing that fuel to produce more than $2^1/_2$ times the amount of electricity generated by coal in 1978—a very doubtful prospect, since present circumstances indicate that even a doubling would be hard to accomplish. If oil were used instead, imports would have to increase by more than 3 million barrels per day, a most unsatisfactory solution in light of the 1978 average daily import total of some 8 million barrels for the nation as a whole.

To sum up, restraints which endanger power supply for the future still continue and, indeed, in some instances have become more binding. These include overlapping and conflicting governmental regulations which impede the siting and timely completion of new facilities; conflicts between environmental goals and energy requirements; lack of timely and adequate rate relief, which adversely affects the ability of utilities to finance needed facilities; impediments to the development of necessary coal and uranium fuels; and lack of stable government policies relative to the supply of electric energy.

The objectives of sound energy policy center on two fundamentals: (1) an energy supply sufficient to provide for needed economic growth and the fulfillment of the basic aspirations of the American people in their desire for a better life; and (2) the capability to supply that essential energy by means which will be unaffected by developments outside the United States. Actions by government should foster, not inhibit, the achievement of these objectives.

The American people, their elected representatives, and their appointed officials must clearly understand these points and test every proposal involving energy against them. If they do this objectively and conscientiously, the right decisions will be made and the appropriate actions will be taken so that energy supply will not become a constraint on America's future.

# INDEX

A-rated bond, 40
Active power, 17
Administrative and general expense, 44
AEC (*see* Atomic Energy Commission)
Aesthetics, 348, 351
  of transmission towers, 352
Air conditioner, 250
Air-conditioning load, 215, 221, 253
  (*See also under* Load)
Air pollution, 347, 356
  first committee on, 350–351
  (*See also under* Clean air)
Air quality control, 352
Alaskan North Slope, 367
All-electric home, 223
Allen, Donald G., 364
Alternating current (ac), 25
Alternative energy forms, 386
American Economic Foundation, 173
American economic system (*see* Economic
  system, American)

American Public Power Association, 286,
  332
Amortization, 43
Ampere (A), 7, 15
Animate energy, 138
Annual revenue, 41
Anthracite coal, 371
Appliances, 213, 221, 328
  saturation index for, 222
  small, 250
Appraising process, 121
Aquatic life, 354
Area coordination of power supply, 69
Area development, 214
Assets, fixed, 37
Association of Edison Illuminating Com-
  panies, Load Research Committee of,
  230, 246
Atomic bomb, 88, 336
Atomic energy (*see* Nuclear energy)
Atomic Energy Act of 1954, 80

Atomic Energy Commission (AEC), now NRC, 64, 79, 337, 358, 362
Atomic Safety and Licensing Board (ASLB), 357
Attitude in employee relations, 311
    measurements of, 312
Automatic meter reading, 327
Automobiles, 384
Average embedded percent return on earnings, 231
Average-load method of cost allocation, 265
Average price of electricity, 96, 97

Balance for return, 44, 109
Balance sheet, 36, 37
Baseload unit, 31
Battery (fuel cell), 327
Bermuda, 375
Bituminous coal, 26, 371
Bituminous Coal Institute, 351
Blackouts, 72
Block rates, 270, 283, 286
Boiler, 11, 78
Boiling-water reactor (BWR), 81, 337
Bonds, 38
    A-rated, 40
    bond rating agencies, 306
    first-mortgage, 40
    interest on, 46
        rate of, 236
    yield of, 41
    (See also Common stock; Preferred stock)
Bonneville Power Administration, 65
Book value of common stock, 38
Breeder reactors, 87, 337–340, 356, 370
    commercial, arguments for, 340
    CRBR, 88, 340
    LMFBR, 87, 339, 340
    LWR (see Light-water reactor)
    PLBR, 340, 342
British plan, 227
British thermal units (Btu), 104
Btu forecast, 381
Bulk energy supply, 72
Bundles (fuel assemblies), 82
Bureau of Mines, 367
Buying power, 142
BWR (boiling-water reactor), 81, 337

Canadian nuclear power reactors, 86
CANDU (Canadian deuterium uranium), 86–87

Capability, power, 51
    exchange of, 58
    reserve, 56, 72
    total, 56, 200
Capacitors, 16
Capacity:
    of electric generator, 12
    firm power (system), 31
    of generating unit (rating), 51
        excess, 52
    standby, 29
    for water pumping, 9
Capital, 131, 144
    return on, 385
Capital consumption, 129
Capital costs, 259
Capital-intensive electric utilities, 91
Capital investment, 161, 204, 257
Capital structure, 289
Capital surplus account, 37
Capitalization of Edison Power Company, 39
Carter, Jimmy, 340, 385
"Cascading" in power interruption, 67, 72
Central dispatching office, 62
Central stations, 288
Certificate of convenience and necessity, 293
Chain reaction in nuclear power, 81
Chicago, Ill., 288
Cisler, Walker, 87, 340
Citizens Advisory Committee on Recreation and Natural Beauty, 350
Civic affairs, 322
Cladding of fuel elements, 83
Class load factors, 22
Clean Air Act of 1970, 352, 354, 363–364
Clean Air Act Amendments (1977), 352, 386
Clean air standards, 353
Clinch River Breeder Reactor (CRBR), 88, 340
Closed cycle in conventional plant, 78
Club of Rome, 197
Coal:
    aesthetics and, 356
    anthracite, 371
    bituminous, 26, 371
    coal alternative, the, 387
    conservation of, 326
    converting coal to oil and gas, 327
    effect of, on environment, 353
    for electric generation, 381
    as fossil fuel, 365
    legislation, surface mining of, 385

Coal (*Cont.*):
  as replacement for oil and gas, 368
  research on getting oil from, 372
  shift to, 371
  solvent refined, 353
  for space heating, 384
  supply of, 355, 366, 377–378
  in transition to fusion/solar power, 344
  volume of, compared with gas and petroleum, 370
Coal mines, 385
Coincident demand, 21
Coincident maximum demand, 22
Cold reserve, generator, 30
Commercial load characteristics, 250
Commercial rates, 283
  schedule of, 272
Commercial sales, 224
Commercial service, 233, 261
Committee on Interior and Insular Affairs, U.S. Senate, 367
Committee on Power Capacity and Pooling, EEI, 65
Common stock, 35
  book value of, 38
  dividends on, 46
  as no-par stock, 38
  stated value of, 38
  stockholders of, 38
  stockholders' equity, 38, 41
    capital surplus account, 37
    earned surplus account, 37
Communicating information, 121
Community affairs, 119
Competition, 146, 245, 274, 275, 287, 288, 292
Computer (*see* Electronic computers)
Condensers, 16, 78
  static, 16
  steam, 78
  synchronous, 16
Condensing water, 78
Conductors, 32
Congress:
  appropriations for TVA by, 296
  Ninety-fifth, and nuclear licensing bill, 386, 387
Conservation:
  and efficient use of energy, 374
  of energy, 197, 225, 377
    marketing and, 212
    National Energy Act (1978) emphasizes, 385
  of environment, 322
    (*See also* Environmental protection)
  of fuels, 350, 368

Construction labor, 99
Construction work in progress (CWIP), 304
Consumer(s):
  education of, 159
  expenditures by, 150
  (*See also* Customers)
Consumer price index, 100
  and electricity component, 101
Control rods, BWR, 81
Controlled thermonuclear reaction (CTR), 343
Conversion:
  to electric energy, 108, 370
  through electric process, 384, 385
  to electricity, 132
  energy, methods of, 333
  methods of, correlations of, 186
Coolant, nuclear reactor, 81
Cooling water, 78, 348
Cooling ponds, 355
Cooling towers, 355, 364
Coordination of power supply, 65, 66, 70, 72
Core of nuclear fuel, 80
Corporate model, 125, 182, 243, 244
  corporate profits tax and, 129
Cost(s):
  of alcoholic products, 243
  of capital, 97, 99
  of clean air legislation, 363
  demand, 263
  of electricity, 225, 239
  embedded, 208, 235, 294, 306
  embedded unit, 267
  energy, 263, 264
  fixed, 236, 257, 262
  of fuels, 110
  of household electricity, 243
  incremental [*see* Incremental cost(s)]
  per kilowatthour sold, 259
  of money, 201, 202, 204, 269, 295, 306
  operating and fuel, 238
  original, 298
  of purchased power, 226
  of service, 245
  of service indexing, 308
  of service pricing, 286
  of tobacco products, 243
  trended present, 300
  unit, 267, 269, 271
  variable, 257, 262
  (*See also* Expenses; Fixed charges)
Cost allocation, 245, 262, 263
  average-load method of, 265
  by customer, 263, 265

Cost allocation (*Cont.*):
  maximum noncoincident method of, 265
  peak-responsibility method of, 265
Cost analyses, 245, 246, 259–260, 262, 263
Cost comparisons, 209, 210
Cost control, 184, 204, 208
Cost curves, 266, 268, 278–280
Cumulative error, 195
Current assets, net, 37
Current dollars, 149
Current liabilities, 37
Current ratio, 37
Customer(s):
  classes of, 260
  costs to serve, 263, 265
  customer accounts expense, 44
  customer-hours of electricity, 102
  customer load factors, 20
  customer-related use, 265
  information to, 214
  service to, 212
  (*See also* Consumers)

Dams, 74–77
  combination, 75
  flood-control, 75
Damsites, 74
Demand, 10, 246, 247, 272, 274, 281, 283
  maximum, 18, 20
  maximum hourly, 12
  peak, 19
  total, 12
Demand charge, 276
Demand cost, 263
Demand-energy rate, 274
  optional, 283
Demand meter, 253, 284
Demand-type rates, 228
Denmark, 375
Department of Energy (DOE), U.S., 64, 289, 330, 333, 354, 358, 372
Depreciation, 36, 45, 289
  taxes and, 44
Depreciation charges, 35
Depreciation reserve (equipment), 45, 299, 300
Detroit Edison Company, The 87
Deuterium, 86, 343, 344
Dike, 77
  (*See also* Dams)
Direct current (dc), 24
Discretionary income, 154
Discrimination, rate, 280, 285, 293
Dispatcher, load, 48

Disposable income, 148
Disposable personal income, 129, 191, 193
Distribution expense, 44
Distribution lines, underground, 352
Distribution systems, 33–34
  facilities of, 33
  investment in, 207
  primary, 34
Distribution (line) transformer, 34
Diversity, 21, 23, 73, 105, 265, 275, 277
Diversity factor, 22
Dividends, stock, 35
Division of income, 156, 170
Division organization of power companies, 53
Domestic energy resources, 339, 370
Domestic oil, 381, 384
Double-A rating on bonds, 40
Dump energy, hydro plant, 59
Duplication and government regulation, 290, 293
Dupree, Walter G., Jr., 377
Duquesne Light Company, 337

Earned surplus account, 37
East Central Area Reliability Coordination Agreement (ECAR), 70
Eastern coal, sulfur content of, 353
Ecological balance, water temperature and, 354
  (*See also* Environmental protection)
Economic characteristics of power companies, 90
  economic climate and, 92
Economic differences among utilities, 291
Economic loading of power plants, 62
Economic Regulatory Administration, 289
Economic system, American, 146, 147, 155
  education and, 152, 155, 173
  facts about, 176
  private property in, 146
  trends in, 108–110
Economies of scale and power industry, 55, 58, 61, 69, 79
  economy energy, exchange of, 58, 59
Edison, Thomas, 324
Edison Electric Institute (EEI), the:
  Committee on Power Capacity and Pooling of, 65
  cooperative research and development and, 79
  *Electric Perspectives* (bimonthly publication), 358

Edison Electric Institute (EEI), the (*Cont.*):
  Electric Power Research Institute and, 330
    funding of, 332
  Electric Research Council and, 328
  electric utility rate design study of, 286
  first air pollution committee of, 350–351
  long-term forecast of, 194
  and nuclear fuel, 337
  purchase and interchange contracts, file kept by, 72–73
  Research Projects Committee of, 328
  research projects undertaken by, 325
  studies on diversity among energy suppliers of, 73
  underground transmission lines in president's environmental program and, 349–350
  wartime shift in use of energy and, 60–61
Edison Power Company:
  annual load chart for, 215
  assistance from EEI, 208
  cost and rate statistics for, 240–241
  cost and sales statistics for, 259
  cost curves for specific years, 279
  distribution of residential customers, 277
  dollar investment per kilowatt of installed capacity, 207
  evolution of power supply and, 53–89
  gross revenue of, 240
  load curves for, summary of, 254–256
  as model for electric energy business, 29–52
  personnel of, 52
  rate design study of, 286
  rate setting and economic climate and, 303
  recordkeeping system of, 298, 299
  total operating expense and fuel cost of, 211
Educational attainment and American economic system, 152
EEI (*see* Edison Electric Institute)
EEI long-term forecast, 194
Efficient use of energy, 374
Einstein, Albert, 334
Electric blanket, 227
Electric energy, 372
  per capita, 367
  efficient use of, 376
  generation of, 194, 195, 379
    breakdown by fuels, 384
    growth in, 102, 134, 194
Electric generator, 11, 29

Electric gross operating revenue, 46
Electric heating, 220, 221
  of houses, 215
Electric net operating revenue, 46
*Electric Perspectives* (EEI), 358
*Electric Power Business, The* (Vennard), 183
Electric Power Research Institute (EPRI), 286, 330, 341
  coordination of, with government agencies, 333
  divisions of, 331
  funding of, 331
  objectives of, 331
  program emphasis of, 331, 333
  summary of activities of, 332
Electric process, conversion through, 384, 385
Electric range (cooking), 246, 247
Electric rates [*see* Rate(s)]
Electric Reliability Council of Texas (ERCOT), 70
Electric Research Council (ERC), 328, 330
Electric systems, 18
  research into, 333
  water analogy and, 5, 9
Electric transportation, 385
Electric utility service (*see* Service, electric utility)
Electric vehicle, the, 228, 327, 351, 385
Electric Vehicle Council (EEI), 351
Electrical systems, 333
*Electrical World*, 378
Electricity:
  industrial use of, 135
  rate of growth of, 102, 134, 194
  rates for [*see* Rate(s)]
  storage of, 91
  in U.S. manufacturing, 226
Electricity component and consumer price index, 101
Electrification of industry, 225
Electromagnetic radiation, 81
Electronic computers, 48, 139
  models, 125, 327
  print outs of, 210
Electrostatic precipitators, 351
Embargo, Middle East oil, 2, 365–366
Embedded cost, 208, 235, 294, 306
Embedded unit cost, 97, 267
Emergency power contracts, 59
Emission standards, 352, 353
Employee relations, 126, 310–314, 316
  employee attitudes and, 311, 312
    surveys of, 315, 320
  employee education and, 172
  employee information and, 312, 318, 320–321

Employee relations (*Cont.*):
  employee opinion and, 157
  personnel for handling, 316
  (*See also* Public relations)
Energy:
  alternative forms of, 386
  animate, 138
  conversion of, methods for, 333
  cost of, 263, 264
  dump, 59
  economy, 58, 59
  efficient use of, 37
  electric (*see* Electric energy)
  from falling water, 134, 365
  fossil fuel, 105, 365, 386
    (*See also specific type*)
  geothermal, 386
  hydroelectric power, 27, 37, 74, 386
  inanimate, 130, 136
  and income, 175
  nuclear (*see* Nuclear energy)
  primary, 370
  solar (*see* Solar energy)
  sources of, 73
  use of: per capita, 174, 187, 198, 367,
    370
    per hour, 140
  use in United States, 375
  useful, 197
  wasteful, 197
Energy crisis, 2, 27, 366
Energy density, 82
Energy policy, 387
  and energy supply, 386, 387
England, 351
  London, 350
Enrico Fermi plant, 87, 338
Environmental protection, 321, 345–347,
  349–351
  concern with, 314
  conservation and, 333
  controls for, 386
    (*See also under* Clean air)
  costs of, 346, 363
  goals of, 387
  overregulation and, 364
  problems of, 99, 352, 353
Environmental Protection Agency (EPA),
  330, 352, 354, 355, 363–364
EPA (*see* Environmental Protection
  Agency)
EPRI (*see* Electric Power Research Insti-
  tute)
Estimated annual revenue (EAR), 231
Excess (generating unit) capacity, 52
Excess power reserves, 31
Exchange of capability, 58

Exchange of economy energy, 58
Expenses, 109, 241
  administrative and general, 44
  distribution, 44
  fixed, 91
  marketing, 44
  operating, 41
  production, 41
  sales, 214
  transmission, 44
  (*See also* Costs; Fixed charges)

Fair return, 245, 257, 280, 294, 295, 297,
  303
Fair value, 267, 289, 294, 297
Falling water, energy from, 134, 365
Family income, 149, 151
  median, 147
Farm market, 223
Farm productivity, 142
Farm service, 261
Fast-flux test facility (FFTF), 341
Federal Energy Regulatory Commission
  (FERC), 210, 289, 298, 308
Federal Power Act, 289
Federal Power Commission (FPC), 279,
  289, 298, 302, 349
Federal power projects, 296
Federal Radiation Council, 359
Federal Reserve Board index of industrial
  production, 191
Federal Water Pollution Control Act, 355,
  363
Feedwater heaters, 78
Felix, Fremont, 174, 187, 367, 373, 375
FERC (*see* Federal Energy Regulatory
  Commission)
Financial ratios, 39
Financing utilities, 35, 306
Firm power (system) capacity, 31
First-mortgage bonds, 40
Fission, 80, 344
  products of, 80
Fixed assets, 37
Fixed charges, 207, 208, 307
  per kilowatt, 237
  on plant, 203
  (*See also* Costs; Expenses)
Fixed costs, 236, 257, 262
Fixed expenses, 91
Flood-control dam, 75
Flue gas desulfurization, 353
Fluorescent lights, 15
Fly ash, 348, 351
Foot-candle (fc), 219
Forced outages, 206

Ford, Henry, 139
Forecasting:
  in kilowatthours, 185, 192, 196, 204
  in kilowatts, 200
  management use of, 119, 183, 184, 199,
    203, 204, 370–373, 381
Foreign oil, 366, 387
Foreign travel, 152
Forward rate base, 203
Fossil-fueled plant, 78
Fossil fuels, 105, 365, 386
Foundation for Economic Education, 173
FPC (see Federal Power Commission)
France, 375
Free market, 145, 295
Free society, 128, 145, 147, 157, 245, 305,
  345
Freedom Foundation, 173
Freezer, 220
Frequency of electricity, 60, 67
Frequency distribution of kilowatthours,
  276
Frequency distribution curve, 276
Fuel(s):
  conservation of, 326
  costs of, 100, 211
    per kilowatthour, 238
  fossil (see Energy, fossil fuel)
  patterns of use of, 26
  prices of, 208
  primary, 100
  raw, 80
  (See also specific type)
Fuel assemblies (bundles), 82
Fuel cell (battery), 327
Fuel clause, 239, 307
Fuel elements, reactor, 81
Fuel pumps, 51
Fuel region blanket, 88
Fuel resources, 367
Functional employee relations,
  311, 316
Functional public relations,
  311, 316
Fusion, 88, 327, 343, 344, 356
Future rate base, 307

Gas, 27, 268, 340, 344, 355, 365, 370,
  373, 377
Gas-cooled reactor, 84
Gas turbines, 79
Gaseous tube lights, 15
Generating plants, 11, 288
  isolated, 292
  on-site, 292
Generating stations, 29

Generation:
  of electric energy, 194, 195, 379
    self-, 105, 107
  by source, 84
  system planning and, 206, 325
Generators, electric, 11, 13, 25, 29
Geothermal energy, 386
Geothermal heat, 26
GNP (see Gross national product)
GNP implicit price deflator, 190
Gompertz curve, 198
Government affairs and public relations,
  317, 321
Government intervention in the economy,
  128, 154, 157, 172–174, 305
Government regulations (see Regulations,
  government)
Governors (devices), 67
Graphite for fuel elements, 84
Great Depression, the, 97, 295
Gross income, 45, 46
Gross national product (GNP), 97, 128,
  129, 147, 191, 367
  implicit price deflator, 190
  U.S. dollar of, 376
Gross revenue, 205, 240
Growth rate, electric energy, 102, 134,
  194

H-frame structure of wood poles, 32
Handy-Whitman index, 94, 300, 301
Heat exchanger, 87
Heat pump, 215, 221
Heating:
  electric, 217, 218, 220, 221
  space, 384
Heavy hydrogen, 343
Heavy water, 86
Heavy-water reactor (HWR), 86
Helium circulator, 86
Hertz (Hz), 25
High voltage in transmitting power, 14
Holding companies, 61, 289
Hope Natural Gas Company case, 302
Horsepower (hp), 7
Horsepower-hour (hp·h), 140
Horsepower-hours per worker-hour, 140
Hot reserve, generator, 30
House-heating rates, 272
Hubbert, M. King, 367, 368
Human behavior, management and, 126,
  154
Hydro-Electric Commission of Ontario, 67
Hydro plant, 29
Hydroelectric power, 27, 73, 74, 386
Hydroelectric reservoir, 74, 78

Hydrogen bomb, 88, 343
Hydrogens, description of, 86

Inanimate energy, 130, 136
Incentives, management use of, 121, 245
Income:
  breakdown of, 164
  per capita, 174
  disposable, 148
  disposable personal, 129, 191, 193
  division of, 156, 170
  gross, 45
  per hour, 140
  national (see National income)
  operating, 45
  personal, 129
  real disposable, 148
  shared, 162, 164
  supernumerary, 152
Incremental cost(s), 235, 237, 280, 294
Incremental unit cost, 306
Incremental unit investment cost, 267
Index of prices, 100
Indirect business taxes, 129
Induction motor, 15
Industrial class, 224
Industrial development, 214
Industrial load curves, 250
Industrial revolution, 117, 132
Industrial service, 233, 262
Industry:
  electrification of, 225
  use of electricity by, 226
Inflation, 1, 96, 98, 167, 190, 303, 346, 386
Informational programs:
  in employee relations, 312, 318, 320–321
  in public relations, 312, 316, 322
Institutional investors, 41, 145
Insulation, 212, 220, 235
Insull, Samuel, 54
Interconnected systems, 30, 54, 56, 66, 67, 72, 104, 292
  and pooling, 326
  vs. single systems, 57
Interconnecting power suppliers, 54
Interim rate increases, 203, 308
Internal cash sources, 35
Interruptible rates, 286
Interruptible service, 228
Investment, 143, 144, 146, 241
  capital, 161, 204, 257
  cost per kilowatt, 98
  dollar, per kilowatt, 207

Investment (Cont.):
  in electric utility plant investor-owned electric utilities, 201
  percent return on (see Percent return on investment)
  plant, 212, 235, 236
  production, 207
Investment ratio, 109
Investor-owned electric utility industry, 296
Investor-owned utilities, 295
Investor relations, 317, 321
Isolated generating plants, 292
Israel, 375

Job accomplishment by management team, 116
Job opportunities, 139, 355
Jobs and living standards, 378
Johnson, Lady Bird, 349
Johnson, Lyndon B., 337, 349
Joint Committee on Atomic Energy, 379
Joppa plant, Illinois, 64

Kellogg, Charles, 60
Kilovolt (kV), 7, 15
Kilovoltampere (kVA), 7, 15, 17
Kilovoltampere capacity, 17
Kilowatt (kW), 4, 7, 12, 15
Kilowatt forecast, 200
Kilowatthour (kWh), 5, 6, 12
  frequency distribution of, 276
  and income, 175
  operating cost per kWh, 238
  percent increase in, 189
  total cost per kWh sold, 239
Kilowatthour forecast, 185, 192, 196, 204
Kilowatts of demand, 204
Krug, J. A., 61

Lapp, Ralph E., 374
Large lighting and power service, 261, 274
Laser-pellet concept, 343
Leadership skills of management, 122
Liabilities, long-term, 37
Licensed nuclear units, 387
Life insurance coverage per family, 153
Lifeline rate, 285
Light-water reactor (LWR), 82, 336–338, 340, 342, 356
Lighting, 213, 217, 219–221, 250
Limestone, boiler use of, 354

"Limits to Growth, The" (Club of Rome), 197n.
Line (distribution) transformer, 34
Liquid-metal fast breeder reactor (LMFBR), 87, 340, 341
Living standards, 127, 130, 141, 355, 366, 375
LMFBR (liquid-metal fast breeder reactor), 87, 340, 341
Load, 10, 47
  characteristics of, 246
  controls on: commercial, 250
    residential, 233
  maximum, 12
  peak, 49
  pricing of, 284
Load centers, 29
Load curves, 48, 231, 247, 250, 265
  composite, 254
  residential, 231
Load dispatcher, 48
Load-duration curve, 288
Load factor:
  of air conditioning, 215, 221, 253
  annual, 21
  average load factor of the class, 268
  cost of furnishing services and, 255
  cost curves of kilowatthours and, 267
  definition of, 17–19
  demand meter and, 284
  diversity in, 21–23
  improvements in, 213–214
    with off-peak service, 225
    trial programs for, 233
  for large light and power service, 274
  load-factor principle in rate schedule, 272, 281, 283
  load management and, 231, 286
  marketing to build, 214–216
  monthly, 20–21
  percent of, and slowing of marketing activity, 237
  and price for electric service, 260
  of range and water heater, 246–247
  of refrigeration, 220
  research on, 230, 326
  rise and fall in, 104
  and storage of electricity, 91
Load management, 205, 212, 213, 221, 224, 231, 285, 286
Load Research Committee of Association of Edison Illuminating Companies, 230, 246
London, 350
Long-run incremental cost pricing (LRIC), 285

Long-term liabilities, 37
Los Angeles, Calif., 349
Losses in the system, 11
Luckiesh, Mathew, 219
LWR (*see* Light-water reactor)

Machinery, use of, 128, 130–131, 136
Magnetohydrodynamics, 327
Main transmission system, 32
Management, 115
  control of business by, 121
  and development of corporation, 117
  development of manager, 122
  forecasting by (*see* Forecasting, management use of)
  objectives of, 118, 124
  planning by, 120, 181
  rewards, use by, 121
  skills of, 199, 122
  staff selection by, 120
Management audit, 209
Management team, tasks of, 116
Manhattan Project, 336
Marketing, 184, 204, 205, 211, 214, 218, 229, 230, 233, 286
Marketing expense, 44
Marx, Karl, 136
Mass transportation, 377
Massachusetts Institute of Technology (MIT), 358
Maximum demand, 18, 20, 22
Maximum hourly demand, 12
Maximum load, 12
Maximum noncoincident method of cost allocation, 265
Median family income, 147
Megawatt (MW), 31
Mid-America Interpool Network (MAIN), 70
Mid-Atlantic Area Council (MAAC), 70
Mid-Continent Area Reliability Coordination Agreement (MARCA), 70
Middle East oil embargo, 2, 365–366
Millirem (unit of measure for radiation), 359
Miscellaneous services, 262
MIT (Massachusetts Institute of Technology), 358
Model system, building a, 66
Moderator in nuclear reactor, 83, 86
Monopoly, public utility, 291–292
  as natural monopoly, 161
Moss, Frank K., 219
Municipal ownership of electric utility systems, 296

Nameplate rating, 51
National Aeronautics and Space Administration (NASA), 333
National Association of Regulatory Utility Commissioners, 286
National Bureau of Standards (NBS), 333
National Economic Research Associates, 364
National Electric Light Association (NELA), 350
National Electric Reliability Council (NERC), 69
National Energy Act (1978), 3, 286, 385, 386
National Energy Plan, 386
National energy policy, 185, 385
National income, 129, 140, 162
  per capita, 367
  per man-hour worked, 140
National Rural Electric Cooperative Association, 286, 332
National Science Foundation, 330
Natural beauty, 350
Natural Gas Pipeline case, 302
Natural monopoly, 161
Nebraska, 289, 296
NELA (National Electric Light Association), 350
Neon lights, 15
Net current assets, 37
Net national product, 129
Net operating revenue, 43, 45
Net revenue, 204, 205, 212
Neutron (subatomic particle), 80
New Mexico, 308
New York City, polluted air in, 350
New York State, formal regulation in, 288
Ninety-fifth Congress and nuclear licensing bill, 386, 387
"No growth" philosophy, 141, 197, 284, 367
No-par stock, common stock as, 38
No significant deterioration amendments, 364
Noncoincident maximum demands, 22
Nondiscrimination clause in legislation, 269
Northeast Power Coordinating Council (NPCC), 70
Northeast power interruption, 67
Northwest Power Pool (NPP), 65
NRC (see Nuclear Regulatory Commission)
Nuclear core of reactor, 81
Nuclear energy, 80, 326, 348, 355, 357, 365, 373, 385
Nuclear fission, 337

Nuclear fuels, 27, 80, 338, 340, 341, 344, 354, 374, 384, 386
Nuclear power, 79, 80, 327
"Nuclear Power: Safe by Any Measure" (Rasmussen), 358
Nuclear power plants, 77, 359, 381, 385
  inspection of, 357
  licensing of, 357
  safety of, 356, 358
Nuclear radiation, 81, 356, 358, 359
Nuclear reactors:
  breeder (see Breeder reactor)
  BWR, 81, 337
  Canadian nuclear power reactors, 86
  core of, 81
  fuel elements in, 81
  gas-cooled, 84
  HWR, 86
  LWR (see Light-water reactor)
  moderator in, 83, 86
  radioactive materials in, 83
  TFTR, 342
  thermal, 83
  water-cooled, 81
Nuclear Regulatory Commission (NRC), formerly AEC, 64, 330, 337, 356, 359
Nuclear research, 334
Nuclear waste, 361–362

Obligation to serve, electric utility, 288
Off-peak periods, 215, 216, 227
Off-peak rates, 228
Off-peak service, 225
Office of War Utilities, 61
Ogive method for distribution computations, 276, 277
Ohio Valley project, 64
Oil:
  Alaskan North Slope, 367
  conservation of, to eliminate importation, 377
  depletion of supply, 355
  domestic, 381, 384
  domestic oil production, 367
  foreign, 366, 387
  as fuel to generate electricity, 216
  transition from, 344
  production of petroleum, 368
    growth rate of, 372
  in short supply, 27
Oil embargo, Middle East, 2, 365–366
Oil imports, 218, 366, 378, 381, 384
Oman, 378

On-site generating plants, 292
Operating cost per kilowatthour, 238
Operating expenses, 41
  total, 211
Operating income, 45
Operating (spinning) reserve,  30, 57
Operating trends, 112
Opinion leaders and public relations, 319
Opinion surveys in public relations, 311,
  319
Optional demand-energy rate, 283
Organizational skills, 120
Original cost, 298
Outage and type of system, 57


Panama, 375
Par value of preferred stock, 37
Parkerson, William, 277
Patterns of energy use, 187, 370
Peak demand, 19
Peak load, 49
Peak-load pricing, 284
Peak-responsibility method of cost alloca-
  tion, 265
Peaks:
  seasonal, 57
    summer, 215, 221
    winter, 221
  system, 12, 265
Pellets of uranium oxide in fuel elements,
  81
Per capita consumer spending, 148, 150
Per capita disposable income, 148
Percent increase in kWh, 189
Percent return on investment, 168, 181,
  184, 200–204, 211, 212, 230, 233,
  244, 262, 298
Perl, Lewis J., 363–364
Personal consumption expenditures, 149
Personal income, 129
Personal outlay, 129
Personal saving, 129
Personal taxes, 129
Petroleum (*see* Oil)
Pickering Generating Station (Canada), 87
PJM (Pennsylvania, New Jersey, Mary-
  land) Interconnection, 65
Plan, corporate model, 5-year, 182
Planning, 120, 181, 184, 204, 206
Plant allocation methods, 265
Plant capacity, 31
Plant investment, 212, 235
  per kilowatt, 236
Plant location, 348

PLBR (prototype large breeder reactor),
  341
Plutonium, 81, 340
Pooling, 68
Population of neutrons, 83
Power, 7, 24
  active, 17
  purchased, 58
  useful, 15
Power Authority of the State of New
  York, 67
Power dam, 75
Power factor, 7, 15
  system, vector diagram of, 16
Power pools, 61
Preferred stock, 37
  dividends on, 46
  par value of, 37
  stockholders of, 38
  (*See also* Common stock)
Pressurized fluidized bed combustion
  (PFBC), 353
Pressurized-water reactor (PWR), 81, 337
Pretesting surveys in public relations, 318
Price of electricity, 188, 190, 202, 208,
  255
  decline in, 188
  increases in, 314
  reductions in, 103
  for residential service, 272
  unit, 94
Pricing policy, 184, 204, 230, 235, 270
Primary coolant pump, 88
Primary distribution, 34
Primary distribution voltage, 34
Primary energy, 370
Primary energy resource processing, 333
Primary fuels, 100, 132, 217
Primary sources of energy, 365
Private ownership of property, 245
Private property in American economic
  system, 146
Production expense, 41
Production investment, 207
Productivity, 99, 107, 128, 131, 143, 147,
  167, 168, 326
Productivity index, 107
Profits, 156, 160, 245
  on all sales, 161
  and losses, 146
Proliferation of nuclear weapons, prevent-
  ing, 341
Property record, continuing, of electric
  utility system, 298
Prototype large breeder reactor (PLBR),
  340–342

Public relations:
    committee for, 316
    government affairs and, 317
    news concerning, 312
    opinion leaders and, 319
    opinion surveys in, 311, 319
    outline of, 316
    personnel for handling, 316
    program for, 314
    public affairs and, 317, 321
    public attitude and, 314, 315
        measurements of, 322
        surveys of, 315
    public attitude and behavior and, re-
        search in, 327
    public information and, 312, 316, 322
    public opinion and, 312
        surveys of, 318
    (*See also* Employee relations)
Public service commissions, 288
Public Utility Act of 1935, 289
Public utility districts, 296
Public Utility Regulatory Policies Act
    (PURPA), 286, 290
Pumped-storage hydroelectric plant, 75
Purchased power, 58
PWR (pressurized-water reactor), 81, 337


R&D (research and development) goals
    and funding, 328, 333
Radiation, nuclear, 81, 356, 359
Radioactive materials in reactor, 83
Radioactive waste, 361–362
Rasmussen, Norman C., 358–361
Rate(s), 260, 288
    block, 270, 283, 286
    change in, proceedings for, 304
    commercial, 283
    design of, 230, 245, 269, 274, 281, 285,
        286, 294
        in future years, 278
    of growth, 102, 134
    of increase, 195
    increase in, 240, 242, 303, 304
    interruptible, 286
    lifeline, 285
    off-peak, 228
    reductions in, 303, 304
    residential, 268, 278, 283
    of return, 306
    room-count, 283
    seasonal, 286
    sliding-scale (*see* Sliding-scale rates)
    summer-winter, 235
    time-differentiated, 286
    valuations and, 297

Rate base:
    forward, 203
    future, 307
Rate cases, 304, 308
    time lag in, 202, 203
Rate making, 246, 262, 267, 270, 280, 301
Rate schedules, 265, 269–271, 274, 306
    commercial, 272
    residential, 271
    schedule of schedules, 279, 281, 306
    upturn, 284
Rating of securities, 39
Ratios, types of, for comparison of com-
    pany efficiencies, 209
    (*See also specific type*)
Raw fuel, 80
REA (Rural Electrification Administra-
    tion) cooperatives, 296
Reactive kilovoltamperes, 17
Reactor (*see* Nuclear reactors)
"Reactor Safety Study: An Assessment of
    Accident Risks in U.S. Commercial
    Nuclear Power Plants" (Rasmussen),
    358
Real disposable income, 148
Recreation, 352, 355
Refrigeration, 220
Refrigerator-freezer, 250
Regional Reliability Councils, NERC, 69
Regulations, government: 94, 97, 146,
    162, 208
    bodies concerned with, 95, 203, 288,
        297, 301, 305, 306, 346, 348, 352
    commissions concerned with, 277,
        286, 294
    duplication and, 290, 293
    legal delays in, 3
    objectives of, 293
    procedures for, 298, 303
    processes of, 385–387
    siting and, 386, 387
Regulatory lag, 302, 307, 308
Reliable service by electric utilities, 69,
    102, 206, 293, 305, 314, 326
Reproduction cost of property, 299
Reproduction cost new of property, 300
Research:
    evolution of, 324, 328
    nuclear, 334
    objectives of, 327
    to remove sulfur from coal, 351
    value of, 323, 324
Research and development (R&D) costs,
    goals, and funding, 328, 333
Research Advisory Committee, EPRI, 332
Research Projects Committee, EEI, 328
Reserve accounts, 37

Reserve capability, 56, 72
Reserves:
  for depreciation (equipment), 45, 299, 300
  excess power, 31
  financial, 37
  operating (spinning) power reserve, 57
  power plant, 29
  uranium, 339
Reservoir, hydroelectric, 74, 78
Residential cost curves, 268
Residential load, 233
Residential rate schedule, 271
Residential rates, 268, 278, 283
Residential service, 261, 276, 283, 284
Resistance heating devices, 15
Return, 241, 263, 288, 294, 302
  on capital, 385
  on investment (*see* Percent return on investment)
Revenue, 41, 109, 161, 241, 276, 277, 279
  for Edison Power Company, 281
  per kilowatthour, 244
  net, 204, 205, 212
  net operating, 43, 45
Revenue requirements, 239, 242, 278
Reversible pump turbine, 77
Rewards, use of, by management, 121
Rockefeller, Laurance, 350
Room-count rate, 283
Roosevelt, Franklin D., 336
Rural electrification, 54
Rural Electrification Administration (REA) cooperatives, 296


Safety:
  of nuclear power plants, 356, 358
  safe use of consumer equipment, 212
  of utility equipment, 212
Sales expense, 214
Saturation index for appliances, 222
Schedule of schedules (rate), 279, 281, 306
Scrubbers for sulfur oxides, 353, 354, 363
Seasonal peaks, 57
Seasonal rates, 286
Secondary distribution voltage, 34
Securities:
  rating of, 39
  value of, 300
Securities and Exchange Commission (SEC), 289
Securities Act, 289
*Seeing* (Luckiesh and Moss), 219
Self-generation of electricity, 105, 107

Service, electric utility, 293
  commercial, 233, 261
  farm, 261
  industrial, 233, 262
  interruptible, 228
  large lighting and power service, 261, 274
  miscellaneous, 262
  obligation to serve, 288
  off-peak, 225
  reliability of (*see* Reliable service by electric utilities)
  residential, 261, 276, 283, 284
  responsibility of, 291
  small power service, 261
  value of, 274, 314
Service wires (services), 34
Shares (*see* Common stock; Preferred stock)
Shippingport, Pa., location of first prototype nuclear power plant, 337
Simpson, John W., 362
Sir Adam Beck Power Plant (Canada), 67
Siting and government regulations, 386, 387
Skills of management, 119, 122
Sliding-scale principle, 272
Sliding-scale rates, 271, 274
  demand or load-factor principle and, 272
Sludge, boiler production of, 354
Small power service, 261
Smog, 349
Smoke, 350, 351
*Smyth* v. *Ames* (1898), 301
Sodium intermediate loop in heat exchanger, 87
Solar energy, 26, 328, 344, 355, 356, 386
Solvent refined coal, 353
South Central Electric Companies (SCEC), 73
Southeast Asia, 365
Southeastern Electric Reliability Council (SERC), 70
Southwest Power Pool (SPP), 60, 65, 70, 73
Space heating, 384
Spacer devices, reactor, 82
Spent core, reactor, 83
Spinning (operating) reserve, 30, 57
Stable system in power plant, 68
Stack emissions, 351
Stack gases, 327
Stacks, 351, 353
Staff selection by management, 120
Standby capacity, 29
Standby contracts, 59

Starr, Chauncey, 330
Stated value of common stock, 38
Static condensers, 16
Stationary motors, 217, 218
Steam engine, 132
Steam turbine, 25
Step-down substations, 32
Step-up transformers, 31
Stockholders (*see* Common stock; Preferred stock)
Stocks (*see* Common stock; Preferred stock)
Storage of electricity, 91
Storage space heater, 227
Storage-water heater, 225
Strip mining, 385
Subatomic particle (neutron), 80
Substation, 31, 32
Sulfur in fuels, 351, 352
    emissions of, 354
    oxides of, 353
Summer peak, 215, 221
Summer-winter rates, 235
Supernumerary income, 152
Supreme Court, U.S., 298, 301, 302
Surface Mining Act (1977), 386
Surveys, employee-attitude, 315, 320
Sweden, 375
Switzerland, 375
Synchronism of generators, 59, 67
Synchronous condensers, 16
System (firm power) capacity, 31
System peak, 12, 265
System planning, 184, 204, 206
System power factor, 16

Tall stacks, 351, 353
Task Force on Atomic Power, EEI, 337
Task Force on Nuclear Power, EEI (1958), 79
Taxes, 44, 111
    and depreciation, 44
    of Edison Power Company, 111
    indirect business, 129
    personal, 129
Tennessee Valley Authority (TVA), 73, 296, 332, 354
Test group in cost comparisons, 210
Thermal efficiency, 104
Thermal reactor, 83
Thorium, 81
Time-differentiated rates, 286
Time lag in rate cases, 202, 203
Time-of-day metering, 228, 284
Time of day rates, 286

Tokamak Fusion Test Reactor (TFTR), 342
Total capability, 56
Total cost per kilowatthour sold, 239
Total demand, 12
Total electric utility industry capability, 200
Total operating expense, 211
Transformer, 17
    distribution, 34
    line, 34
    step-up, 31
Transmission expenses, 44
Transmission grid, 44
Transmission lines, 32
    underground, 349
Transmission system(s), 31
    high voltage in, 14
    investment in, 36, 207
    main, 32
    objectives for, 327
    plans for, 206
    reducing voltage at load centers, 32
Transmission towers, 14, 351
    aesthetics of, 352
Transmission voltage:
    high, 14
    increased, 325
    reducing, at load centers, 32
Transportation, 217, 218, 384
    mass, 377
Trended present cost, 300
Trended value, 301
Tritium, 343
Turbine:
    blades of, construction of, 55
    reversible pump, 77
    steam, 25
Turbine-driven circulator, 86
Turbine generators, 29, 78
Turbo-generators, 97
TVA (Tennessee Valley Authority), 73, 296, 332, 354

Underground cables, 327
Underground distribution lines, 352
Underground transmission lines, 349
Unemployment and Clean Air Act Amendments, 386
Unit (turbine generator), 29
Unit cost, 267, 269, 271
Unit embedded cost, 97
Unit incremental cost, 97
Unit price, 94
United States, energy use in, 375

U.S. Department of Energy [*see* Department of Energy (DOE), U.S.]
U.S. Department of the Interior, 332, 367
U.S. dollar of GNP, 376
U.S. oil, 381, 384
U.S. Treasury, sale of revenue bonds to, 296
Unregulated business operation, 290
Upturn rate schedule, 284
Uranium, 81, 343, 356, 366, 368, 370, 387
  $U^{235}$, 337
  $U^{238}$, 337
Uranium oxide, 81
Uranium reserves, 339
Useful energy, 197
Useful power, 15
Useless current, 17
Utility responsibility, 291


"Valleys" between seasonal peaks, 57
Valuations, 297
  (*See also* Rates)
Value of service, 274, 314
Variable costs, 257, 262
Vector diagram of power factor, 16
Volt (V), 15
Voltage:
  as measure of pressure in electric system, 7, 13
  primary distribution, 34
  secondary distribution, 34
Voltampere (VA), 7

Washington, State of, 296
"Waste heat" plants, 65
Waste recycling, 348
Wasteful energy, 197
Water:
  analogy with use of electricity, 5, 9
  cooling, 78, 348
  pollution of, 348
  purification of, 348
  quality of, 354
  storage of, 9
Water-cooled reactors, 81
Water heater, 247, 272
  storage-, 225
Water systems, 9
Watt, James, 117, 132
Watt (W), 7, 8
West, James A., 377
Western coal, use of, by Eastern utilities, 353
Western Systems Coordinating Council (WSCC), 70
White House Conference on Natural Beauty, 349, 350
Wholesale price index, 92, 93
Winter peak, 221
Wisconsin, 288
"Wobble effect" in power forecasts, 194
Worker-hour, 147
Workweek, decline in hours of, 143
World energy, trends and patterns in, 373–376
World nuclear energy, table, 85

Yields, bond, 41